INTELLIGENT MULTIMEDIA MULTI-AGENT SYSTEMS

A Human-Centered Approach

THE KLUWER INTERNATIONAL SERIES IN ENGINEERING AND COMPUTER SCIENCE

INTELLIGENT MULTIMEDIA MULTI-AGENT SYSTEMS

A Human-Centered Approach

Rajiv Khosla

Ishwar K. Sethi

Ernesto Damiani

Kluwer Academic Publishers
Boston/Dordrecht/London

ISBN 978-1-4419-5008-6

Distributors for North, Central and South America:
Kluwer Academic Publishers
101 Philip Drive
Assinippi Park
Norwell, Massachusetts 02061 USA
Telephone (781) 871-6600
Fax (781) 871-6528
E-Mail <kluwer@wkap.com>

Distributors for all other countries:
Kluwer Academic Publishers Group
Distribution Centre
Post Office Box 322
3300 AH Dordrecht, THE NETHERLANDS
Telephone 31 78 6392 392
Fax 31 78 6546 474
E-Mail services@wkap.nl>

Electronic Services <http://www.wkap.nl>

Library of Congress Cataloging-in-Publication

Khosla, Rajiv, 1959-
Intelligent multimedia multi-agent systems : a human-centered approach / Rajiv Khosla, Ishwar K. Sethi, Ernesto Damiani.
p. cm. -- (The Kluwer international series in engineering and computer science ; SECS 582)
Includes bibliographical references and index.

1. Multimedia systems. 2. Intelligent agents (Computer software) 3. Human-computer interaction. I. Sethi, Ishwar K., 1948- II. Damiani, Ernesto. III. Title. IV. Series.

QA76.575 .K52 2000
006.7--dc21

00-135009

Printed on acid-free paper.

Printed in the United States of America

TABLE OF CONTENTS

PART II: (Chapters 6-8)
HCVM Applications in Health Informatics, Face Detection, Net Euchre and Sales Recruitment

PREFACE

This book is about developing intelligent multimedia multi-agent systems based on a human-centered approach. It is relevant to practitioners and researchers in areas like human-centered systems, intelligent systems, multimedia, image processing, internet and e-commerce, e-health and health informatics, software engineering (especially those using patterns and adapter technologies) and cognitive science. The book is an outcome of research done in areas like work-centered design, situated and distributed cognition, activity theory, multimedia, problem solving ontologies and pattern and adapter design concepts. It is instructive here to clarify the title of the book. The term "intelligent" in the title represents the hard and soft computing technologies and their hybrid configurations used for modeling complex tasks. The term "Multimedia" in the title represents two contexts of multimedia. Firstly, multimedia artifacts have been used for human-centered modeling of data in order to improve the representational efficiency and effectiveness of the human-computer interface. Secondly, it is used in the context of multimedia information management and retrieval issues on the internet and the role of human or user-centered models in addressing these issues. The third term "Multi-agent" represents the software artifact used to define among other aspects, the intelligent technology artifacts and multimedia artifacts at the computational level of system development. Finally, the term "system" in the title represents the fact that we have used a system (consisting of a number of components) as the unit of analysis for developing computer-based artifacts.

The motivation for writing this book has come from the work done by Donald A. Norman ("The Invisible Computer," 1998; "The Psychology of Everyday Things," 1988) and by the contributions made by several important researchers in the Human-Centered Systems workshop held by NSF in February 1997 in Crystal Gateway Marriott Hotel, Arlington, VA. The book does not claim to have succeeded fully in developing all facets of human-centered systems. However, as the subtitle of the book suggests we have adopted a human-centered approach for software development. In this approach humans are the prime drivers in the system development process and technology is a primitive that is used based on its conformity to the needs of people in the field of practice.

The book consists of three parts:

Part 1: provides the motivation behind the book, introduces various intelligent technologies used for intelligent systems development. This introduction is followed by a discussion on traditional theories as well as recent ones from philosophy, cognitive science, psychology, and the work place which have contributed towards a human-centered software development. These theories are used as a foundation for developing a human-centered system development framework and a human-centered virtual machine.

Part II: describes the applications of the human-centered vitual machine in health informatics, image processing, internet games and sales recruitment

Part III: introduces the area of intelligent multimedia information management and electronic commerce. It describes applications of human-centered virtual machine in electronic commerce and web-based medical image retrieval.

Part I is described through chapters 1, 2, 3, 4 and 5 respectively.

Chapter 1: primarily outlines the difference between a technology-centered approach and a human-centered approach to software development. It discusses the problems associated with the technology-centered approach and describes the criteria for human-centered software systems.

Chapter 2: describes various intelligent, multimedia and internet technologies that are used as software modeling and computational primitives in intelligent multimedia multi-agent systems. It covers a range of intelligent technologies like neural networks, fuzzy logic, genetic algorithms and expert systems, software development technologies, like object-oriented and agents, multimedia artifacts and XML-the emerging internet standard.

Chapter 3: discusses how areas, like intelligent systems, software engineering, multimedia databases, data mining and human-computer interaction, are evolving towards human-centeredness. We follow these pragmatic developments with enabling theories for human-centered system development in philosophy, cognitive science, psychology and workplace.

Chapters 4 and 5: build on the foundations laid down in chapter 3. These two chapters describe four components of the human-centered system development framework. The four components are activity-centered analysis component, problem solving ontology component, transformation agent component, and multimedia interpretation component.

At the computational level, a Human-Centered Virtual machine (HCVM) is realized through integration of activity-centered analysis component, problem solving ontology component and the multimedia interpretation component with technology based models like, intelligent technology model, object-oriented model, agent model, distributed process model and XML/XTL (eXtensible Transformation Language) model.

HCVM consists of four transformation agent layers and one object layer. The five layers represent a component based approach for building intelligent multimedia multi-agent systems.

Part II is covered through chapters 5, 6,7 and 8 respectively

Chapter 6: describes a detailed application of four components of HCVM in medical diagnosis and treatment decision support. It defines a range of problem solving agent intelligent agents and multimedia agents.It looks into the role of external representations in this domain and how multimedia artifacts can be used for effective symptomatic data gathering. .

Chapter 7: describes a face detection and annotation application. It primarily focuses on the application of the problem-solving ontology component of the HCVM

Chapter 8: describes modeling of human dynamics and breakdowns through an internet based Net Euchre card game application and a sales recruitment and benchmarking application respectively. It describes the role of intelligent distributed back up support game agents in human-computer interaction, and intelligent agents for modeling human breakdowns in the sales recruitment decision making situations.

Part III is described through chapters 9,10 11 and 12:

Chapter 9: introduces the reader to the area of intelligent multimedia information management. It covers the three evolution stages of multimedia information management and retrieval (the latest one being on the internet) and discusses problems and issues that need to be addressed in this area.

Chapter 10: discusses the need for human or user-centered architectures for electronic commerce. It describes the application of HCVM in the hardware adapter domain. The chapter also outlines the new internet language XML and describes the integration of the problem-solving ontology component with XML in the hardware adapter domain.

Chapter 11: describes a web-based medical image retrieval application of the problem-solving ontology component of HCVM.

Chapter 12: concludes the book by highlighting some of the contributions made by it in various areas.

RAJIV KHOSLA
ISHWAR K. SETHI
ERNESTO DAMIANI

ACKNOWLEDGEMENTS

A number of people have contributed to this work. The authors wish to especially acknowledge the support of their research students Somkiat Kitjongthawonkul, Damian Phillips, Daniela Stan, Gang Wei and Avi. The authors would also like to thank Department of Computer Science at Wayne State University, Detroit, Michigan, USA for the use of their facilities and resources in the production of the work.

1 SUCCESSFUL SYSTEMS OR SUCCESSFUL TECHNOLOGIES?

1.1. Introduction

Since the advent of computers, technology has been used by the software industry as a means of developing a competitive edge. On the other hand research institutions and universities have pursued technology as a means of innovation and novelty. As a result today, we have a range of technologies and techniques in intelligent systems, software engineering, databases and other areas. At the receiving end of this technology -based competition and innovation are the computer users. These users (majority of whom are nonspecialists) today have to cope with not only the range of technologies but also pace of technological change. They are used as guinea pigs for every technological innovation. More so, most technologies are presented to the users as a panacea of all ills. This approach has lead to a number of problems ranging from ineffective use of computers to loss of human lives (Preece et al 1997). These problems have given birth to new type of systems, namely, human-centered systems in the last decade (Norman 1993,97; Clancey 1997; Flanagan et. al. 1997).

This introductory chapter discusses the need for human-centered systems and outlines the criteria for their development. The criteria are used to develop human-centered system development framework in this book.

1.2. Technology-Centeredness vs Human-Centeredness

The technology-centered approach can be analyzed from a research perspective as well as from a marketing perspective. Most studies in computer science (and other science disciplines) can be seen as following technology-centered motto "Science Finds, Industry Applies and Man Conforms" (Chicago World Fair 1933). In other words, science invents new technologies, industry applies them for solving various problems and people or users are expected to comply with the nuances of the technology. This approach is also captured in Figure 1.1.

Thus in this technology–centered approach, technology is the prime driver in the system development process as shown in Figure 1.1. A technologist or a software developer armed with one or more technologies (e.g. object-oriented, neural network, etc.) models a software solution to a human problem.

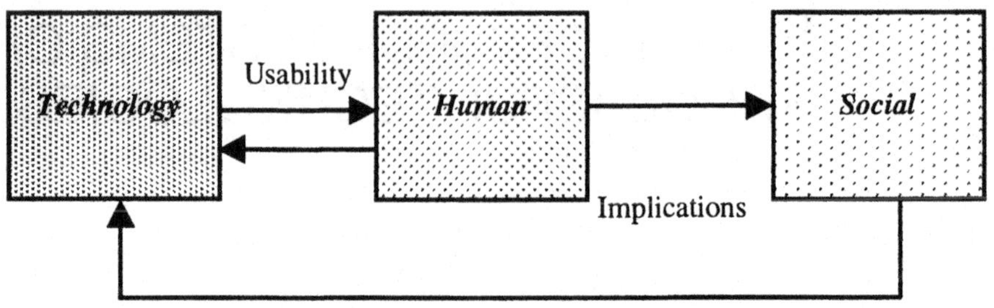

Figure 1.1: Technology-Centered Approach

The conceptualization of a problem domain in the technology-centered approach is largely based on system designer's perspective rather than the user's perspective (Figure 1.2). Once the technology-centered artifact is created it is connected to its users through user interfaces developed using usability engineering and human-computer interaction techniques. At this stage human-computer interaction and usability specialists play an active part in system interpretation. The technology-centered software system is handed over to the users who are provided with manuals (consisting of hundreds of pages) to get acquainted with its use. Over a period of time feedback from users (related to usability) and the social scientists (on social consequences of technology use) is employed to improve the future use of technology.

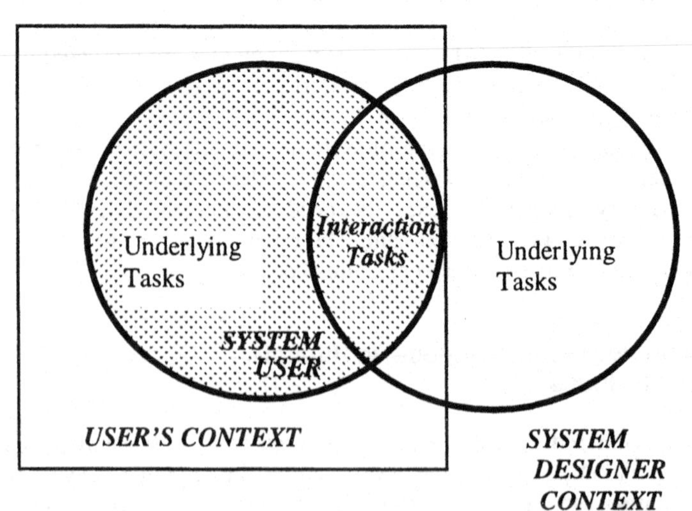

Figure 1.2: Context

Although the benefits of technology can by no means be underestimated, empirical studies on the impact of new technology on actual practitioner cognition and performance has revealed that new systems often have surprising consequences or even fail (Norman 1988; Sarter, Woods and Billings 1997). These surprising consequences and failures can be appreciated in the context that computerization/automation changes the nature of activity (and the tasks) in a field of practice, roles people play and strategies they employ to accomplish various tasks in that activity. The mismatch between the automated activity and how it was being done previously (i.e. before being automated) results in breakdowns in human-computer interaction (Norman 1988; Preece et al. 1997). These breakdowns occur at two levels, namely, the interaction level and the underlying task level. The interaction level determines the presentation efficiency of a computerized system. On the other hand, the underlying task level determines how involved the user is going to be with the system. As an analogy, the interaction level can be seen to represent a façade or the exterior features of a home. The underlying task level can be seen to represent the floor plan that determines the configuration and organization of various rooms (e.g., study, living, bedroom, etc.). The quality of the first one invites one into the home. The quality of the second one provides an immersive environment for one to live in.

The mismatch in these two levels manifests in different kinds of breakdowns in human-computer interaction. These include sheer complexity and bewilderment, cognitive overload, error and failure and others (Norman 1988). However more serious consequences can also result from the mismatch. Studies done of major accidents in power plants, aircraft and other complex systems (e.g. ambulance dispatch systems) found that 60 to 80 percent of accidents were blamed on operator error and incorrect actions by people (Perrow 1984; Preece et al 1997). Given the human limitations, the problem could have easily have lied with system modeling and design.

The remedial measures in most technology-centered systems are mainly applied at the interaction-level (through usability engineering and HCI methods) because the underlying technology and the system model are difficult to change once a software system has been developed.

As indicated in the introduction, the motivation for technology-centered approach also comes from the way the computer industry functions and markets itself. As Norman (1998) has put it

The computer industry is still in its rebellious adolescent stage. It is mature enough that its technology, functions and reliability should be taken for granted, but it still has a good deal of immaturity. It keeps trying to grow bigger, faster, more powerful. The rest of us wish it would just quiet down and behave. Enough already. Grow up. Settle down and provide good, quiet, competent service without all the fuss and bother. Ah, but to make this change from youth to maturity is to cross the chasm between technological excitement of youth and the staid utility of maturity. It is a difficult chasm to bridge.

This chasm is shown in Figure 1.3. It represents the transition point in the technology life cycle of products in the computer industry. Figure 1.3 shows the transition of a product from a technology-centered infant to a human-centered adult. In early stages of the technology life cycle (left side of the curve in Figure 1.3), technology is used to fill basic unfilled technology needs of customers. For example, video-on demand, and learning-on demand are new technologies on the internet and are in the early stages of their technology life cycle. Their customers generally are early adopters (innovators, technology enthusiasts and visionaries) who are ready to pay any price for the new technology. They have good technical abilities and are prepared to live with the idiosyncrasies of the new technology. Based on these early adopters, the computer industry labels the technology as a successful technology.

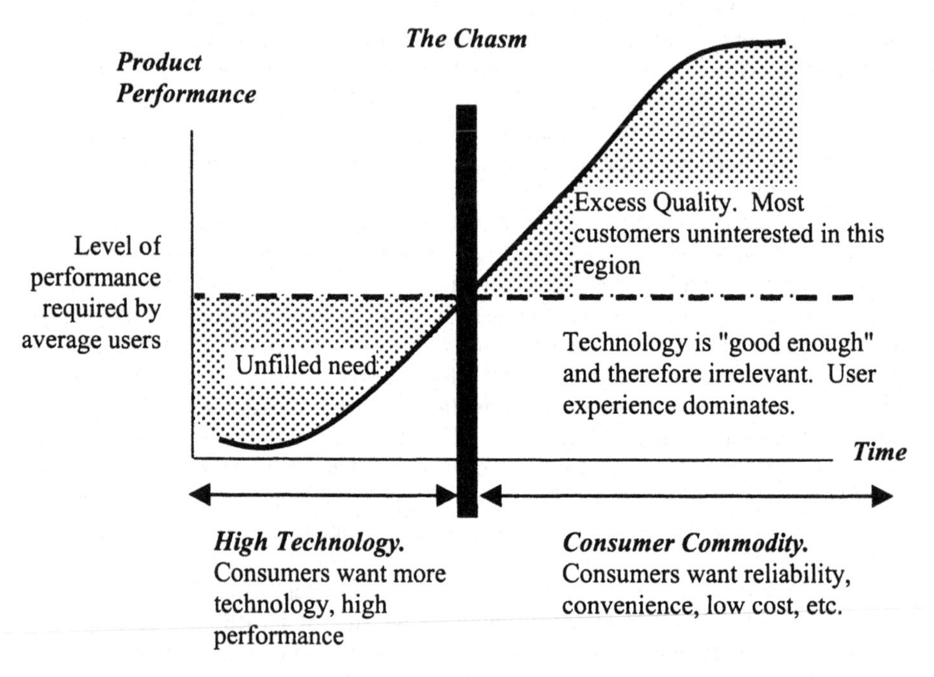

Figure 1.3: Technology Life Cycle of Products in Computer Industry (adapted from Norman (1998))

However, these early adopters who thrive on technological superiority, represent a small part of the potential marketplace. The majority of customers are late adopters and are represented on the right side of the chasm in Figure 1.3. These customers wait till the technology has established itself in the marketplace and has matured. This chasm or transition is reached once the basic technology needs have been satisfied. In this phase of the life cycle, the customers see the technology-based product as a commodity where user experience, cost and reliability dominate the buying decision. The technological aspects of the product are taken for granted and are no longer relevant.

Most of the time the technology-centered companies react to this change by loading their products with still more features or excess quality (represented by the shaded upper right hand part of the curve in Figure 1.3). Actually, what is required is a change in the product development philosophy that targets the human user rather than the technology. In other words, a more human-centered approach to product development is required. In this approach, technology is just one component that has to adapt to other system components. In the next section we outline the criteria for human-centered product development.

1.3. Human-Centered Approach

Human-centered development is about achieving synergy between the human and the machine. This synergism goes (as outlined in the preceding section) beyond human-computer interaction concepts, people in the loop philosophy and other interpretations given to human-centeredness. Although most systems are designed with some consideration of its human users, most are far from human-centered. The informal theme of the recently held NSF workshop on human-centered systems (1997) was people propose, science studies, and technology conforms. In other words, humans are the centerpiece of human-centered research and design (as shown in Figure 1.4). They are the prime drivers and technology is a primitive that is used -based on its conformity to the needs of people in a field of practice. The three criteria laid down in the workshop for human-centered system development are:

1. Human-Centered research and design is problem/need driven as against abstraction driven (although there is an overlap)
2. Human-Centered research and design is activity centered
3. Human-Centered research design is context bound.

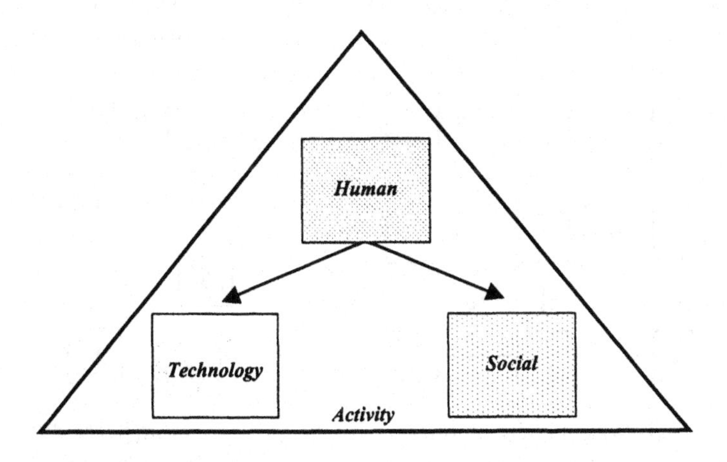

Figure 1.4: A Human-Centered Approach Leading to Successful Systems

The first criteria outlines a need for developing software systems that are modeled based on how people use various artifacts to solve problems in a field of practice. The modeling should include not only the normal or repetitive tasks but also exceptional and challenging situations. These exceptional and challenging situations can also be likened to breakdowns in problem solving. For example in the sales management area, a sales (or customer service) manager while recruiting new salespersons (customer service representatives) is faced with an exceptional situation of determining a benchmark for recruiting the new salespersons (customer service representatives). This benchmark should represent the type of salespersons who are successful in a their organization and culture. Another challenging situation is when they have to decide between two candidates who (as the sales manger perceives) are equally good for the job. On the other hand, a fresh sales recruit faces an exceptional situation when they are unable to understand or explain why a particular customer does not respond or "freezes up" during their interaction with them.

Further, this criteria also suggests that generic problem solving abstractions should be extracted from problem solving situations as people perceive and solve them, rather than employ abstract theories like graph theory or logic or other domain theories to solve problems in various fields.

The second criteria emphasizes system development based on practitioners or users goals and tasks rather than system designers goals and tasks. In other words, this criteria emphasizes the need for maximizing the overlap between a user's model of the problem domain and a system's model of the domain. The focus is on how well the computer serves as an effective tool for accomplishing user's goals and tasks.

Finally, the third criteria emphasizes that human cognition, collaboration and performance is dependent upon context. It particularly looks at the representational context. That is, how the problem is represented influences the cognitive work needed to solve the problem (see Figure 1.5). Problem solving is distributed across external and internal representations. Software systems based only on internal representations or models of a problem domain are likely to put a higher cognitive load on their users as against systems that are based on external or perceptual representations. Other contexts that need to be taken into account are social/organizational context and task context (as outlined in the second criteria).

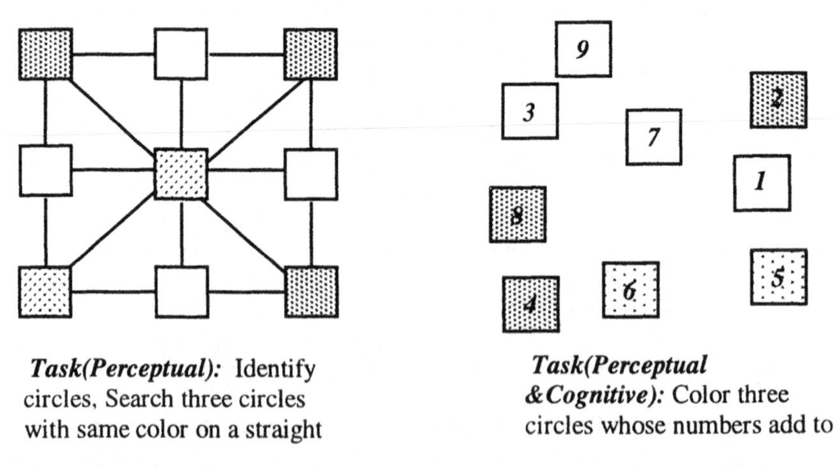

Figure 1.5: Two Representations of Tic-Tac-Toe

The human-centered framework and its applications described in this book are based on the above three human-centered criteria. A bird's eye view of the applications developed in this book is shown in Figure 1.6. The central image represents the human-centered system development framework.

1.4. Summary

This chapter highlights the need for human-centered systems. It analyzes the need from a system development perspective, as well as the marketing perspective. It outlines three human-centered criteria which are used as guidelines for developing the human-centered system development framework and the human-centered virtual machine.

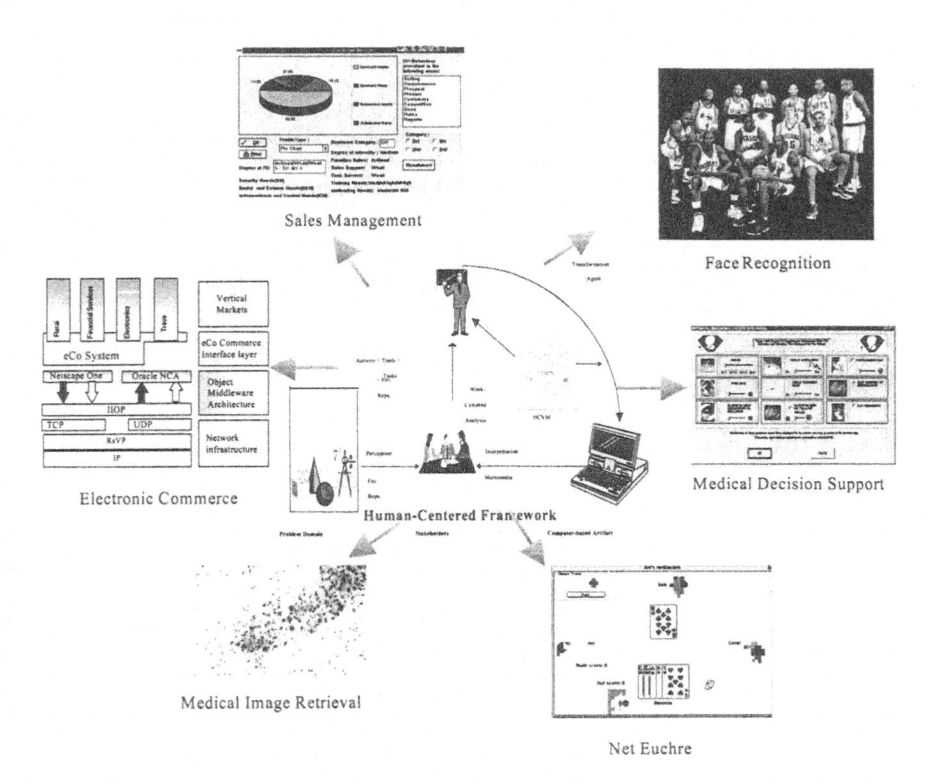

Figure 1.6: Overview of the Human-Centered Framework and its Applications

References

Clancey W.J. (1989). "The Knowledge Level Reconsidered: Modeling How Systems Interact" in *Machine Learning* 4, pp 285-92.

Clancey, W.J. (1993) "Situated Action: A Neuropsychological Interpretation (Response to Vera and Simon)" in *Cognitive Science*, 17, 87-116.

Flanagan, J.,Huang, T et al. (1997). "Human-Centered Systems: Information, Interactivity, and Intelligence," *Final report NSF Workshop on Human-Centered Systems, February.*

Norman, D. A. (1988). The Psychology of Everyday Things. Basic Books: New York

Norman, D. A. (1993). Things That Make Us Smart. Reading: Addison-Wesley

Norman, D. A. (1998). *The Invisible Computer,* Massachusetts: MIT Press.

NSF Workshop on *Human-Centered Systems, February 1997. Final report.*

Flanagan, J.,Huang, T et al. (1997). Human-Centered Systems: Information, Interactivity, and Intelligence.

Perrow, C., *Normal accidents: living with high-risk technologies*. 1984, NY: Basic Books.
Preece, J., et al (1997), *Human-Computer Interaction*, Massachusetts: Addison-Wesley Pub.
Sarter, N., Woods, D.D. and Billings, C. (1997), "Automation Surprises," in G. Slavendy, (ed.), *Handbook of Human Factors/Ergonomics*, second edition, Wiley,
Zhang, J., Norman, D. A. (1994), "Distributed Cognitive Tasks", *Cognitive Science*, pp. 84-120

2 TECHNOLOGIES

2.1 Introduction

The applications in this book employ a number of technologies. These technologies can be grouped under intelligent systems, software engineering, agents, multimedia and the internet/electronic commerce. This chapter introduces the reader to these technologies. The various technologies covered in this chapter are:

- Expert Systems
- Case based Reasoning
- Artificial Neural Networks
- Fuzzy Systems
- Genetic Algorithms
- Intelligent Fusion, Transformation and Combination
- Object-Oriented Technology
- Agents and Agent Architectures
- Multimedia
- XML – the new internet standard

2.2. Expert Systems

Expert Systems handle a whole array of interesting tasks that require a great deal of specialized knowledge and these tasks can only be performed by experts who have accumulated the required knowledge. These specialized tasks are performed in a variety of areas including Diagnosis, Classification, Prediction, Scheduling and Decision Support (Hayes-Roth et al 1983). The various expert systems developed in these areas can be broadly grouped under four architectures.

- Rule Based Architecture
- Rule and Frame Based Architecture
- Model Based Architecture
- Blackboard Architecture

These four architectures use a number of symbolic knowledge representation formalisms developed in the last thirty years. Thus, before describing these architectures, the symbolic knowledge representations are briefly described.

2.2.1. Symbolic Knowledge Representation

Symbolic Artificial Intelligence (AI) has developed a rich set of representational formalisms that have enabled cognitive scientists to characterize human cognition. The symbolic representational power has in fact been seen for a long time as an advantage over the connectionist representations for human problem solving. The real reason according to Chandrasekaran (1990), for loss of interest in the perceptrons by Minsky and Papert (1969), was not due to limitations of the single layer perceptrons, but for the lack of powerful representational and representation manipulation tools. These AI knowledge representational formalisms are briefly overviewed.

Knowledge representation schemes have been used to represent the semantic content of natural language concepts, as well as to represent psychologically plausible memory models. These schemes facilitate representation of semantic and episodic memory.

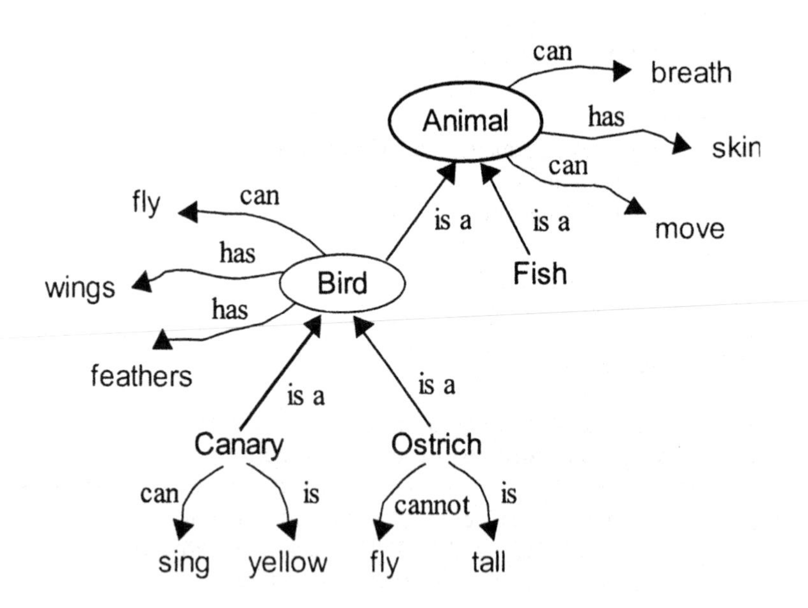

Figure 2.1. A Semantic Network

Human semantic memory is the memory of facts we know, arranged in some kind of hierarchical network (Quillian 1968; Kolodner 1984). For example, in a semantic memory "stool" may be defined as a type of "chair", in turn defined as an instance of "furniture". Properties and relations are handled within the overall hierarchical framework. Semantic networks are a means of representing semantic memory in which any concept (e.g. Bird in Figure 2.1) is represented by a set of properties, which in turn consists of pointers to other concepts as shown in Figure 2.1 (Quillian 1968). The properties are made up of attribute value pairs. Semantic networks also introduce property inheritance as a means to establish hierarchies and a form of default reasoning. Whereas, semantic nets provide a form of default reasoning, first order predicate calculus and production systems (Newell 1977) provide a means of deductive reasoning observed in humans. First order predicate calculus and production systems are both a representational and processing formalism. A predicate calculus expression like:

$$\forall X \, \exists Y \, (student(X) \,)\, AI\_subject(Y) \wedge likes(X; Y))$$

provides a clear semantics for the symbols and expressions formed from objects and relations. It also provides a means for representing connectives, variables, and universal and existential quantifiers (like forall ($\forall$) and forsome ($\exists$)). The existential and universal quantifiers provide a powerful mechanism for generalization that is difficult to model in semantic networks.

Knowledge in production systems is represented as condition-action pairs called production rules (e.g., "if it has a long neck and brown blotches, infer that it is a giraffe").

If semantic memory encodes facts, then episodic memory encodes experience. An episode is a record of an experienced event like visiting a restaurant or a diagnostic consultation. Information in episodic memory is defined and organized in accordance with its intended uses in different situations or operations. Frames and scripts (Schank 1977; Minsky 1981) which are extensions of semantic networks are used to represent complex events (e.g., like going to a restaurant) in terms of structured units with specific slots (e.g., being seated, ordering), with possible default values (e.g., ordering from a menu), and with a range of possible values associated with any slot. These values are either given or computed with help of demons (procedures) installed in slots. Schank's (1972) earlier work in this direction on conceptual dependencies, involves the notion of representing different actions or verbs in terms of language independent primitives (e.g., object transfer, idea transfer). The idea was to be able to represent all the paraphrases of a single idea with the same representation (e.g., Mary gave the ball to John; John got the ball from Mary).

Object-Oriented representation, a recent knowledge representational formalism from research in artificial intelligent and software engineering has some similarities with frames as shown in Figure 2.2. It is a highly attractive idea, as it does both development from the theory of programming languages, and knowledge representation. The object-oriented representational formalism identifies the real-world objects relevant to a problem as humans do, the attributes of those objects, and the processing operations (methods) in which they participate. Some similarities with frame-based class hierarchies and various procedures (methods in objects) attached to the slots are evident. However, demons or procedures in frames are embedded in the slots, whereas in objects procedures or methods and attributes are represented separately. This delineation of methods from attributes provides them with strong encapsulation properties which makes them attractive from a software implementation viewpoint.

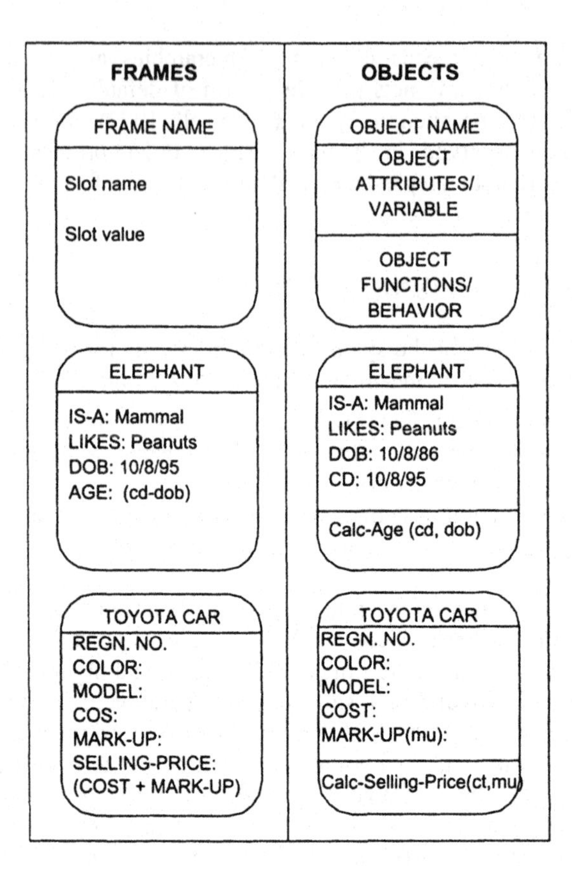

Figure 2.2. Frames and Objects

The four expert system architectures are now briefly described in the rest of the subsections.

The basic components of a rule based expert system are shown in Figure 2.3. The knowledge base contains a set of production rules (that is, IF . . . THEN 8 rules). The IF part of the rule refers to the antecedent or condition part and the THEN part refers to the consequent or action part.

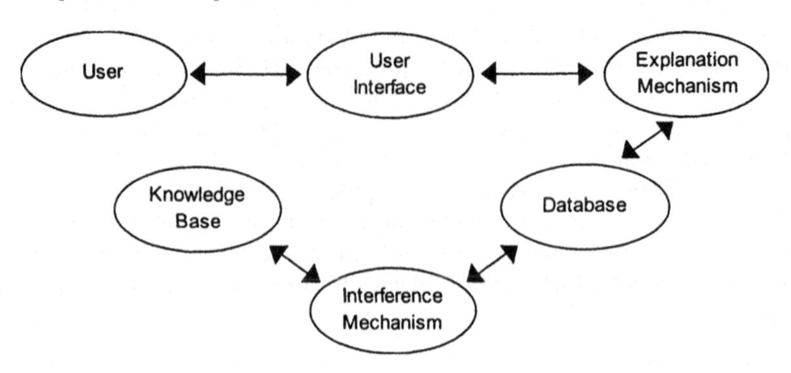

Figure 2.3. Components of a Rule Based Expert System

2.2.2. *Rule Based Architecture*

The database component is a repository for data required by the expert system to reach its conclusion/s based on the expertise contained in its knowledge base. That is, in certain types of expert systems, the knowledge base, though endowed with expertise, cannot function unless it can relate to a particular situation in the problem domain. For example, data on an applicant's present credit rating in a loan analysis expert system needs to be provided by the user to enable the expert system to transform the information contained in the knowledge base into advice. This information is stored in a data base.

The inference mechanism in a rule based system compares the data in the database with the rules in the knowledge base and decides which rules in the knowledge base apply to the data. Two types of inference mechanisms are generally used in expert systems, namely, forward chaining (data driven) and backward chaining (goal driven). In forward chaining, if a rule is found whose antecedent matches the information in the database, the rule fires, that is, the rule's THEN part is asserted as the new fact in the database. In backward chaining on the other hand, the system forms a hypothesis that corresponds to the THEN part of a rule and then attempts to justify it by searching the data base or querying the user to establish the facts appearing in the IF part of the rule or rules. If successful, the hypothesis is established, otherwise another hypothesis is formed and the process is repeated. An important component of any expert system is the explanation mechanism. It uses a trace of rules fired to provide reasons for the decisions reached by the expert system.

The user of an expert system can be a novice user or an expert user depending upon the purpose of the expert system. That is, if the expert system is aimed at training operators in a control center or for education, it is meant for novice users, whereas if it is aimed at improving the quality of decision making in recruitment of salespersons, it is meant for an expert user. User characteristics determines to a large extent the type of explanation mechanism or I/O (Input/Output) interface required for a particular expert system. The user interface shown in Figure 2.3 permits the user to communicate with the system in a more natural way by permitting the use of simple selection menus or the use of a restricted language which is close to a natural language.

Many successful expert systems using the rule based architecture have been built, including MYCIN, a system for diagnosing infectious diseases (Shortliffe 1976), XCON, a system developed for Digital Equipment Corp. for configuring computer systems (Kraft et al. 1984), and numerous others.

2.2.3. *Rule and Frame (Object) Based Architecture*

In contrast to rule base systems certain expert systems consists of both heuristic knowledge and relational knowledge. The relational knowledge describes explicit or implicit relationships among various objects/entities or events. Such knowledge is usually represented by semantic nets, frames or by objects. Figure 2.4 shows an object-oriented network for cars.

The is-a link is known as the inheritance or generalization-specialization link and is-a-*PART-OF* is known as the compositional or whole-part link. These two relational links give lot of expressive power to object-oriented representation. They help to express the hierarchies (classes) and aggregations existing in the domain. That is, VOLKSWAGEN and MAZDA 131 objects are of type SMALL CAR. Such hybrid systems use both deductive and default (inheritance) reasoning strategies for inferencing with knowledge.

Expert systems with hybrid architectures like CENTAUR (Aitkins 1983) - a system for diagnosing lung diseases use both deductive and default reasoning for inferencing with knowledge.

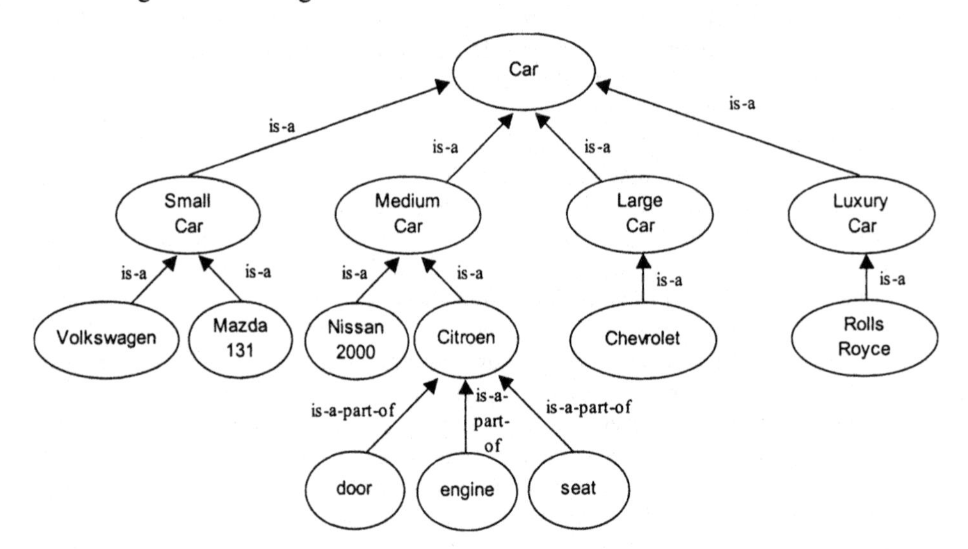

Figure 2.4. An Object-Oriented Network

2.2.4. *Model Based Architecture*

The previous two architectures use heuristic or shallow knowledge in the form of rules. The model based architecture on the other hand, uses an additional deep model (derived from first principles) which gives them an understanding of the complete search space over which the heuristics operate. This makes two kinds of reasoning possible: a) the application of heuristic rules in a style similar to the classical expert systems; b) the generation and examination of a deeper search space following classical search techniques, beyond the range of heuristics.

Various models can be used to describe and reason about a technical system. These include anatomical models, geometric models, functional models and causal models (Steels 1989). Anatomical models focus on the various components and their part-whole relations whereas geometrical models focus on the general layout of the geometrical relations between components. Functional models predict the behavior of a device based on the functioning of its components. On the other hand, a causal model knows about the causal connections between various properties of the components but unlike a functional model does not know how each of the components actually works. Causal models represent an alternative, more abstract, view of a device which is particularly effective for diagnosis in cooperation with a functional model.

First generation expert systems are brittle, in the sense that as soon as situations occur which fall outside the scope of the heuristic rules, they are unable to function at all. In such a situation, second generation systems fall back on search that is not knowledge-driven and therefore potentially very inefficient. However, because these traditional search techniques can theoretically solve a wider class of problems, there is a gradual degradation of performance.

Model based reasoning strategies have been applied on various problems, e.g., XDE (Hamscher 1990), a system for diagnosing devices with heuristic structure and known component failure modes, HS-DAG (Ng 1991), a model based system for diagnosing multiple fault cases in continuous physical devices, DIAGNOSE (Wang and Dillon 1992), a system for diagnosing power system faults, etc.

2.2.5. Blackboard Architecture

Blackboard systems are a class of systems that can include all different representational and reasoning paradigms discussed in this section. They are composed of three functional components, namely, knowledge sources component, blackboard component, and control information component, respectively.

The knowledge sources component represents separate and independent sets of coded knowledge, each of which is a specialist in some limited area needed to solve a given subset of problems. The blackboard component, a globally accessible data structure, contains the current problem state and information needed by the knowledge sources (input data, partial solutions, control data, alternatives, final solutions). The knowledge sources make changes to the blackboard data that incrementally lead to a solution. T he control information component may be contained within the knowledge sources, on the blackboard, or possibly in a separate module. The control knowledge monitors the changes to the blackboard and determines what the immediate focus of attention should be in solving the problem. HEARSAY-II (Erman et al 1980) and HEARSAY III (Balzer 1980; Erman et al 1981) -a speech understanding project at Stanford University is a well known example of blackboard architecture.

2.2.6. Some Limitations of Expert System Architectures

Rule based expert systems suffer from several limitations. Among them is the limitation that they are too hard-wired to process incomplete and incorrect information. For this reason they are sometimes branded as 'sudden death' systems. This limits their application especially in the real-time systems where incomplete information, incorrect information, temporal reasoning, etc. are major system requirements. Further, knowledge acquisition in the form of extraction of rules from a domain expert is known to be a long and tedious process.

Model based systems overcome the major limitations of rule based systems. However, model based systems are slow as may involve exhaustive search. The 12 response time deteriorates further in systems that require temporal reasoning and consist of noisy data. Also, it may not be always possible to build one. Blackboard systems, which combine disparate knowledge sources, try to maximize the benefits of rule based and model-based systems. The major problem, however, with these systems lies in developing an effective communication medium between disparate knowledge sources. Further, given the use multiple knowledge sources it is not easy to keep track of the global state of the system.

Overall, these architectures have difficulty handling complex problems where the numbers of combinatorial possibilities are large and/or where the solution has a non-deterministic nature, and mathematical models do not exist. Artificial neural networks have been successfully used for these types of problems.

2.3. Case Based Reasoning Systems

In some domains (e.g. law), it is either not easy or possible to represent the knowledge using rules or objects. In these domains one may need to go back to records of individual cases that record primary user experience. Case based reasoning is a subset of the field of artificial intelligence that deals with storing past experiences or cases and retrieving relevant ones to aid in solving a current problem. In order to facilitate retrieval of the cases relevant to the current problem a method of indexing these cases must be designed. There are two main components of a case based reasoning system, namely, the case base where the cases are stored and the case based reasoner. The case based reasoner consists of two major parts:

- mechanism for relevant case retrieval
- mechanism for case adaptation.

Thus given a specification of the present case, the case based reasoning system searches through the database and retrieves cases that are closest to the current specification.

The case adapter notes the differences between the specification of the retrieved cases and the specification of the current case, and suggests alternatives to the retrieved cases so that the current situation is best met.

Case based reasoners can be used in open textured domains such as the law or design problems. They reduce the need for intensive knowledge acquisition and try to use past experiences directly.

2.4. Artificial Neural Networks

The research in artificial neural networks has been largely motivated by the studies on the function and operation of the human brain. It has assumed prominence because of the development of parallel computers and, as stated in the previous chapter, the less than satisfactory performance of symbolic AI systems in pattern recognition problems like speech and vision.

The word 'neural' or 'neuron' is derived from the neural system of the brain. The goal of neural computing is that by modeling the major features of the brain and its operation, we can produce computers that can exhibit many of the useful properties of the brain. The useful properties of brain include parallelism, high level of interconnection, self-organization, learning, distributed processing, fault tolerance and graceful degradation. Neural network computational models developed to realize these properties are broadly grouped under two categories, namely, supervised learning and unsupervised learning. In both types of learning a representative training data set of the problem domain is required. In supervised learning the training data set is composed of input and target output patterns. The target output pattern acts like an external "teacher" to the network in order to evaluate its behavior and direct its subsequent modifications. On the other hand in unsupervised learning the training data set is composed solely of input patterns. Hence, during learning no comparison with predetermined target responses can be performed to direct the network for its subsequent modifications. The network learns the underlying features of the input data and reflects them in its output. There are other categories like rote learning which are also used for categorization of neural networks.

Although, numerous neural network models have been developed in these categories, we will limit our discussion to following well known and popularly used ones.

- Perceptron (Supervised)
- Multilayer Perceptron (Supervised)
- Kohonen nets (Unsupervised)
- Radial Basis Function Nets (Unsupervised and Supervised).

Here again, before outlining the neural network architectures, the knowledge representation in neural networks is briefly overviewed.

2.4.1. Perceptron

The Perceptron was the first attempt to model the biological neuron shown in Figure 2.5. It dates back to 1943 and was developed by McCulloch and Pitts. Thus as a starting point, it is useful to understand the basic function of the biological neuron which in fact reflects the underlying mechanism of all neural models.

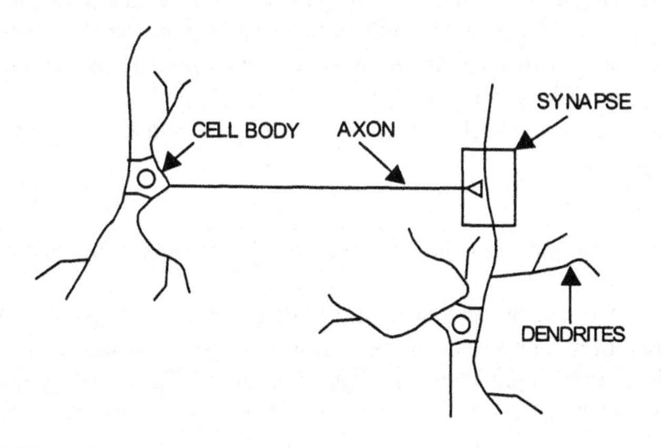

Figure 2.5. A Biological Neuron

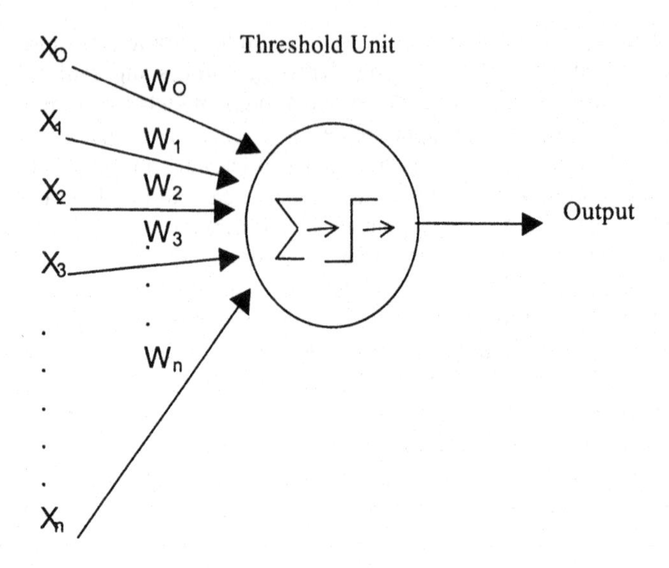

Figure 2.6. The Perceptron

The basic function of a biological neuron is to add up its inputs and produce an output if this sum is greater than some value, known as the threshold value. The inputs to the neuron arrive along the dendrites, which are connected to the outputs from other neurons by specialized junctions called synapses. These junctions alter the effectiveness with which the signal is transmitted. Some synapses are good junctions, and pass a large signal across, whilst others are very poor, and allow very little through. The cell body receives all these inputs, and fires if the total input exceeds the threshold value.

The perceptron shown in Figure 2.6 models the features of the biological neuron as follows:

a. The efficiency of the synapses at coupling the incoming signal into the cell body is modeled by having a weight associated with each input to the neuron. A more efficient synapse, which transmits more of the signal, has a correspondingly larger weight, whilst a weak synapse has a small weight.
b. The input to the neuron is determined by the weighted sum of its inputs

$$\sum_{i=0}^{n} w_i x_i$$

where x_i is the ith input to the neuron and w_i is its corresponding weight.

c. The output of the neuron, which is on (1) or off (0), is represented by a step or heaveside function. The effect of the threshold value is achieved by biasing the neuron with an extra input x0 which is always on (1). The equation describing the output can then be written as

$$y = fh[\sum_{i=0}^{n} w_i x_i]$$

The learning rule in perceptron is a variant on that proposed in 1949 by Donald Hebb, and is therefore called Hebbian learning. It can be summarized as follows:

set the weights and thresholds randomly
present an input
calculate the actual output by taking the thresholded value of the weighted sum of the inputs
alter the weights to reinforce correct decisions and discourage incorrect decisions, i.e. reduce the error.

This is the basic perceptron learning algorithm. Modifications to this algorithm have been proposed by the well-known Widrow and Hoff's (1960) delta rule. They realized that it would be best to change the weights by a lot when the weighted sum is a long way from the desired value, whilst altering them only slightly when the weighted sum is close to that required. They proposed a learning rule known as the Widrow-Hoff delta rule, which calculates the difference between the weighted sum and the required output, and calls that the error. The learning algorithm basically remains the same except for step 4, which is replaced as follows:

4. Adapt weights - Widrow-Hoff delta rule

$$\Delta = d(t) - y(t)$$

$$w_i(t + 1) = w_i(t) + \eta \Delta x_i(t)$$

$$d(t) = \begin{cases} 1, \text{ if input from class A} \\ 0, \text{ if input from class B} \end{cases}$$

where Δ is the error term, $d(t)$ is the desired response and $y(t)$ is the actual response of the system. Also $0 \leq \eta \leq$ is a positive gain function that controls the adaptation rate.

The delta rule uses the difference between the weighted sum and the required output to gradually adapt the weights for achieving the desired output value. This means that during the learning process, the output from the unit is not passed through the step function, although the actual classification is effected by the step function.

The perceptron learning rule or algorithm implemented on a single layer network, guarantees convergence to a solution whenever the problem to be solved is linearly separable. However, for the class of problems, which are not linearly separable, the algorithm does not converge. Minsky and Papert first demonstrated this in 1969 in their influential book, Perceptrons using the well-known XOR example. This in fact dealt a mortal blow to the area, and sent it into hibernation for the next seventeen years till the development of multilayer perceptrons (popularly known as 'backpropagation') by Rumelhart et al. (1986). If the McCulloch-Pitts neuron was the father of modern neural computing, then Rumelhart's multilayer perceptron is its child prodigy.

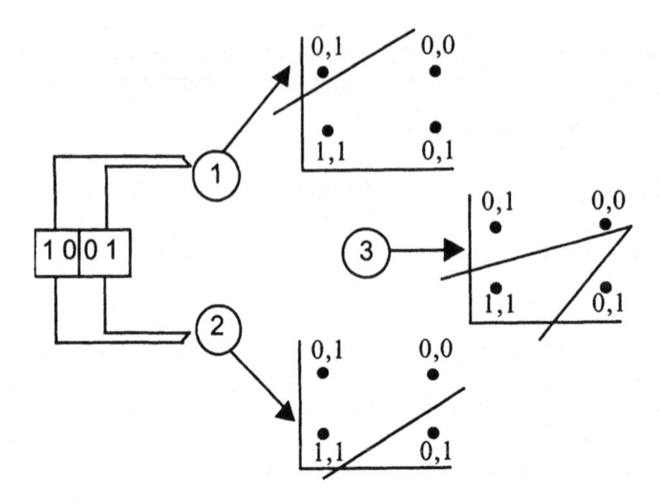

Figure 2.7. Combination of Perceptrons to Solve XOR Problem

2.4.2. Multilayer Perceptrons

An initial approach to solve problems that are not linearly separable problems was to use more than one perceptron, each set up to identify small, linearly separable sections of the outputs into another perceptron, then combining their outputs into another perceptron, which would produce a final indication of the class to which the input belongs. The problem with this approach is that the perceptrons in the second layer do not know which of the real inputs were on or not. They are only aware of the inputs from the first layer. Since learning involves strengthening the connections between active inputs and active units (Hebb 1949), it is impossible to strengthen the correct parts of the network, since the actual inputs are masked by the intermediate (first) layer. Further, the output of neuron being on (1) or off (0), gives no indication of the scale by which the weights need to be adjusted. This is also known as the credit assignment problem.

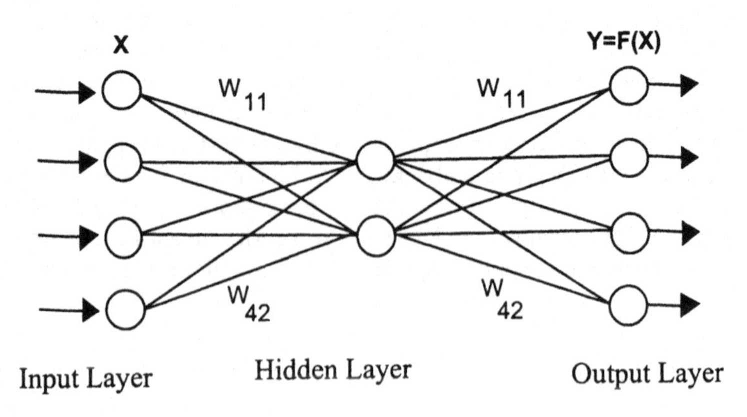

Figure 2.8. Multilayer Perceptron

In order to overcome this problem of linear inseparability and credit assignment, a new learning rule was developed by Rumelhart et al. (1986). They used a three layer multipreceptron model/network shown in Figure 2.8. The network has three layers; an input layer, an output layer, and a layer in between, the so-called hidden layer. Each unit in the hidden layer and the output layer is like a perceptron unit, except that the thresholding function is a non-linear sigmoidal function (shown in Figure 2.9),

$f(net) = 1/(^{1+e\text{-}knet})$

where k is a positive constant that controls the "spread" of the function. Large values of k squash the function until as $k \rightarrow \infty$ $f(net) \rightarrow$ *Heavside function.*

It is a continuously differentiable i.e. smooth everywhere and has a simple derivative. The output from the non-linear threshold sigmoid function is not 1 or 0 but lies in a range, although at the extremes it approximates the step function. The non-linear differentiable sigmoid function enables one to overcome the credit assignment problem by providing enough information about the output to the units in the earlier layers to allow them to adjust the weights in such a way as to enable convergence of the network to a desired solution state.

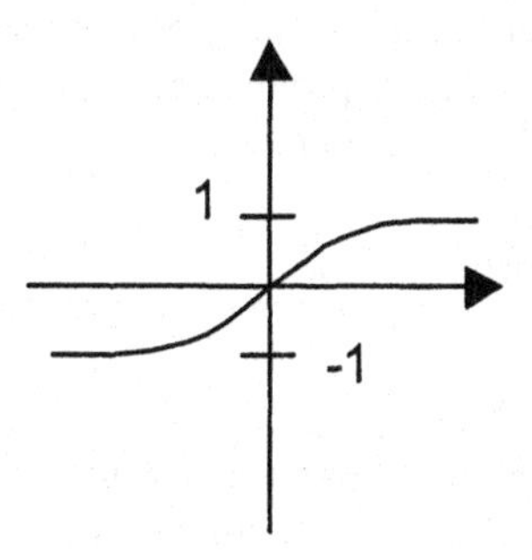

Figure 2.9. Sigmoidal Function

The learning rule, which enables the multilayer perceptron to learn complex non-linear problems, is called the generalized delta rule or the 'backpropagation' rule. In order to learn successfully, the value of the error function has to be continuously reduced for the actual output of the network to approach the desired output. The error surface is analogous to a valley or a deep well, and the bottom of the well corresponds to the point of minimum energy/error. This is achieved by adjusting the weights on the links between units in the direction of steepest downward descent (known as the gradient descent method). The generalized delta rule (McClelland, et al. 1986) does this by calculating the value of the error function for a particular input, and then back-propagating (hence the name) the error from one layer to the previous one. The error term delta for each hidden unit is used to alter the weight linkages in the three layer network to reinforce the correct decisions and discourage incorrect decisions, i.e. reduce the error.

The learning algorithm is as follows:

- Initialize weights and thresholds. Set all the weights and thresholds to small random values.
- Present input and target output patterns. Present input $X_p = x_0; x_1; x_2; ::::; x_n - 1$ and target output $T_p = t_o; t_1; ::::; t_{m-1}$ where n is the number of input nodes and m is the number of output nodes. Set w_0 to be - theta, the bias and x_0 to be always 1. For classification, T_p is set to zero except for one element set to 1 that corresponds to the class that X_p is in.
- Calculate actual output. Each layer calculates

$$y_{pj} = f[\sum_{i=0}^{n-1} w_{ij} x_i] = f(net_{pj}) = 1/(1+e^{-knet_{pj}})$$

  where w_{ij} represents weight from unit i to unit j, net_{pj} is the net input to unit j for pattern p, and y_{pj} is the sigmoidal output or activation corresponding to pattern p at unit j and is passed as input to the next layer (i.e. the output layer). The output layer outputs values o_{pj}, which is the actual output at unit j of pattern p.
- Calculate the error function E_p for all the patterns to be learnt

$$E_p = 1{=}2[\sum_j (t_{pj} - o_{pj})^2]$$

  where t_{pj} is the target output at unit j of pattern p.
- Starting from the output layer, project the error backwards to each layer and compute the error term δ for each unit.

  For output units

$$\delta_{pj} = ko_{pj}(1 - o_{pj})(t_{pj} - o_{pj})$$

  where δ_{pj} is an error term for pattern p on output unit j

  For hidden units

$$\delta_{pj} = ko_{pj}(1 - o_{pj} \sum r \, \delta_{pr} w_{jr}$$

  where the sum is over the r units above unit j, and δ_{pj} is an error term for pattern p on unit j which is not an output unit.
- Adapt weights for output hidden units

$$w_{ij}(t+1) = w_{ij}(t) + \eta \, \delta_{pj} o_{pj}$$

  where w_{ij} represents the weights from unit i to unit j at time t and j is a gain term or the learning rate.

2.4.3. Radial Basis Function Net

The Radial Basis Function (RBF) net is a 3-layer feed-forward network consisting of an input layer, hidden layer and output layer as shown in Figure 2.10. The mapping of the input vectors to the hidden vectors is non-linear, whereas the mapping from hidden layer to output layer is linear. There are no weights associated with the connections from input layer to hidden layer. In the radial basis function net, the N activation functions g_j of the hidden units correspond to a set of radial basis functions that span the input space and map. Each function $g_j(\|x - c_j)\|)$ is centered about some point c_j of the input space and transforms the input vector x according to its Euclidean distance, denoted by $\| \|$, from the center c_j. Therefore the function g_j has its maximum value at c_j. It has further an associated receptive field that decreases with the distance between input and center and which could overlap that of the functions of the neighboring neurons of the hidden layer.

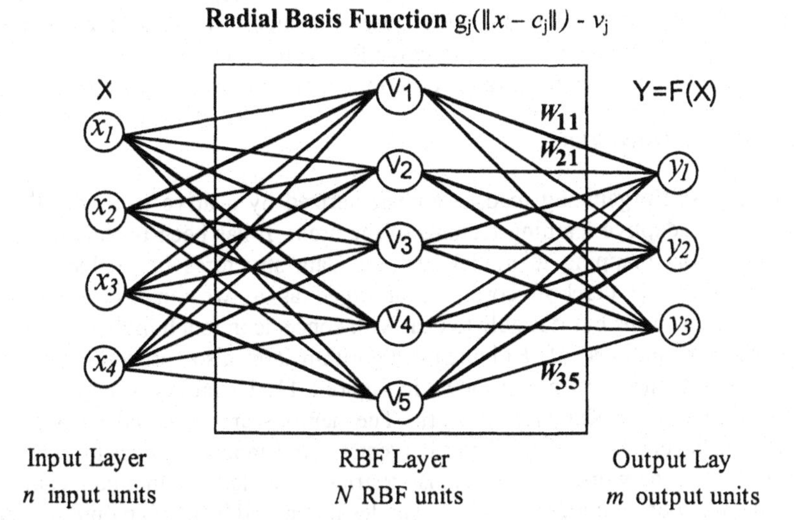

Figure 2.10. Radial Basis Function Net

The hidden units are fully connected to each output unit y_i with weights w_{ij}. The outputs y_i are thus linear combinations of the radial basis functions i.e.

$$y_i = \sum_{j=1}^{N} w_{ij} g_j(\|x - c_j\|)$$

One such set of radial basis functions that is frequently used are Gaussian Activation functions centered on the mean value c_j and with a receptive field whose size is proportional to the variances fixed for all units:

$$g_j(\|x - c_j\|) = exp(\|x - c_j\|^2/4\sigma^2)$$

where:

$\sigma = d = \sqrt{(2N)}$ with N number of (hidden) RBF units and d the maximum distance between the chosen centers. One could use an unsupervised learning approach to determine the centers c_j and the width σ of the receptive fields, see Haykin (1994). One could then use the delta learning rule to determine the weights between the hidden units and the output units. Since the first layer can be said to be trained using an unsupervised approach and the second using a supervised approach one could consider such a net as a hybrid net. The RBF network is an important approximation tool because like spline functions it provides a quantifiable optimal solution to a multi-dimensional function approximation problem under certain regularization constraints concerning the smoothness of the class of approximating RBF functions.

The applications of multilayer perceptrons can be found in many areas including natural language processing (NETalk- Sejnowski and Rosenberg 1987), prediction (airlines seat booking, stock market predictions, bond rating, etc.) and fault diagnosis.

The use of neural networks is dependent upon availability of large amounts of data. In some cases, both input and output patterns are available or known, and we can use supervised learning techniques like backpropagation, whereas in other cases the output patterns are not known and the network has to independently learn the class structure of the input data. In such cases, unsupervised learning techniques characterized by Kohonen nets, and Adaptive Resonance Theory are used. The more commonly used Kohonen networks are described in the following section.

2.4.4. Kohonen Networks

Kohonen's self-organizing maps are characterized by a drive to model the self-organizing and adaptive learning features of the brain. It has been postulated that the brain uses spatial mapping to model complex data structures internally (Kohonen 1990). Much of the cerebral cortex is arranged as a two-dimensional plane of interconnected neurons but it is able to deal with concepts in much higher dimensions.

The implementations of Kohonen's algorithm are also predominantly two-dimensional. A typical network is shown in Figure 2.11. The network shown is a one-layer two-dimensional Kohonen network. The neurons are arranged on a flat grid rather than in layers as in a multilayer perceptron. All inputs connect to every node (neuron) in the network. Feedback is restricted to lateral interconnections to immediate neighboring nodes. Each of the nodes in the grid is itself an output node.

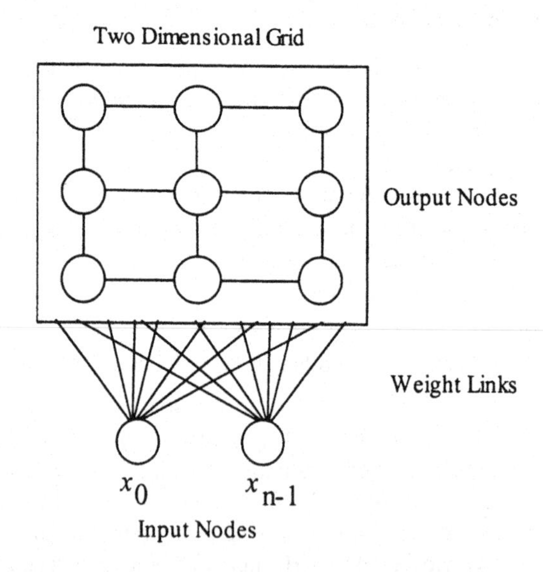

Figure 2.11. Kohonen's Self Organizing Feature Map

The learning algorithm organizes the nodes in the grid into local neighborhoods or clusters (shown in Figure 2.12) that act as feature classifiers on the input data. The biological justification for that is the cells physically close to the active cell have the strongest links. No training response is specified for any training input. In short, the learning involves finding the closest matching node to a training input and increasing the similarity of this node, and those in the neighboring proximity, to the input. The advantage of developing neighborhoods is that vectors that are close spatially to the training values will still be classified correctly even though the network has not seen them before, thus providing for generalization.

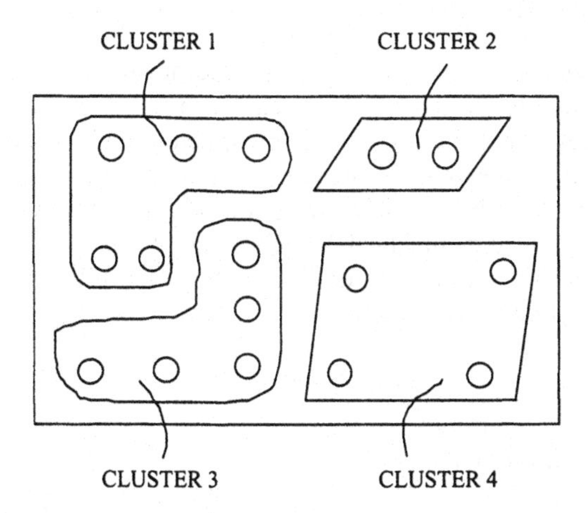

Figure 2.12. Kohonen Network Clusters

The learning algorithm is notionally simpler than the backpropagation algorithm as it does not involve any derivatives.

1. Initialize network

Define $w_{ij}(t)$ $(0 \leq i \leq n - 1)$ to be the weight from input i to node(unit) j. Initialize weights from the n inputs to the nodes to small random values. Set the initial radius of the neighborhood around *node j*, $Nj(0)$, to be large.

2. Present input

Present input $x_0(t); x_1(t); x_2(t); ::::::; x_{n-1}(t)$, where $x_i(t)$ is the input to node i at time t.

3. Calculate distances

Compute the distance dj between the input and each output node j, given by

$d_j = \Sigma_{I=}0(x_i(t) - w_{ij}(t))^2$

4. Select minimum distance using the Euclidean distance measure

Designate the output node with minimum d_j to be j^* .

5. Update weights

Update weights for node j^* and its neighbors, defined by the neighborhood size $N_j{}^*(t)$. New Weights are

$w_{ij}(t+1) = w_{ij}(t) + \eta(t)(x_i(t) - w_{ij}(t))$

For j in $N_{j^*}(t)$,hspace1cm $0 \leq i \leq n - 1$

The term $\eta(t)$ is a gain term $(0 < \eta(t) < 1)$ that decreases in time, so slowing the weight adaptation. The neighborhood $N_{j^*}(t)$ decreases in size as time goes on, thus localizing the area of maximum activity.

6. Repeat steps 2 to 5.

The most well known application of self-organizing Kohonen networks is the Phonetic typewriter (Kohonen 1990) used for classification of phonemes in real time. Other applications include evaluating the dynamic security of power systems (Neibur et al. 1991).

2.5. Fuzzy Systems

Fuzzy systems provide a means of dealing with inexactness, imprecision as well as ambiguity in everyday life. In fuzzy systems relationships between imprecise concepts like "hot," "big," and "fat," "apply strong force," and "high angular velocity" are evaluated instead of mathematical equations. Although not the ultimate problem solver, fuzzy systems have been found useful in handling control or decision making problems not easily defined by practical mathematical models.

This section provides introduction to the reader on the following basic concepts used for construction of a fuzzy system:

- Fuzzy Sets
- Fuzzification of inputs
- Fuzzy Inferencing and Rule Evaluation
- Defuzzification of Fuzzy Outputs

2.5.1. Fuzzy Sets

Although fuzzy systems have become popular in the last decade, fuzzy set and fuzzy logic theories have been developed for more than 25 years. In 1965 Zadeh wrote the original paper formulating fuzzy set and fuzzy logic theory. The need for fuzzy sets has emerged from the problems in the classical set theory. In classical set theory, one can specify a set either by enumerating its elements or by specifying a function f(x) such that if it is true for x, then x is an element of the set S. The latter specification is based on two-valued logic. If f(x) is true, x is an element of the set, otherwise it is not. An example of such a set, S, is:

$S = x : weightofpersonx > 90kg$

Thus all persons with weight greater than 90 kg would belong to the set S. Such sets are referred to as crisp sets.

Let us consider the set of "fat" persons. It is clear that it is more difficult to define a function such that if it is true, then the person belongs to the set of fat people, otherwise s/he does not. The transition between a fat and not-fat person is more gradual. The membership function describing the relationship between weight and being fat is characterized by a function of the form given in Figure 2.13.

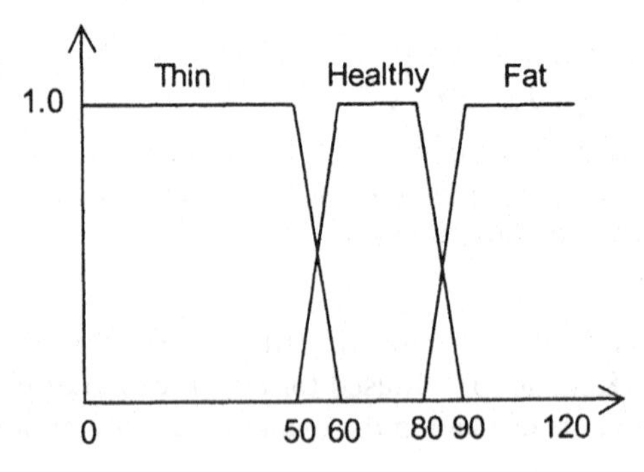

Figure 2.13. Fuzzy Membership Function

Such sets where the membership along the boundary is not clear-cut but progressively alters are called fuzzy sets. The membership function defines the degree of membership of the set x : x fat. Note this function varies from 0 (not a member) to 1 (definitely a member).

From the above, one can see that the truth value of a statement, *person X is fat*, varies from 0 to 1. Thus in fuzzy logic, the truth value can take any value in this range, noting that a value of 0 indicates that the statement is false and a value of 1 indicates that it is totally true. A value less than 1 but greater than 0 indicates that the statement is partially true. This contrasts with the situation in two-valued logic where the truth value can only be 0 (false) or 1 (true).

2.5.2. Fuzzification of Inputs

Each fuzzy system is associated with a set of inputs that can be described by linguistic terms like "high," "medium" and "small" called fuzzy sets. Fuzzification is the process of determining a value to represent an input's degree of membership in each of its fuzzy sets. The two steps involved in determination of fuzzy value are:

- membership functions
- computation of fuzzy value from the membership function

Membership functions are generally determined by the system designer or domain expert based on their intuition or experience. The process of defining the membership function primarily involves:

- **defining the Universe of Discourse (UoD):** UoD covers the entire range of input values of an input variable. For example, the UoD for an input variable Person Weight covers the weight range of 0 to 120 Kilograms
- **partitioning the UoD into different fuzzy sets:** A person's weight can be partitioned into three fuzzy sets and three ranges, i.e. 0-60, 50-90 and 80-120 respectively
- **labeling fuzzy sets with linguistic terms:** The three fuzzy sets 0-60, 50-90, and 80-120 can be linguistically labeled as "thin," "healthy" and "fat" respectively
- **allocating shape to each fuzzy set membership function:** Several different shapes are used to represent a fuzzy set. These include piecewise linear, triangle, bell shaped, trapezoidal (see Figure 2.14), and others. The shape is said to represent the fuzzy membership function of the fuzzy set.

Once the fuzzy membership function has been determined, the next step is to use it for computing the fuzzy value of a system input variable value. Figure 2.14 shows how the degree of membership or fuzzy value of a given system input variable X with value Z can be computed.

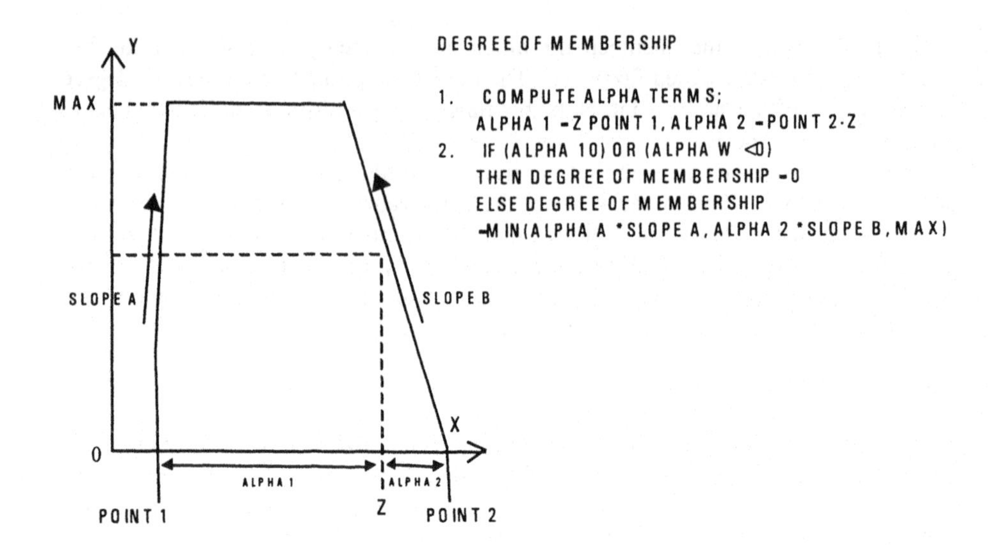

Figure 2.14. Fuzzification of Inputs

2.5.3. *Fuzzy Inferencing and Rule Evaluation*

In order to express relationships between imprecise concepts and model the system's behavior, a fuzzy system designer develops a set of fuzzy IF-THEN rules in consultation with the domain expert. The IF part of a fuzzy rule is known as the antecedent and the THEN part is known as the consequent. The antecedent or antecedents of a fuzzy rule contain the degrees of membership (fuzzy inputs) calculated during the fuzzification of inputs process. For example consider the Fuzzy Rule 1:

IF share_price_is_decreasing AND trading_volume_is heavy THEN market_order_ is_sell

Here, the two antecedents share_price_is_decreasing and trading_volume_is_heavy are the rule's fuzzy antecedents. Further share price and trading volume are fuzzy inputs with fuzzy sets "decreasing," "stable" and "increasing," and "light," "moderate" and "heavy" respectively.

The consequent of the fuzzy rule is represented by the THEN part which in this case is market_order_is_sell. Generally, more than one fuzzy rule has the same fuzzy output. For example, Fuzzy Rule 2 can be :

IF share_price_ is_decreasing AND trading_volume_is_moderate THEN market_order_is_sell.

In order to evaluate a fuzzy rule, rule strengths are computed based on the antecedent values and then assigned to a rule's fuzzy outputs. The antecedent values are computed based on the degree of membership of an input variable. For example, the fuzzy value of the antecedent share price is decreasing will correspond to the degree of membership of the "decreasing" fuzzy set of share price input variable. The most commonly used fuzzy inferencing method is the maxmin method. In this inferencing method minimum operation is applied making the rule strength equal to the least-true or weakest antecedent value.

For example, say, for share price of $50, and trading volume of 1000 contracts, the degree of membership values for fuzzy sets "decreasing," "heavy" and "moderate" are 0.7, 0.2, and 0.4 respectively. Rule strength of Fuzzy Rule 1 and 2 can be computed as follows:

Rule (firing) Strength of Fuzzy Rule 1 = min(0:7; 0:2) = 0:2
Rule (firing) Strength of Fuzzy Rule 2 = min(0:7; 0:4) = 0:4

The fuzzy output of selling shares is carried out to a degree reflected by the rule's strength. Since two rules apply to the same fuzzy output, the strongest rule strength is used in this case. This is done by applying the max operation as follows:

Market_order_is_sell = max(min(0:7; 0:2*); min*(0:7; 0:4)) = 0:4

2.5.4. Defuzzification of Outputs

Defuzzification is required to determine crisp values for fuzzy outputs or actions like *market_order_is_sell*. Another reason for defuzzification is to resolve conflict among competing outputs or action. For example, competing output market order is hold can also be triggered based with a rule strength of say, 0.6 on the share price of $50, and trading volume of 1000 contracts with a Fuzzy Rule 3:

IF share_price_is_stable AND trading_volume_is_small THEN market_order_is_hold

In order to resolve the conflict between competing outputs one of the common defuzzification techniques used is the center-of-gravity method.

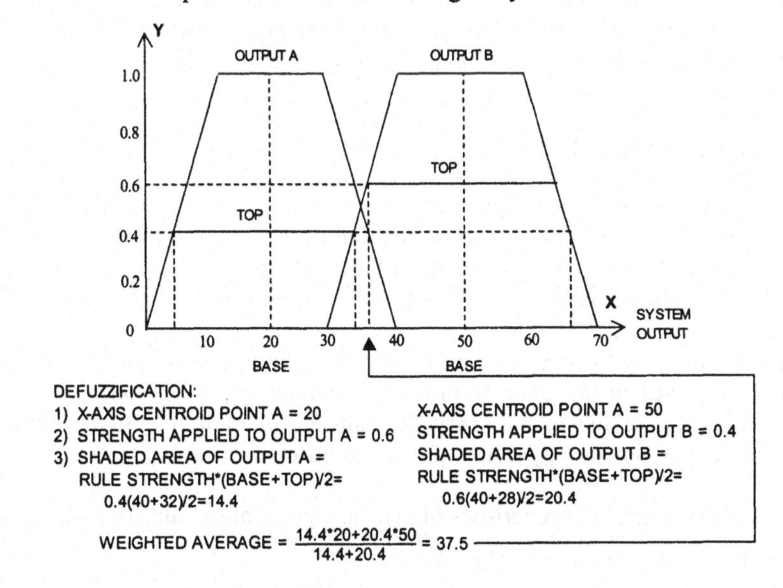

Figure 2.15. Defuzzification of Outputs

For a general case, the following steps as shown in Figure 2.15 are taken for defuzzification:

- determine the centroid point on the X-axis for each competing output membership function. In Figure 2.15 it is 20 and 50 for the competing outputs A and B respectively
- compute new output membership trapezoidal areas based on rule strengths of 0.6 and 0.4 for competing outputs A and B respectively as shown in Figure 2.15
- compute the defuzzified output by taking a weighted average of X-axis centroid points and the output membership areas (for rule strengths 0.6 and 0.4 respectively).

Fuzzy systems have been applied in a number of industrial control applications under the following conditions:

- When the system is highly nonlinear and mathematical models do not exist or are difficult to build.
- When parameters of the system change.
- Where sensor accuracy is a problem.
- Where the system may receive conflicting or uncertain input but must still make correct control decisions.

The satisfaction of these conditions provides number of advantages to systems modeled on fuzzy logic. These include increased efficiency of control cycles which may result in reduced power consumption or higher stability, requirement of less expensive and less accurate sensors because of the ability to model imprecision in sensor data, and ability to extract knowledge from an expert using everyday language. In spite of a number of advantages, fuzzy systems also suffer from some disadvantages. These include problems with learning of control process, tuning/optimization of membership functions, and optimization of rules. Genetic algorithms that are good at optimization problems are described in the next section.

2.6. Genetic Algorithms

Genetic algorithms refer to a class of adaptive search procedures based on principles derived from the dynamics of natural population genetics (Grefenstette 1990). These search procedures exploit the mechanics of natural selection and natural genetics. Survival of the fittest is thus combined with mechanisms for generation of new candidates to form search algorithms.

John Holland (1975) first developed genetic algorithms. His aim was to abstract and rigorously explain the adaptive process of natural systems. From this, software systems that capture the evolutionary mechanisms of natural systems have been developed. These genetic algorithms provide robust search procedures for complex spaces.

Some of the general characteristics of genetic algorithms include (Goldberg 1989):

- working with a coding of the parameter set
- searching from a population of points, rather than a single point
- using a pay-off function
- using probabilistic transition rules.

We will now discuss these characteristics. First, the parameter set used in a problem is coded as a finite length string. To explain this, consider the following:

- Problem : determine the best days for making interstate phone calls.
- In this case, the parameters are the seven days of the week. Therefore, to code the parameter set as a finite length string, a string of length 7 with each element representing a different day of the week is chosen.

In general, the parameter set is coded into:

$A = a_1\ a_2{:}{:}{:}a_n;$

where $a_1\ a_2{:}{:}{:}a_n$ represents the individual parameters in the set of size n. The individual parameters are binary, with values of either 1 or 0. This encoding is required by the genetic algorithm for several reasons, including the flexibility to apply the three operators of reproduction, crossover, and mutation on the strings, as discussed below. New strings are created by applying these three string operators. Genetic algorithms actually work by processing and manipulating a group of coded strings. As these coded strings are used in determining the solution, genetic algorithms are regarded as working with a group of points simultaneously. To search effectively through the strings, some form of pay-off value is required for each string. These pay-off values are objective function values used to guide the search through the strings (Goldberg 1989).

The coding of the parameter set is usually over an extended alphabet and not just over 0 and 1. This extended alphabet includes a hash symbol #, representing a don't care or wild card symbol (Booker et al. 1990; Goldberg 1989). Thus, the # matches either a 0 or 1, as illustrated below:

Suppose the following strings are in the population being considered:

0110011 1111111
0010010 0111010

- The string ##1##1#, represented in the extended alphabet, matches each string in the population above. To match this string, a string must have a 1 in both positions 3 and 6; the other elements are not relevant.

The flexibility of this notation is apparent, as large groups of patterns can be stated in a concise form. Thus, the notation greatly simplifies the analysis of genetic algorithms because it explicitly recognizes all the possible similarities in a population of strings.

There are three important operators commonly used in genetic algorithms; reproduction, crossover, and mutation. The first operator, reproduction, is a process in which individual strings are copied according to their objective function values (Goldberg 1989). These objective function values represent a measure of goodness that needs to be maximized. Copying strings with higher fitness values gives those strings a higher probability of contributing one or more of their offspring to the next generation. In the natural world, this is equivalent to natural selection, resulting in the survival of the fittest. Crossover occurs when members of the newly reproduced strings are mated at random. This is the second operator used by genetic algorithms. Goldberg (1989) argues that much of the power of genetic algorithms derives from the combination of reproduction and crossover which provides structured, though randomized, information exchange. Mutation, the third operator, usually plays a less significant role.

To illustrate these operators, let string 1 be 11011 with an objective function value of 0.25, and string 2 be 11101 with an objective function value of 0.50.

- Applying the reproduction operator to the two strings results in string 2 being reproduced because it has a higher objective function value.
- Applying the crossover operator, assuming the crossover point in the string is the third position, results in the following:

original strings:	11011	gives new strings:	11111
and	11101		11001

- Applying the mutation operator to the two strings may result in the following (mutation being the random inverting of a bit in the string):

original strings:	11011	gives new strings:	11010
and	11101		11001

Comprehension of a genetic algorithm's manipulation of building blocks is crucial (Booker et al. 1990). Building blocks are high performance schemata that are combined to form strings with expected higher performance. The schemata or similarity templates are strings over the extended alphabet 0,1,# (Booker et al. 1990; Goldberg 1989). According to Booker et al. (1990):

A genetic algorithm rapidly explores the space of schemas, a very large space, implicitly rating and exploiting the schemas according to the strengths of the rules employing them.

Two useful schemata properties are (Goldberg 1989):

- order - the number of fixed positions present in each schema
- length - the defining length of each schema; that is, the distance between the first and last specific string positions.

The generating procedure samples the schemata with above average instances more regularly. This results in further confirmation of the useful schemata and degradation of the less useful. The overall average strength is thus increased. This leads to an ever-increasing criterion that a schema must meet to be above average. Also the generating procedure uses the distribution of instances rather than the most recent, best schemata. This helps the system to be more robust and to overcome local minima which misdirect development. The better than average building blocks are rapidly accumulated by the generating procedure. The regularities and interactions in the environment (Booker et al. 1990) determine the strengths.

2.7. Intelligent Fusion, Transformation and Combination

Intelligent technologies such as Expert Systems, Neural Networks and Generic Algorithms have also been used in various hybrid configurations. These hybrid configurations have been grouped into four classes, namely, fusion, transformation, combination and associative systems. The structure of these configurations is described in chapter 3. These classes of systems, and their knowledge and task modeling issues stages, have been described in Khosla and Dillon (1997).

2.8. Object-Oriented Software Engineering

The recent research in object-oriented software engineering and databases has shown objects provide a powerful and comprehensive mechanism for capturing the relationships between concepts not only in terms of their structure but also their behavior (Cox 1986; Booch 1986; Kim et al. 1988; Myer 1988; Unland et al. 1989; Coad et al. 1990, 92; Rumbaugh 1990; Dillon and Tan 1993, and others). Computationally, they offer more powerful encapsulation than other knowledge representation formalisms like frames (Dillon and Tan 1993). They also have special features like message passing and polymorphism which make them attractive computationally than other symbolic knowledge representation formalisms semantic networks and frames.

As a result of research in these two communities and in artificial intelligence, a common set of characteristics which define the general object-oriented model unique object identifier, data and behavior (i.e. operation/method/procedure) abstraction or encapsulation, inheritance, composition, message passing and polymorphism.

Object-oriented methodology has been used in this book in the context of its knowledge modeling features like inheritance and composability, and software implementation features like encapsulation, message passing and polymorphism. These features are now briefly described in the following subsections.

2.8.1. Inheritance and Composability

The structured representation of real world concepts includes various kinds of relationships. Inheritance and composition are two constructs which have given expressive power to knowledge representation formalisms for relating real world concepts in a meaningful fashion. These two relational constructs are also important features of the object-oriented methodology.

Inheritance allows real-world objects to be organized into classes so that objects of the same class can have similar properties, and more specific classes or subclasses may inherit properties of the more general classes. In fact inheritance is also used as a mechanism for default reasoning in computer applications where knowledge is organized hierarchically. Put another way inheritance makes extensive use of abstraction. The classes at the higher levels in the hierarchy are expressed as generalizations of the lower level classes (which are their specializations). The relationship is expressed through an *IS-A* link. The two basic modes in which inheritance can be realized are single and multiple inheritance. In single inheritance, there is a single *IS-A* or *INSTANCE-OF* link from a class or instance at a lower level to class at a higher level, whereas in multiple inheritance there can be multiple *IS-A* links from a class at a lower hierarchical level and classes at higher level/s. Aggregation or composition is another common form of relationship observed in hierarchical structures. It represents the whole-part concept used in our everyday life

(Britannica 1986) and is incorporated in the object-oriented model. A composite object consists of a collection of two or more heterogeneous, related objects referred to as component objects. The component objects have a *PART-OF* relationship to the composite object. Each component object may, in turn, be a composite object, thus resulting in a 'component-of' hierarchy. There are other forms of non-hierarchical relationships like *ASSOCIATION* (Dillon and Tan 1993) and others obtained from entity-relationship models (Coad and Yourdon 1990; Hawryszkiewyz 1991) which are also modeled as extensions to the object-oriented methodology.

2.8.2. Encapsulation

Encapsulation is a property of object-oriented models by which all the information (i.e. data and behavior) of a object is captured under one name, that is the object name. For example a real world object like *Chair* encapsulates attribute values that define the *Chair*, methods that are applied to change the attributes of *Chair*, and other related information. This notion of encapsulating information related to a particular concept does not distinguish between the type of attributes or methods used to define that concept. In other words, it is a useful software implementation methodology for realizing heterogeneous architectures involving more than one intelligent methodology.

2.8.3. Message Passing

When integrating two fundamentally different paradigms it becomes important to ascertain the communication mechanism between the two from a computational point of view. The object-oriented model provides a uniform communication mechanism between objects. In order to communicate, objects pass messages. A message defines the interface between the object and its environment. Essentially, a message consists of the name of the receiving object, a message name or method selector and arguments of the selected method.

2.8.4. Polymorphism

In large-scale domains, genericity is an important element to promote comprehensibility and intelligibility of the domain. Polymorphism (Pressman 1992; Dillon and Tan 1993) is another feature of the object-oriented models that brings about the genericity in terms of the behavior of different objects or concepts in the domain. It is one of the key features of object-oriented programming (Blair et. al. 1989; Pressman 1992; Dillon and Tan 1993). It allows object-oriented systems to separate a generic function from its implementation. These generic functions or virtual functions (as they are called sometimes) provide the ability to carry out function overloading (Berry 1988; Dillon and Tan 1993).

2.9. Agents and Agent Architectures

Intelligent agents and multi-agent systems are one of the most important emerging technologies in computer science today. A dictionary definition of the term "agent" is:

An entity authorized to act on another's behalf. The definitional term "another's behalf" in this book refers to a problem solver or a user. The definitional term "entity" refers to the agent which maps percepts (e.g. inputs) to actions for achieving a set of tasks or goals assigned to it in a largely nondeterministic environment. The four ingredients Percept, Action, Goal and Environment (PAGE) of an agent have been derived from Russell and Norvig (1995). Table 2.1 provides a PAGE description of systems modeled as agent types. In fact Russell and Norvig (1995) define an agent as consisting of an architecture and a program *(agent = architecture + program).* As a software program it maps percepts to actions and in the process exhibits the following characteristics:

- **Autonomy:** An agent should be able to exercise a degree of autonomy in its operations. It should be able to take initiative and exercise a non-trivial degree of control over its own actions.
- **Collaboration:** An agent should have the ability to collaborate and exchange information with other agents in the environment to assist other agents in improving their quality of decision making as well as its own.
- **Flexibility and Versatility:** An agent should be able to dynamically choose which actions to invoke, and in what sequence, in response to the state of its external environment. Besides, an agent should have a suite of problem solving methods from which it can formulate its actions and action sequences. This facility provides versatility as well as more flexibility to respond to new situations and new contexts.
- **Temporal History:** An agent should be able to keep a record of its beliefs and internal state and other information about the state of its continuously changing environment. The record of its internal state helps it to achieve its goals as well as revise its previous decisions in light of new data from the environment.
- **Adaptation and Learning:** An agent should have the capability to adapt to new situations in its environment. This includes the capability to learn from new situations and not repeat its mistakes.
- **Knowledge Representation:** In order to support its actions and goals with an agent should have the capabilities and constructs to properly model structural and relational of the problem domain and its environment.
- **Communication:** An agent should be able to engage in complex communication with other agents, including human agents, in order to obtain information or request for their help in accomplishing its goals.
- **Distributed and Continuous Operation:** An agent should be capable of distributed and continuous operation (even without human intervention) in one machine as well as across different machine for accomplishing its goals.

An agent program with above characteristics can be a single agent system or a multi-agent system. Multi-agent systems are concerned with coordinating problem solving behavior amongst a collection of agents. Each agent in a multi-agent system represents a specific set of problem solving skills and experience. The intention is to coordinate the skills, knowledge, plans and experience of different agents to pursue a common high-level system goal.

Table 2.1: PAGE Description of Agent Types

Agent Type	Percepts	Actions	Goals	Environment
Fruit Storage Control System	Temperature, Humidity Reading	Control Fruit Weight Loss	Retain Fruit Freshness	Controlled Storage
		Control Fruit Disease		
Oil Dewaxing System	Oil Type, Oil Inflow Rate, Tank Oil Level	Adjust Oil Inflow Rate Valve, Adjust Oil Outflow Rate Valve	High Quality Dewaxed Lubricant Oil	Petroleum Plant
Inventory control System	Sales Forecast Existing Stock	Stockpile, Liquidate, Replenish	Minimize Storage Cost	Inventory and Sales Databases, User

An agent program describes the behavior of an agent in the sense that for a given set of percepts or inputs a particular action is performed. A number of agent programs can be found to assist an user in e-mail filtering, on line news management, and in various other manufacturing and business areas (Maes et al. 1994; Dinh 1995; Lee 1996). Agent applications can also be found in the areas of air-traffic control, network resource allocation, and user-interface design. On the other hand, an agent architecture outlines how the job of generating actions from percepts to actions is organized (Russell and Norvig 1995). Maes (1994) has provided a more elaborate definition. Maes defines an agent architecture as a particular methodology for building agents. It specifies how the agent can be decomposed into the construction of a set of component modules and how these modules should be made to interact. The total set of modules and their interactions has to provide an answer to the question of how the sensor data and the current internal state of the agent determine its actions and future internal state of the agent. Architecture encompasses techniques and algorithms that support this methodology.

2.10. Multimedia

Media can exist in various forms, namely, text, video, sound and music. The term multimedia is typically applied to use of some sort of interaction across media (or carriers) and concerns are focussed on integrating carriers (image, text, video and audio). The media characteristics shown in Table 2.2 are used for mapping media to the data characteristics and information content to be communicated to the user. The temporal dimension defines the permanent (Perm in Table 2.2) or static and transient (Tans in Table 2.2) or dynamic nature of the media. The granularity is indicative of the continuous or discrete form of the media. On the other hand, baggage characteristics reflect the level or interpretation associated with the media. These and other characteristics (described in chapter 4) are used for designing multimedia as a means for interpreting the computer-based artifact. We look at these aspects in more detail in chapter 5.

Table 2.2: Media Characteristics

Medium	Carrier Dimension	Temporal Dimension	Granularity	Medium Type	Default Delectability	Baggage
Map	2D	Perm	Continuous	Visual	Low	High
Picture	2D	Perm	Continuous	Visual	Low	High
Table	2D	Perm	Discrete	Visual	Low	High
Form	2D	Perm	Discrete	Visual	Low	High
Graph	2D	Perm	Continuous	Visual	Low	High
Ordered List	1D	Perm	Discrete	Visual	Low	Low
Sliding Scale	1D	Perm	Continuous	Visual	Low	Low
Written Sentence	1D	Perm	Continuous	Visual	Low	Low
Spoken Sentence	1D	Perm	Continuous	Aural	Mhigh	Low
Animation	2D	Trans	Continuous	Visual	High	High
Music	1D	Trans	Continuous	Aural	Mhigh	Low

Further, modeling data using media artifacts involves a number of terms. Some of the terms that are used in this book are outlined here:

Consumer: a person interpreting a communication.

Medium: a single mechanism by which to express information, e.g. spoken and written natural language, diagrams, sketches, graphs, tables, pictures.

Exhibit: a complex exhibit is a collection or composition of several simple exhibits. A simple exhibit is that which is produced by one invocation of one medium, e.g. a diagram, computer beep.

Substrate: is a background to a simple exhibit. It establishes to the consumer the physical or temporal relationship and the semantic context within which new information is presented to the information consumer. For example, a piece of paper or screen (on which information may be drawn or presented) or a grid (on which a marker might indicate the position of an entity).

Information Carrier: is that part of the simple exhibit which, to the consumer, communicates the principal piece of information requested or relevant in the current communicative context, e.g. a marker on a map substrate, prep phrase within sentence predicate substrate.

Channel: the total number of channels gives the total number of independent pieces (dimensions) of information the carrier can convey, e.g., a single mark or icon (say, ship icon) can convey information by its shape, color, position and orientation in relation to a background map.

2.11. The eXtensible Markup Language

The eXtensible Markup Language (XML) is a document mark-up *metalanguage* originally designed to enable the use of the *Standard Generalized Mark-up Language* (SGML) on the World Wide Web and later standardized by the World Wide Web Consortium (World Wide Web Consortium 1998). Today, a huge amount of information is made available in XML format, both on corporate Intranets and on the global Net. In this section, we shall give an outline of the main features of XML; the interested reader may refer to (Pardi 1999) for a comprehensive description of the language and its applications.

While HTML is defined by means of SGML (*Standard Generalized Mark-up Language*: ISO 8879), XML is a sophisticated subset of SGML, and is designed to describe the data using arbitrary tags. One of the goals of XML is to be suitable for the use on the Web; thus to provide a general mechanism for extending HTML. As its name implies, the *extensibility* is a key feature of XML; users or applications are free to declare and use their own tags and attributes. Therefore, XML mark-up ensures that both the logical structure and content of semantics-rich information is retained.

Other approaches, especially from academia, suggest that first order logic or special purpose formal languages such as KQML (Finin et al. 1994) would allow for more precise specification of content. While XML is currently becoming so widespread that it could be chosen based on this criterion alone, it should be noted that expressing semantics in syntax rather than in first-order logic leads to a simpler evaluation function while needing no agreement on an associated ontology. XML is widely accepted in the Web community now, and current applications of XML include *MathML* (Mathematical Mark-up Language) to describe mathematical notation, *CDF* (Channel Data Format) for push technologies, *OFX* (Open Financial Exchange) to describe financial transactions and *OSD* (Open Software Distribution) for the software distribution on the Net.

XML focuses on the description of information structure and content as distinct from its presentation. The data structure and its syntax are defined in a *Document Type Definition* (DTD) specification, which is a derivative from SGML and defines a series of tags and their constraints. In contrast to information structure, the presentation issues are addressed by *XSL* (*XML Style Language*) (Adler 1998), which is also a W3C standard for expressing how XML-based data should be rendered. XSL is based on *DSSSL* (*Document Style Semantics and Specification Language* ISO/IEC 10179) and interoperable with CSS (Cascading Style Sheet), which was originally a style definition language specific to HTML.

In addition to XML and XSL, XLL (*XML Linking Language*) (World Wide Web Consortium 1998) is a specification to define anchors and links within XML documents. Moreover, the *Extensible Forms Description Language* (XFDL) (Blair and Boyer 1999), developed by Tim Bray (Bray et al 1998), is an application of XML that allows organizations to move their paper-based forms systems to the internet while maintaining the necessary attributes of paper-based transaction records. XFDL was designed for implementation in business-to-business electronic commerce and intra-organizational information transactions.

As such, XML has a great potential as an exchange format for general structured data and increases the productivity to author and maintain, together with style sheet and linking mechanism, while remaining the feature that HTML has provided.

XML documents can be classified into two categories: *well formed* and *valid*. An XML document is well formed if it obeys the syntax of XML (e.g., non-empty tags must be properly nested, each non-empty start tag must have the corresponding end tag). A well-formed document is valid if it conforms to a proper DTD. A DTD is a file (external, included directly in the XML document or both) that contains a formal definition of a particular type of XML documents. A DTD states what names can be used for element types, where they may occur, how each element relates to the others, and what attributes an element may have.

As shown in Figure 2.17, an XML DTD may include four kinds of declarations: element declarations, attribute list declarations, entity declarations, and notation declarations. Element declarations specify the names of elements and their content. Attribute declarations specify the attributes of each element, indicating their name, type, and default value. Attributes that must necessarily appear are said to be required (#REQUIRED). Entities allow for incorporating text and/or binary data into a document. There are two kinds of entities: internal entities are used to introduce special character in the document or as shorthand for some text frequently mentioned, external entities are external files containing either text or binary data. Notation declarations specify what to do with the binary entities.

```
<! --Sample DTD-->
<! ELEMENT DOCUMENT (HEAD?,INTRO, ALEAF*)>
<! ELEMENT INTRO (BT+)>
<! ELEMENT HEAD (LINK)>
<! ELEMENT ALEAF (BT, (%content)*)>
<! ELEMENT BT (PCDATA)>
<! ENTITY %content "(BT|FIGURE)">
<! ELEMENT LINK (EMPTY)>
<! ELEMENT FIGURE (FIGREF, FIGNUM)
<! ELEMENT FIGREF EMPTY>
<! ELEMENT FIGNUM (PCDATA)>
<! NOTATION tiff SYSTEM "viewer.exe">
<! NOTATION bmp SYSTEM "viewer.exe">
<! NOTATION eps SYSTEM "viewer.exe">
<! ATTLIST FIGREF
    SRC CDATA #REQUIRED
    TYPE NOTATION (tiff|bmp|eps) "tiff">
<!ATTLIST LINK
    REL CDATA  #REQUIRED
    HREF  CDATA #REQUIRED)
```

Figure 2.16. Sample XML DTD

The set of declarations defines the vocabulary that can be used in tagging a document. XML vocabularies can be *open* or *closed*, the former allowing for using additional tags beyond what is declared in the base DTD. When XML documents are shared between applications, an open vocabulary can be extended, with the receiving application determining how to interpret extended elements and attributes. Depending on the application, unrecognized extensions to a vocabulary can often be ignored. As far as links are concerned, there are two types of links: *simple links* (like the one used in Figure 2.19) which are similar to the HTML links, and *extended links*, which allow expressing relationships between more than two resources. In Figure 2.18 the LINK element is defined whose attributes allow for simple link definition. Elements and attributes declaration have associated the cardinality with which they can appear:

character '*' indicates zero or more occurrences, character '+' indicates one or more occurrences, character '?' indicates zero or one occurrence, and no label indicates exactly one occurrence.

A sample valid document for the above DTD is shown in Figure 2.17.

```
<? xml version="1.0">
<! DOCTYPE DOCUMENT SYSTEM "http://127.0.0.1/document.dtd">
<DOCUMENT>
<HEAD> <LINK REL="0"
HREF="http://127.0.0.1/style.xsl"/></HEAD>
<INTRO>
<BT> This is a sample XML document </BT>
</INTRO>
<ALEAF>
<BT> This is a leaf that contains a paragraph (this one)
and a figure</BT>
<FIGURE> <FIGREF SRC="image.tif"
TYPE="tiff"></FIGREF><FIGNUM>110.2<?FIGNUM></FIGURE>
</ALEAF>
</DOCUMENT>
```

Figure 2.17 Sample XML Document

It should be noted that the element LINK is used in Figure 2.17 to identify the XSL style sheet that contains presentation information for this document. Again as shown in Figure 2.19, declarations that form a standardized XML DTD are usually stored in separate files, which can be referenced, as an XML external subset through the Uniform Resource Locator that its author has assigned to a publicly available copy of the data. Alternatively, if public access is to be restricted, the document type definition can be stored as the internal subset within the document type definition sent with the message.

```
<!ENTITY % address SYSTEM "http://www.sample.org/XML/address.xml" >
<!ENTITY % items SYSTEM "http://www.sample.org/XML/items.xml">
<!ENTITY % data "(#PCDATA)">
<!ELEMENT order (deliverylocation, invoicing, order-no, item+) >
<!ELEMENT deliverylocation (address) >
<!ELEMENT invoicing (address) >
<!--Import standard address class-->
%address;
<!ELEMENT order-no %data; >
<!--Import standard item class-->
%items;
```

Figure 2.18. A DTD Fragment

Where DTD is based on classes of information shared by more than one message, each class of information can be defined in a separate file, known in XML as an external entity.

For example, an XML DTD could have the form shown in Figure 2.18.

DTD1

```
<!ELEMENT list-manuf
  (manufacurer+)>
<!ELEMENT manufacturer (mn-
  name,year,model+)>
<!ELEMENT mn-name #PCDATA>
<!ELEMENT year #PCDATA>
<!ELEMENT model
(mo-name,front-rating,side-
  rating,rank)>
<!ELEMENT mo-name #PCDATA>
<!ELEMENT front-rating #PCDATA>
<!ELEMENT side-rating #PCDATA>
<!ELEMENT rank #PCDATA>
```

DTD2

```
<!ELEMENT list-vehicles (vehicle+)>
<!ELEMENT vehicle(vendor,(make|reference),
model,year,color,option*,price?)>
<!ELEMENT vendor #PCDATA>
<!ELEMENT make #PCDATA>
<!ELEMENT reference EMPTY>
<!ATTLIST reference manufactured-by
  IDREF>
<!ELEMENT model #PCDATA>
<!ELEMENT year #PCDATA>
<!ELEMENT color #PCDATA>
<!ELEMENT option #PCDATA>
<!ATTLIST option opt PCDATA
  #REQUIRED>
<!ELEMENT price #PCDATA>
<!ELEMENT company (name,address)>
<!ATTLIST company id ID #REQUIRED>
<!ELEMENT name #PCDATA>
<!ELEMENT address #PCDATA>
```

Figure 2.19. Two DTDs about cars

INSTANCE1

```
<list-manuf>
<manufacturer>
<mn-name>Mercury</mn-name>
<year>1998</year>
<model>
<mo-name>Sable LT</mo-name>
<front-rating>3.84</front-rating>
<side-rating>2.14</side-rating>
<rank>9</rank>
</model>
<model>
<mo-name>Sable LG</mo-name>
<front-rating>3.75</front-rating>
<side-rating>2.76</side-rating>
<rank>8</rank>
</model>
</manufacturer>
<manufacturer>
<mn-name>...</mn-name>
<year>1997</year>
<model>
<mo-name>...</mo-name>
<front-rating>3.05</front-rating>
<side-rating>2.00</side-rating>
<rank>11</rank>
</model>
</manufacturer>
</list-manuf>
```

INSTANCE2

```
<list-vehicle>
<vehicle>
<vendor>Scott Thomason</vendor>
<make>Mercury</make>
<model>Sable LT</model>
<year>1999</year>
<color>metallic blue</color>
<option opt="sunroof"/>
<option opt="M">A/C</option>
<price>26800</price>
</vehicle>
<vehicle>
<vendor>Scott Thomason</vendor>
<reference
  manufactured_by="C1"></reference>
<model>Sable LG</model>
<year>1999</year>
<color>metallic gray</color>
<option opt="SR">8</option>
<option opt="SF">ABS</option>
<price>27500</price>
</vehicle>
</list-vehicle>
<company id="C1">
<name>Mercury</name>
<address>Chicago</address>
</company>
```

Figure 2.20. Two XML documents

This DTD fragment defines two external and one internal parameter entity; moreover, it declares four locally defined elements and contains two parameter entity references (%address; and %items;) that call in the contents of the external entities at appropriate points in the definition. Both of the parameter entity references are preceded by explanatory comments.

Note that the source of each class of information is identified not in the call to the class itself (%address;) but within a formal definition of the data storage entities required to process the class definition references (the first two lines of the DTD). This technique allows files to be moved without having to change the main definitions of the DTD.

Of course, different applications may represent the same kind of information using completely different DTDs. As the focus on XML is shifting from document formatting to knowledge representation issues, this situation is becoming more and more common. As an example, we show two DTDs and two document fragments describing cars: in the former, the manufacturer company reports result data for the NHSC crash-safety test, in the latter auto dealers and brokers list their prices.

2.11.1. XML Namespaces

The above example should have highlighted the problem of compatibility between related DTDs at the level of tag and attribute names. Fortunately, XML provides *namespaces* (World Wide Web Consortium 1999), a simple and elegant mechanism for DTD extensibility. This technique leverages the Net's *Uniform Resource Identifier* (URI) namespace to allow arbitrary attributes and elements to be added to an existing XML vocabulary, and can be best illustrated through an example. Consider the following XML fragment:

```
<order orderno="33666">
    <vendor vendno="5573" />
    <part partno="4463" />
    <part partno="2930" />
</order>
```

This fragment indicates that vendor number 5573 is being ordered items 4463 and 2930; of course, for this document to be useful part numbers need to be shared between the vendor's and the customer's information system. Suppose now that the organization placing the order needs to annotate this message with additional information, adding an identifier that associates the order with a larger transaction. At first sight, simply adding an attribute as follows could make it:

```
<order orderno="33666" transid="12345">
<vendor vendno="5573" />
    <part partno="4463" />
    <part partno="2930" />
</order>
```

However, several problems arise, due to the fact that in general the application receiving the document at the vendor's site will have been developed independently from the sending application at the customer's site. If a closed vocabulary is used, the receiver may not recognize the additional elements/attributes added to the message.

Even if an open XML vocabulary is employed, the problem of *ambiguity* remains. This problem arises when both the sender and the receiver extend the vocabulary in the same way (e.g. adding independently two `transid` attributes with different semantics).

Namespaces were designed to relieve this problem, inasmuch as they allow attributes and elements to be scoped by a URI. The following XML fragment illustrates how XML namespaces can be used to unambiguously add the `transid` attribute to the order request:

```
<order orderno="33666"
       xmlns:acme="http://acme.org/trans/ns"
       acme:transid="55291" >
<vendor vendno="5573" />
    <part partno="4463" />
    <part partno="2930" />
</order>
```

This notation allows the vendor's application to detect that the `transid` attribute is scoped by the namespace http://acme.org/trans/ns and is not the same as the `transid` attribute used at its site (which would have a different namespace URI, e.g. `http://hop.org/trans/ns`). The following fragment illustrates how the request can be made completely unambiguous:

```
<order orderno="33666"
       xmlns:acme="http://acme.org/trans/ns"
       xmlns:hop ="http://hop.org/trans/ns "
        acme:transid="55291" >
         hop:transid="46722" >
<vendor vendno="5573" />
    <part partno="4463" />
    <part partno="2930" />
</order>
```

There are currently several initiatives to standardize domain-specific XML vocabularies, though it is unlikely that any of these standards will achieve 100 percent penetration in a particular application domain.

2.11.2. XML-based Agent Systems Development

XML, which was originally designed as a document standard for adding extensions to HTML, is becoming widely used as a means for autonomous agents to interoperate. As we have seen, the core XML specification is extremely simple as it only lays down the syntactic ground rules for forming valid XML messages. In other words, XML defines the minimal shared representation for data and message interchange needed to ensure that software agents can communicate. Moreover, XML neither mandates a type representation technique nor depends on a particular operating system, programming language, or hardware architecture. As long as two agents can exchange XML messages, they can potentially interoperate despite their differences.

Moreover, XML is easy to understand, author and process. Unlike binary-wire protocols like DCOM, CORBA/MASIF, or Aglets-Java/RMI, XML allows agents to easily create messages using standard string manipulation functions in the programming language of choice. The text-based nature of XML also makes it easier to debug and monitor distributed agent-based applications, as all agent-to-agent messages are readable to us when using a network debugging tool.

Due to the use of open vocabularies and namespaces, XML can support weakly typed communications. While strong typing has many benefits, it is extremely easy to build weakly typed systems using XML. This makes XML extremely adaptable to generic application frameworks, data-driven applications, and rapid development scenarios, such as disposable or transient Web-based applications. We shall elaborate on this subject in the following chapters.

2.12. Summary

A number of technologies have been used in this book. These include expert systems, artificial neural networks, fuzzy systems, genetic algorithms, object-oriented and agents and agent architectures. Expert systems, artificial neural networks, fuzzy systems and genetic algorithms are the four most widely used intelligent technologies. Expert systems can be broadly grouped under four architectures, namely, rule based, rule and frame (object) based, model based, and blackboard architectures. These architectures employ a number of representation formalisms including, predicate calculus, production rules, semantic networks, frames and objects. Case based reasoning methods are a requirement of expert systems and are used in domains like law where it is not possible to represent the knowledge using rules or objects. However, these architectures suffer from a number of limitations including combinatorial explosion of rules, inability to handle problems of non-deterministic or fuzzy nature, and others. Artificial neural networks that are used for problems of random and non-deterministic character can be grouped under supervised and unsupervised learning. Although there are a number of supervised and unsupervised neural network based learning algorithms, the most commonly used ones are backpropagation and self organizing Kohonen maps. Radial basis function networks which incorporate both unsupervised and supervised characteristics are also used. Fuzzy systems which, like artificial neural networks involve approximate reasoning have been widely used in various industrial applications. Fuzzy system construction involves determination of fuzzy sets and fuzzy membership functions, fuzzification of inputs, fuzzy inferencing and rule evaluation, and defuzzification of outputs.

Although genetic algorithms have not been exclusively applied in this book, they form an important class of adaptive search procedures based on principles derived from the dynamics of natural population genetics. The general characteristics of genetic algorithms include working with a coding of the parameter set, searching from a population of points, rather than a single point, using a pay-off or fitness function, and using probabilistic transition rules.

Object-oriented software engineering methodology today is one of the premium software engineering methodologies for building software systems. Besides its ability to structure data through inheritance and composability relationships and other non-hierarchical relationships, its encapsulation, and polymorphic properties make it attractive from a software implementation viewpoint. Agents and agent architectures are one of the most important emerging technologies in computer science today. They provide a means to map percepts to actions and in the process incorporate very sophisticated characteristics in a software program, namely, autonomy, collaboration, flexibility and versatility, adaptation and learning, complex communication, and others.

Multimedia technologies are being used for enhancing the effectiveness of use of computer-based artifacts. XML is a new internet standard for defining the declarative semantics of web documents.

References

Adler S. (1998), "Initial Proposal for XSL", available from: http://www.w3.org/TR/NOTE-XSL.html

Aitkins, J. (1983), "Prototypical Knowledge for Expert Systems," *Artificial Intelligence*, vol. 20, pp. 163-210.

Balzer, R., Erman, L. D., London, P. E. and Williams, C. (1980), "Hearsay-III:A Domain-Independent Framework for Expert Systems," in *First National Conference on Artificial Intelligence* (AAAI), pp. 108-110

Berry, J. T. (1988), *C++ Programming, Howard W. Sams and company*, Indianapolis, Indiana, USA.

Blair B. and Boyer J. (1999), "XFDL: Creating Electronic Commerce Transaction Records Using XML", *Proceedings of the WWW8 Intl. Conference,* Toronto, Canada, pp. 533-544

Bray T. et al. (ed.) (1998), "Extensible Markup Language (XML) 1.0", available at http://www.w3.org/TR/1998/REC-xml-19980210

Encyclopedia Britanica, (1986), Articles on "Behaviour, Animal," "Classification Theory," and "Mood," Encyclopedia Britanica, Inc.

Chandrasekaran, B. (1990), *What Kind of Information Processing is Intelligence,* The Foundations of AI: A Sourcebook, Cambridge, UK: Cambridge University Press, pp. 14-46.

Coad, P. and Yourdon, E. (1990), *Object-Oriented Analysis*, Prentice Hall, Englewood Cliffs, NJ, USA.

Coad, P. and Yourdon, E. (1991), *Object-Oriented Analysis and Design*, Prentice Hall, Englewood Cliffs, NJ, USA.

Coad, P. and Yourdon, E. (1992), *Object-Oriented Design*, Prentice Hall, Englewood Cliffs, NJ, USA.

Cox, B. J. (1986), *Object-Oriented Programming*, Addison-Wesley.

Dillon, T. and Tan, P. L. (1993), *Object-Oriented Conceptual Modeling*, Prentice Hall, Sydney, Australia.

Erman, L. D., Hayes-Roth, F., Lesser, V. R. and Reddy, D. R. (1980), "The Hearsay-II Speech-Understanding System: Integrating Knowledge to Resolve Uncertainty," *ACM Computing Surveys*, vol. 12, no. 2, June, pp. 213-53.

Erman, L. D., London, P. E. and Fickas, S. F. (1981), "The Design and an Example Use of Hearsay-III," in *Seventh International Joint Conference on Artificial Intelligence*, pp. 409-15.

Finin T., Fritzson R., MacKay D. and MacEntire R. (1994), "KQML as an Agent Communication Language, *Proceedings of the Third International Conference on Information and Knowledge Management*, pp.112-124

Greffenstette, J.J. (1990) "Genetic Algorithms and their Applications" *Encyclopedia of Computer Science and Technology,* vol. 21, eds. A. Kent and J. G. William, AIC-90-006, Navla Research laboratory, Washington DC, pp. 139-52.

Goldberg, D.E. (1989), *Genetic Algorithms in Search, Optimization and Machine Learning*, Addison-Wesley, Reading, MA, pp. 217-307.

Hamscher, W. (1990), "XDE: Diagnosing Devices with Hierarchic Structure and Known Failure Modes," *Sixth Conference of Artificial Intelligence Applications,* California, pp. 48-54.

Hawryszkiewyz, I. T. (1991), *Introduction to System Analysis and Design*, Prentice Hall, Sydney, Australia.

Hayes-Roth, F., Waterman, D. A. and Lenat, D. B. (1983), *Building Expert Systems,* Addison-Wesley.

Haykin, 1994 *Neural Networks: A comprehensive foundation.* IEEE Press, New York.

Hebb, D. (1949), The Organisation of Behaviour, Wiley, New York.

Holland, J. (1975), *Adaptation in Neural and Artificial Systems*, University of Michigan Press, Ann Arbor, Michigan, USA.

Inmon, W.H., and Kelley, C., (1993), Rdb/VMS, *Developing the Data Warehouse,* QED, Publication Group, Boston, USA.

Kim, Ballou, Chou, Garza and Woelk (1988), "Integrating an Object-Oriented Programming System with a Database System," *ACM OOPSLA Proceedings*, October.

Kohonen, T. (1990), *Self Organisation and Associative Memory*, Springer-Verlag.

Kolodner, J. L. (1984), "Towards an Understanding of the Role of Experience in the Evolution from Novice to Expert," *Developments in Expert Systems,* London: Academic Press.

Kraft, A. (1984), "XCON: An Expert Configuration System at Digital Equipment Corporation," *The AI Business: Commercial Uses of Artificial Intelligence*, Cambridge, MA: MIT Press.

McClelland, J. L., Rumelhart, D. E. and Hinton, G.E. (1986), "The Appeal of Parallel Distributed Processing," *Parallel Distributed Processing,* vol. 1, Cambridge, MA: The MIT Press, pp. 3-40,

Minsky, M. and Papert, S. (1969), *Perceptrons*, MIT press.

Minsky, M. (1981) "A Framework for representing Knowledge," *Mind Design*, Cambridge, MA: the MIT Press, pp. 95-128.

Myer, B. (1988), *Object-Oriented Software Construction,* Prentice Hall.

Neibur, D. and Germond, A. J. (1992) "Power System Static Security Assessment Using The Kohonen Neural Network Classifier," *IEEE Transactions on Power Systems*, May, vol. 7, no. 2, pp. 865-72.

Newell, A. (1977), "On Analysis of Human Problem solving," *Thinking: Readings in Cognitive Science,* Cambridge UK: Cambridge University Press.

Ng, H.T., 1991, "Model-Based, Multiple-Fault Diagnosis of Dynamic, Continuous Physical Devices," *IEEE Expert*, pp. 38-43.

Pardi W. J. (1999), *XML In Action*, Microsoft Press

Pressman, R. S. (1992), *Software Engineering: A Practioner's Approach*, McGraw Hill International, Singapore.

Quillian, M. R. (1968), "Semantic Memory," *Semantic Information Processing*, Cambridge, MA: The MIT Press, pp. 227-270.

Rumbaugh, J. et al. (1990), *Object-Oriented Modeling and Design*, PrenticeHall, Englewood Cliffs, NJ, USA.

Rumelhart, D. E., Hinton, G. E. and Williams, R. J. (1986), "Learning Internal Representations by Error Propagation," *Parallel Distributed Processing,* vol. 1, Cambridge, MA:The MIT Press, pp. 318-362.

Russell, S., and Norvig, P. (1995), *Artificial Intelligence - A Modern Approach*, Prentice Hall, New Jersey, USA, pp. 788-790.

Schank, R. C. (1972), "Conceptual Dependency," *Cognitive Psychology,* vol. 3,pp. 552-631.

Schank, R. C. and Abelson, R. P. (1977), *Scripts, Plans, Goals and Understanding*, Hillsdale, NJ: Lawerence Erlbaum.42

Sejnowski, T.J., and Rosenberg, C.R. (1987), "Parallel Networks that Learn to Pronounce English Text," *Complex Systems*, pp, 145-168.

Shortliffe, E.H. (1976), *Computer-based Medical Consultation: MYCIN*, New York: American Elsevier.
Smolensky, P. (1990), "connectionism and Foundations of AI", *The Foundations of AI: A Sourcebook*, Cambridge, UK: Cambridge University.
Steels, L. (1989), "Artificial Intelligence and Complex Dynamics," *Concepts and Characteristics of Knowledge Based Systems*, Eds., M. Tokoro, et al., North Holland, pp. 369-404.
Unland, R, and Schlageter, G. (1989), "An Object-Oriented Programming Environment for Advanced Database Applications," *Journal of Object-Oriented Programming*, May/June.
Wang, X. and Dillon, T. S. (1992), "A Second Generation Expert System for Fault Diagnosis," in *Journal of Electrical Power and Energy Systems*, April/June, 14 (2/3), pp. 212-16.
Widrow, B., and Hoff, M.E. (1960), "Adaptive Switching Circuits," *IRE WESCON Convention Record*, Part 4, pp. 96-104.
World Wide Web Consortium (1998), "Extensible Markup Language (XML) 1.0" (W3C Recommendation) http://www.w3.org/TR/1998/REC-xml-19980210
World Wide Web Consortium (1999), "Namespaces in XML" (W3C Recommendation) http://www.w3.org/TR/1999/REC-xml-names-19990114/
Zadeh, L.A. (1965), "Fuzzy sets," *Information and Control*, vol. 8, pp. 338-353.

43 Notes

1. This term is used interchangeably with connectionist networks, connectionist models, connectionist systems or simply neural networks/nets throughout the thesis

2. data is used here interchangeably with attribute

3 PRAGMATIC CONSIDERATIONS AND ENABLING THEORIES

3.1. Introduction

Human-centered system development is not a revolutionary concept in computer science and information systems but an evolutionary and enabling one. In this chapter we look at how some areas in computer science and information systems are evolving or moving towards human-centeredness. These areas include intelligent systems, electronic commerce, software engineering, multimedia databases, data mining, enterprise modeling and human-computer interaction. This evolution is based on the need for addressing pragmatic issues in these areas. We follow these pragmatic issues with enabling theories in philosophy, cognitive science, psychology and work-oriented design for human-centered system development framework. These theories are described and discussed in terms of their contributions toward human-centered system development framework. We conclude the chapter with a discussion section that outlines the foundations of the human-centered system development framework described in the chapter.

3.2. Pragmatic Considerations for Human-Centered Software Development

In chapter 1 we quoted Norman (1997) as saying that the computer industry is still in its rebellious adolescent stage where technology provides all the excitement of youth as compared to the staid utility of maturity. Pragmatic considerations are about how various information technologies areas are evolving towards bridging this chasm between youth and maturity. The pragmatic considerations represent the practical problems associated with use of various technologies. These practical problems are underpinned in epistemological limitations (and strengths) which human and computers have, human vs. technology mismatch, relationship between technology and people and the impact of this relationship on the use of technology, social, organizational and task context in which technology is used, and some others.

The various areas looked into are:

- Intelligent Systems
- Software Engineering,
- Multimedia Databases
- Electronic Commerce
- Data Mining

- Enterprise Modeling, and
- Human-Computer Interaction

3.2.1. Intelligent Systems and Human-Centeredness

The four most commonly used intelligent methodologies in the 90's are symbolic knowledge based systems (e.g. expert systems), artificial neural networks, fuzzy systems and genetic algorithms.

Symbolic knowledge based systems have served varied purposes in industry and commerce during the last three decades. The most widely used versions being expert systems which have found their way into industry and commerce, including manufacturing, planning, scheduling, design, diagnosis, sales and finance. In these applications, the unitary architecture of production rules, normally enhanced by frames or objects, has been used to capture human expertise and to solve different problems. The different knowledge representation techniques like semantic networks, frames, scripts and objects have been able to capture some of the ways in which humans utilize knowledge. However, practitioners have also identified some of the limitations of symbolic knowledge based systems. These include among others, slow and constricted knowledge acquisition processes, inability to properly deal with imprecision in data, inability to process incomplete information, combinatorial explosion of rules, retrieval problems in recovering relevant past cases, and inability to reason under time constraints on occasions.

People deal every day with imprecision and fuzziness in data. This imprecision may be represented by linguistic statements. A number of fuzzy systems have been built based on fuzzy concepts and imprecise reasoning. Fuzzy systems have been used in a number of areas including control of trains in Japan, sales predictions, and stock market risk analysis. A major disadvantage of fuzzy systems and expert systems is their heavy reliance on human experts for knowledge acquisition. This knowledge may be in the form of rules used to solve a problem and/or the shape of the membership functions used for modeling a fuzzy concept. Besides the knowledge acquisition problem, these systems are restricted in terms of their adaptive and learning capabilities.

The limitations in knowledge based systems and fuzzy systems have been primarily responsible for the resurgence of artificial neural networks. In the financial sector, neural networks are used for prediction and modeling of markets, signature analysis, automatic reading of handwritten characters (checks), assessment of credit worthiness and selection of investments. In the telecommunication sector, applications can be found in signal analysis, noise elimination and data compression. Similarly, in the environment sector, neural networks have been used for risk evaluation, chemical analysis, weather forecasting and resource management. Other applications can be found in quality control, production planning and load forecasting in power systems. In these applications, the inherent parallelism in artificial neural networks and their capacity to learn, process incomplete information and generalize have been exploited. However, the stand-alone approaches of artificial neural networks have exposed some limitations such as the problems associated with lack of structured knowledge representation, inability to interact with conventional symbolic databases and inability to explain the reasons for conclusions reached. Their inability to explain their conclusions has limited their applicability to high-risk domains (e.g. real-time alarm processing). Another major limitation associated with neural networks is the problem of scalability. For large and complex problems, difficulties exist in training the networks and also in assessing their generalization capabilities.

Optimization of manufacturing processes is another area where intelligent methodologies like artificial neural networks and genetic algorithms have been used. Genetic algorithms are being used for solving scheduling and control problems in industry. They have also been successfully used for optimization of symbolic, fuzzy and neural network based intelligent systems because of their modeling convenience. One of the problems associated with genetic algorithms is that they are computationally expensive, which can restrict their on-line use in real-time systems where time and space are at a premium.

In fact, real-time systems add another dimension to the problems associated with the various intelligent methodologies. These problems are largely associated with the time and space constraints of real-time systems. Some examples of real-time systems are command and control systems, process control systems, flight control and alarm processing systems.

These computational and practical issues associated with the four intelligent methodologies have made the practitioners and researchers look at ways of hybridizing the different intelligent methodologies from an applications viewpoint. However, the evolution of hybrid systems is not only an outcome of the practical problems encountered by these intelligent methodologies but is also an outcome of deliberative, fuzzy, reactive, self-organizing and evolutionary aspects of the human information processing system (Bezdek 1994).

Intelligent hybrid systems can be grouped into three classes, namely, fusion systems, transformation systems, combination systems (Khosla et al. 1997b). In fusion systems (Edelman 1992; Fu & Fu 1990; Hinton 1990; Sethi 1990; Sun 1994), the representation and/or information processing features of intelligent methodology A are fused into the representation structure of another intelligent methodology B. In this way, the intelligent methodology B augments its information processing in a manner which can cope with different levels of intelligence and information processing. From a practical viewpoint, this augmentation can be seen as a way by which an intelligent methodology addresses its weaknesses and exploits its existing strengths to solve a particular real-world problem. The hybrid systems based on the fusion approach revolve around artificial neural networks and genetic algorithms. In artificial neural network based fusion systems, representation and/or information processing features of other intelligent methodologies like symbolic knowledge based systems and fuzzy systems are fused into artificial neural networks. Genetic algorithm based fusion systems involve fusion of intelligent methodologies like knowledge based systems, fuzzy systems, and artificial neural networks.

Transformation systems (Gallant 1988; Ishibuchi et al. 1994) are used to transform one form of representation into another. They are used to alleviate the knowledge acquisition problem by transforming distributed or continuous representations into discrete representations. From a practical perspective, they are used in situations where knowledge required to accomplish the task is not available and one intelligent methodology depends upon another intelligent methodology for its reasoning or processing. For example, neural nets are used for transforming numerical/continuous data into symbolic rules which can then be used by a symbolic knowledge based system for further processing. Transformation systems have also been used for knowledge discovery and data mining (Khosla et al. 1997b).

Combination systems (Chiaberage et al. 1995; Fukuda et al. 1995; Hamada et al. 1995; Srinivasan et al. 1994) involve explicit hybridization. Instead of fusion, they model the different levels of information processing and intelligence by using intelligent methodologies that best model a particular level. Intelligent combination systems, unlike fusion systems, retain the separate identity of each intelligent methodology within a module. These systems involve a modular arrangement of two or more intelligent methodologies to solve real-world problems.

These three different classes of intelligent hybrid systems and their industrial applications have been researched and reported in Khosla et al. (1997b).

The concepts of fusion, transformation, and combination have been used in different situations or tasks, and by applying a top-down and/or bottom-up knowledge engineering strategy. All these hybrid architectures have a number of advantages in that the hybrid arrangement is able to successfully accomplish tasks in various situations. However, these hybrid architectures also suffer from some drawbacks. These drawbacks can be explained in terms of the quality of solution and range of tasks covered as shown in Figure 3.1. Fusion and transformation architectures on their own do not capture all aspects of human cognition related to problem solving. For example, fusion architectures result in conversion of explicit knowledge into implicit knowledge, and as a result lose on the declarative aspects of problem solving. Thus, they are restricted in terms of the range of tasks covered by them. The transformation architectures with bottom-up strategy get into problems with increasing task complexity. Therefore the quality of solution suffers when there is heavy overlap between variables, where the rules are very complicated, the quality of data is poor, or data is noisy. Also, because they lack explicit reasoning, the range of tasks covered by them becomes restricted. The combination architectures cover a range of tasks because of their inherent flexibility in terms of selection of two or more intelligent methodologies. However, because of lack of (or minimal) knowledge transfer among different modules the quality of solution suffers for the very reasons the fusion and transformation architectures are used.

As fusion, transformation, and combination architectures have been motivated by and developed for different problem solving tasks/situations, it is useful to associate these architectures in a manner so as to maximize the quality as well as range of tasks that can be covered. These class of systems are called associative systems (or associative hybrid systems) as shown in Figure 3.1.

The groundwork related to associative hybrid systems has been reported and explored in a book by Khosla et al. (1997b) and other publications (Khosla 1997c-f; Main et. al 1995; Tang et al. 1996 1995). As may be apparent from Figure 3.1, associative systems consider the four intelligent methodologies and their hybrid configurations, namely, fusion, transformation, and combination as technological primitives that are used to accomplish tasks. The selection of these technological primitives is contingent upon satisfaction of task constraints (e.g. presence/absence of domain knowledge, noisy incomplete data, learning, adaptation, etc.) which in Figure 3.1 have been grouped under the quality of solution dimension.

In summary, it can seen from the discussion in this section that intelligent associative systems have evolved from a technology-centered approach to intelligent systems where standalone intelligent technologies (e.g. neural networks, fuzzy logic, etc.) have been used for building intelligent systems to a task-centered approach where various intelligent technologies are used as primitives rather than prime drivers for building intelligent systems. The task-centered approach intends not only to model user/stakeholder tasks and capture deliberate, fuzzy, reactive, self-organizing and evolutionary aspects of human information processing through use of various technological primitives but also account for epistemological limitations which humans and computers have through satisfaction of various pragmatic task constraints.

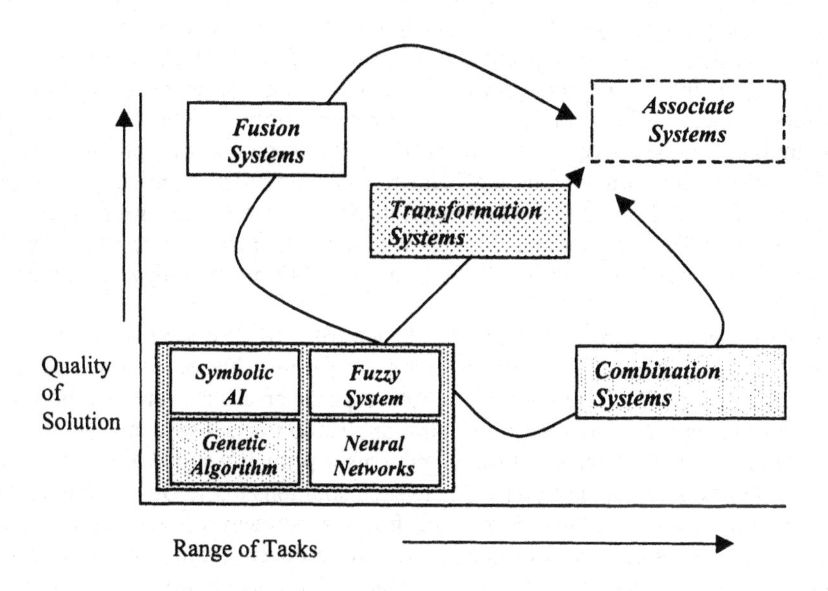

Figure 3.1: Intelligent Task-Centered Associative Systems

3.2.2. Software Engineering and Human-Centeredness

Software engineering can be defined as a layered technology that encompasses a definition phase, a development phase, and a maintenance phase. The development phase involves analysis, design, implementation and testing phases, respectively. In the past two decades it has evolved through three software engineering methodologies, namely, structured design or data flow methodology (Yourdon 1978), object-oriented methodology (Coad and Yourdon 1991; Jacobson 1995; Gamma et al. 1995; Pree 1995), and agent methodology (Wooldridge and Jennings 1994; Weilingaet.al. 1993).

Analysis and design phases are the main focus of the three software engineering methodologies. Analysis involves modeling of the information, functional, and behavioral aspects of a domain. On the other hand, software design is the technical kernel of software engineering. It is based on design characteristics like modularity, high cohesion, and low coupling. It is a multi-step process that involves data, architecture, procedure and interface design of a domain.

The traditional structured analysis and design methodology involves functional modeling using data flow methodology. Transformation and transaction flow methods are used to convert the data flow analysis into structured (architectural) design. One of the problems with the structured design methodology is the non-isomorphic transition between analysis and design phases. It can result in loss of information between analysis and design phases. The object-oriented methodology alleviates this problem by retaining the same vocabulary of objects and classes in the analysis and design phases respectively. Unlike the functional characteristics of the data flow methodology, object-oriented methodology intuitively captures the structural aspects of a domain. By definition, from a software design and programming perspective, the object-oriented methodology provides strong encapsulation and information hiding characteristics. In order to enrich the functional modeling aspects of the O-O methodology, some researchers (e.g. Rumbaugh 1991) have integrated the data flow methodology with the O-O methodology. More recently design patterns have been added to the armor of O-O methodology (Gamma et al. 1995; Pree 1995; and others).

A design pattern names, abstracts, and identifies the key aspects of a common design structure that makes it useful for creating a reusable object-oriented design (Gamma et. al. 1995). However these key aspects or abstractions are primitives, which are not expected to design an entire application or subsystem (Gamma et al. 1995). Further, the vocabulary of design patterns is primarily suited to meet the needs of software designers and not users. Many such additions to the O-O methodology in the last decade are much like "add more features" strategy of technology-centered products. As a result, today we have more than a dozen definitions of objects and classes, many of which (unlike the earlier stages of this paradigm) are not intuitive or user-centered for system modeling purposes. Moreover, from problem solving and human-centered perspectives, O-O paradigm does not provide an intuitive means of modeling goals and tasks.

Recently questions have been raised on the appropriateness of considering software analysis and design as two distinct processes. In fact the development of software design pattern construct is being seen as a means of alleviating this problem associated with software analysis and design process. The problem is primarily related to the fact that software engineering development process is not necessarily analogous to product development process in more mature traditional engineering disciplines. Thus, software design patterns which abstract software design structure from existing applications to be reused in a new one are being seen as means of ensuring software quality which is the main aim of the software engineering process. Although, the design patterns are a useful construct, their present state-of-the-art is still limited to the component level rather than an entire application. Further, their definition suggests a technology-based focus (i.e., they are likely to carry the limitations of the technology along with them) which makes them somewhat unsuitable for human-centered systems.

Unlike O-O methodology, agent methodology is primarily driven by problem solving, tasks and task-based behavior (Jennings et al. 1996; Khosla and Dillon 1997; and others). In fact, agents have transformed the internet's character and mission. They are being used for searching and retrieving information for users from distributed sources on the internet in a collaborative manner. However, besides their collaborative and task characteristics, they lack the structural characteristics of the O-O methodology. Traditionally, agent modeling is embedded in laws of thought, that is, classical AI logic. Most of the agent based systems on the internet and in the field model tasks based on logic. This is at cross roads with the human-centered objectives, like activity-centeredness, focus on practitioner's goals and tasks, and the need to model tasks based on how users accomplish them rather than force fit a particular technology (like AI logic) on to the tasks.

Thus although, the three distinct approaches discussed in this section have made significant contributions towards achieving the primary goal of software engineering, namely, software quality, it is apparent they have not contributed in the same vein towards human-centeredness. The technology-centeredness of these approaches constrains them to model all aspects of a domain using a particular technology. This undermines to some extent the syntactic and semantic quality of a computer based artifact (software system) from a human-centered viewpoint. The syntactic quality determines the intuitiveness of the constructs used to model a domain. That is, how close are the constructs used by a particular technology to those used by humans (i.e. users/stakeholders and not system designers). On the other hand, semantic quality determines how people use various artifacts to solve problems. That is, how close is the software design of a human problem to the human solution of that problem.

3.2.3. Multimedia Databases and Human-Centeredness

In recent years, large amounts of data in structured (e.g., relational/object-oriented), unstructured (e.g., image) and sequential (e.g., audio) formats has been collected and stored in thousands of repositories. These repositories which exist in organizations as well on the internet (e.g., World Wide Web) are used for locating and accessing multimedia data.

The progress made in locating and accessing data related to a single media type (e.g. image, audio, video or text) has been significant. Most of the recent techniques (Grosky 1994; Anderson et al. 1994; Chen et al. 1994; Glavitsch et al. 1994; Kashyap, Shah and Seth 1995; Jain 1996; Jain et al. 1994, and others) for searching and accessing digital data have employed the concept of metadata. Metadata represents implicit information about the data in individual databases and can be seen as an extension of the concept of schema in structured databases (Kashyap, Shah and Seth 1995). It can be classified as content-dependent metadata, content–descriptive metadata, and content-independent metadata (Kashyap Shah and Seth 1995). Content-dependent metadata depends only on the content of the original data. For example, in an image database color, texture, and shape are content-dependent metadata where information is determined by the content, e.g., hue and saturation values (for color).

Content-dependent data can be automatically extracted from the contents. Content-descriptive metadata on the other hand, is determined exclusively by looking at the content, employs cognitive processes and can be domain-dependent or domain-independent. Domain-dependent metadata employs domain-specific concepts like retrieving all mammals or tigers from an animal kingdom multimedia database. Domain-independent metadata would be the one which captures the structure of a multimedia document (Bohm and Rakow 1994).

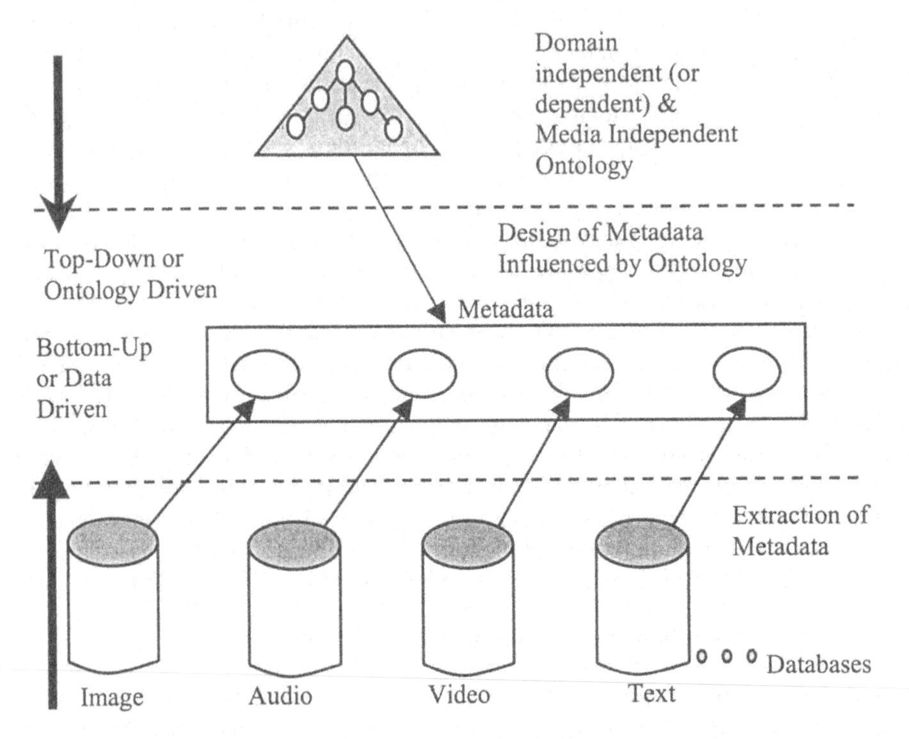

Figure 3.2: Multimedia Databases, Metadata and Ontology Levels:

However, as emphasized by Kashyap, Shah and Sheth (1995) and Sheth. (1996) the above classifications are not sufficient to capture the semantic correlation[1] (as done by humans) for problems which involve more than one media (e.g., image, video and audio).. Even for a single media type like image, a number of models can be used for image indexing and retrieval. A user is constrained to use one model for querying, indexing and retrieval which may be entirely based on content-dependent and content-independent data. A user in actual practice may be employing more than one model for querying, indexing and retrieval and may actually be using a mixture of content-descriptive, content-dependent and other metadata. Here again, from a human-centered perspective there is a scarcity of models or architectures that address this problem. Thus from a human-centered and semantic correlation perspectives there is a need for defining another level, namely, the ontological level above the metadata level as shown in Figure 3.2.

[1] Semantic correlation is defined as meaning and usage of data.

An ontology is a representation vocabulary, typically specialized to some technology, domain or subject matter. However, here we are dealing with upper ontology, i.e., ontology that describes generic knowledge that holds across many domains. Further, we are dealing with problem solving knowledge (e.g. tasks) about problem solving. As shown in Figure 3.2 the ontology can be domain independent and media independent or domain dependent and media independent. The domain independent and media independent ontology is based on generic tasks that are mapped on to the domain tasks and associated conceptual data structures (e.g., classes and objects). The domain dependent ontology on the other hand is based on specific domain tasks and associated conceptual data structures. Given the ontology level, one can adopt a top-down or bottom-up strategy for designing the metadata. The top-down strategy for designing metadata is also called ontology driven metadata design strategy because it will be influenced by the problem tasks in a domain understudy. However, the bottom-up strategy will be primarily media data driven rather than task driven.

3.2.4. Electronic Commerce and Human-Centeredness

Electronic Commerce (eCo) can be broadly seen as the application of information technology and telecommunications to create *trading networks* where goods and services are sold and purchased, thus increasing the efficiency and the effectiveness of traditional commerce (Figure 3.3).

The scenario in which vendors, brokers and buyers interact on private networks or, more frequently, on the global Net to execute commercial transactions has become increasingly familiar to the general public. Many software architectures have been proposed for supporting such networks in the past few years; in Hands, et. al., (1998) some of them are presented, addressing issues such as sales, ordering and delivery of products in the framework of the global internet.

Indeed, internet based electronic commerce is currently a driving force behind the evolution of many *Web-based technologies* such as HTTP, HTML, Java, CGI and others (Hamilton 1997), all of which were originally conceived for different applications.

A further step toward open markets standardization has been envisioned in *eCo* (refer Figure 3.3), a reference architecture for electronic commerce proposed by the CommerceNet Consortium (including Actra, Bank of America, Visigenic, World Wide Web Consortium, Mitsubishi, NEC and Oki), which exploits Web-based technologies such as HTML, and Java/CORBA (Tenenbaum et. al. 1998).

The eCo platform was originally a framework of reusable software components based on CORBA middleware standard, (Orfali, R. and Harkey, D. 1997) that can be used to build electronic commerce applications.

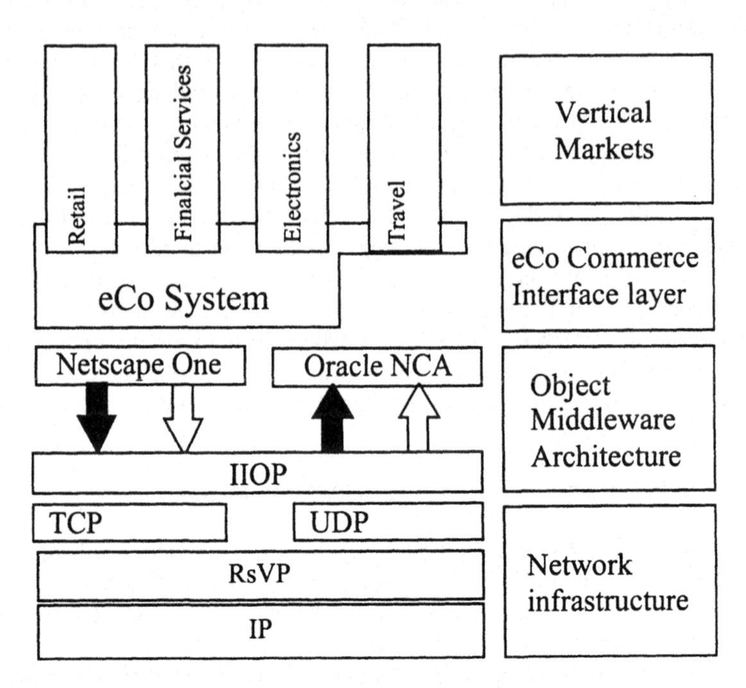

Figure 3.3: Electronic Commerce Framework

It includes a high-level domain specific language, *Common Business Language (CBL)* allowing software modules to communicate much like humans involved in commercial transactions, but exchanging *EDI-compliant object documents* instead of traditional paper documents.

Industry-wide standardization of eCo objects should allow companies to build *open markets* for business-to-business electronic commerce. Objects can be created in all the main proprietary environments, including IBM, Oracle, JavaSoft and Netscape.

In 1997, eCo-system was entirely recast on an XML foundation, due to XML adoption by all key vendors of the CommerceNet Consortium.

The eCo system framework overcomes a long-standing barrier to the development of electronic commerce, as XML documents provide, at least in principle, an incremental path to business automation, whereby browser-based tasks are gradually transferred to software agents. This development might allow traditional supply chains to evolve into open markets, while agents interact with business services through object documents.

However, traditional commerce, the existing electronic commerce architectures on the internet are supplier-centered. XML's human readability, while an advantage over CORBA [Glushko 1999], does not eliminate the risk of a supply-side market model, where the structure and content of metadata are modeled w.r.t. the needs of vendors and distributors, leaving it to the brokers to transform them into a form more suitable for the buyers.

In this setting, nearly every electronic commerce purchase is preceded by a *network search* or *product brokering* phase (Tenenbaum et al. 1998). when the customer navigates the trading network looking for the needed products or services.

However, the avalanche of on-line suppliers and multimedia information about goods and services currently available, makes it difficult to locate, purchase and obtain the desired products at the best prices. General-purpose internet search engines seem wholly unfit for this task.

A solution to this problem is the definition of a human or consumer-centered brokerage architecture which will locate all the vendors carrying a specific product or service, then query them in parallel to locate the best deals. In chapter 10 we apply consumer-centered brokerage architecture in a hardware adapter electronic commerce domain using XML.

3.2.5. Data Mining and Human-Centeredness

An important aspect of data mining is to impart meaningfulness (or meaningful knowledge) to the mined patterns from large databases (Khosla et al 1997b pp. 150-185).

One way of imparting meaningfulness is to integrate the task-based problem solving model of the user with the data mining process. That is, the problem solving model can be used to provide a priori knowledge to data mining techniques like summarization, clustering, association, prediction, and classification and enhance the quality and applicability of results of the computing mechanisms (e.g., data warehouse, object-oriented, neural networks, etc.) employed by these techniques. Further, this way data mining can become an integral part of the decision-making processes of stakeholders/user). The task based modeling process will also enable one to account for different user perspectives associated with data in the data mining process. For example, information related to a customer in an electric distribution utility is viewed by forecasting and pricing manager in terms of their energy consumption (for forecasting) and credit rating (for pricing) and other perspectives depending the task and the task model.

3.2.6. Enterprise Modeling and Human-Centeredness

The rapid growth of the use of computers in organizations in the past twenty years has reflected the important role played by information technology in a business enterprise. Most organizations, including businesses, government agencies, industrial firms, and hospitals, now depend on computers as an integral part of their operations.

Whatever form of information technology is utilized, the fact is that the management of business relies heavily on information throughout the business process where data, information, and knowledge are the three main resources to support that business process. In terms of organizational levels, the demand for data is very high at the operational level as shown in Figure 3.3. The demand for information and knowledge increases as we move up from operational level to the strategic level. This is because degree of unstructuredness of problems increases as we move up from operational level to the strategic level. Further, like information systems exist at all levels, there is also evidence that intelligent systems exist at all levels. The evidence can be seen with the development of intelligent systems such as intelligent air line reservation systems at the operational level (Nwana and Ndumu 1997), intelligent e-mail and news management systems at knowledge work level (Maes 1994), intelligent production scheduling systems at the management level (Hamada et al. 1995), and intelligent forecasting and prediction systems at the strategic level (Khosla and Dillon 1997b).

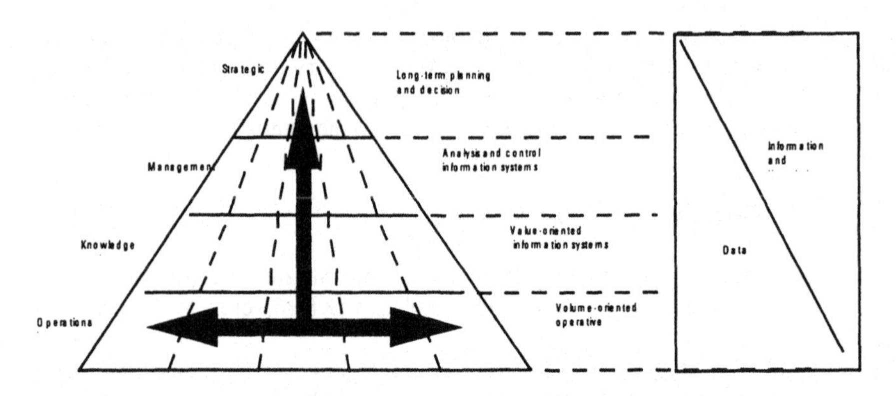

Figure 3.4: Applicability of Data, Information and Knowledge w.r.t. Organizational Level

Like other areas covered in this chapter, technology-centered perspective has also dominated the development of enterprise systems at the operational, management and strategic levels in the last two decades. Although, database systems, information and intelligent systems represent different stages of evolution of enterprise systems, the distinctions between these systems to some extent today are technology-centered. Database systems are modeled using data abstraction technologies like entity – relationship diagrams, information systems rely on technologies based on functional abstraction like data flow diagrams, and intelligent systems rely on technologies like expert systems, fuzzy-logic, neural networks, and genetic algorithms which model different aspects of human-cognition, brain and evolution. As outlined in the previous paragraph, these three types of systems exist at all levels of an enterprise. This has created an obvious need for integration and interoperability of these systems. However, the technology mismatch is the one of the major obstacles today for their integration and interoperability. More so, fields like knowledge discovery and data mining have demonstrated that one type of system (i.e. a standard database system with DBMS capabilities) can evolve into another type of system (i.e. an intelligent system) to provide sophisticated intelligent decision support. Thus there is a need to build enterprise systems which are problem driven and which have capabilities to evolve with time. Such enterprise systems will need to have architectures which facilitate use of range of technologies for different tasks and needs which evolve with time. Thus in this scenario, technologies are more likely to be used based on their intuitive modeling strengths.

On the other hand, from a human-centered perspective (as outlined in chapter 1), in the past decade there has been an increasing emphasis on modeling of complex software systems which are based on synergy between human and the machine

(Perrow 1984; Norman 1993). In the 70's and 80's information technology has been primarily used by organizations for automation (and enhancing the bottom line) without looking into its psychological and social side effects and the revolutionary impact it has had on the overall nature of workplace activity. In the 90's the disruptive effects of information technology have become all too visible and have forced the organizations to adopt a more balanced view where information technology and computers have to coexist (rather than necessarily replace) with people and their activities. In the 90's computers and information technology are being deployed based on the incentives they offer to workers in terms of their personal goals as well as the organizational goals. The computers and information technology in the 90's are seen as tools that assist people in their day-to-day activities and in breakdown situations rather than as prime drivers which redefine workplace activities and tasks in an organization.

3.2.7. Human-Computer Interaction and Human-Centeredness

Human-Computer Interaction (HCI) is about designing computer systems that support people so that they can carry out their activities productively and safely (Preece, et al. 1997). Human factors engineering (which deals with factoring of human characteristics like limited attention span, faulty memory, etc. into the design of computer system) and usability engineering (which is concerned with making systems easy to learn and easy to use) are among a number of areas contributing to HCI. A lot has been said in the literature about making user interface human or user-centered. Most of the initial work done in the HCI has been based on interaction tasks. In this scenario, the user interface has been treated as a distinct entity detached from the underlying system design and/or model. As a result, the initial response of the software industry to the usability problems was to add more features to the user-interface and hope for the problems to go away. More recently, however, the HCI community has looked into areas like artificial intelligence, linguistics sociology, anthropology, sociology, design, engineering, social and organizational psychology and cognitive psychology to alleviate some of the problems associated with user-interface design. These areas have highlighted the need for considering user-interface design as being tightly integrated with overall system design. Thus today among other aspects, researchers are working on designing task-oriented interfaces, incorporating aspects related to industrial design in user-interface design, and integrating multimedia as means for reducing the cognitive load on users. In the task-oriented interfaces specifically look into integration of the interaction tasks with the underlying problem domain tasks. However, complex software systems that clearly demonstrate such integration have still not emerged out of lab settings.

3.3. Enabling Theories for Human-Centered Systems

In the last section we have looked at how pragmatic considerations have resulted in the evolution towards human-centeredness of various areas of information technology. In this section, we discuss some theories from philosophy, cognitive science, psychology, and workplace that have influenced research and design of human-centered systems. These are:

- Semiotic Theory
- Cognitive Science Theories
- Activity Theory
- Work-oriented Design Theory

The rest of this section will describe these theories and their implications for design of human-centered systems.

3.3.1. Semiotic Theory – Language of Signs

The aim of this section is to establish the theoretical foundations for development of human-centered intelligent systems. For that matter, firstly theoretical aspects related to understanding of human intelligence from human science perspective (i.e. semiotic theory) and computer science perspective (artificial intelligence and computational intelligence) are outlined.

Human intelligence has always been of interest and curiosity in the scientific world. The understanding of human intelligence before the advent of computers was primarily rooted in human sciences and philosophy (Pierce 1960). After the advent of computers, the developments in this area have evolved under two fields, namely, artificial intelligence, and computational intelligence. The field of artificial intelligence is grounded mainly in symbolic logic and the physical symbol system (Newell 1980). The physical symbol hypothesis has led to development of class of intelligent agents embodied in symbols. The symbols represent knowledge at a higher level (also called the knowledge level) compared to ordinary computer programs. A number of knowledge level models of general intelligence were developed as a consequence including KL-ONE (Brachman et al. 1985), SOAR (Laird et al. 1987; Norman 1991) and ACT* and PUPS (Anderson 1989). An important aspect of these symbolic models has been the concept of inference and inference patterns.

While all computable problems can, in principle, be represented in symbolic terms as demonstrated by Turing's work on universal Turing machines, there is no reason to believe that aspects of the macrostructure of cognition (as opposed to the microstructure) are amenable to a purely symbolic treatment. This point is made clearly by McClelland, Rumelhart & Hinton (1986:12):

In general, from the PDP [parallel distributed processing] point of view, the objects referred to in macrostructural models of cognitive processing are seen as approximate descriptions of emergent properties of the microstructure. Sometimes these approximate descriptions may be sufficiently accurate to capture a process or mechanism well enough; but many times ... they fail to provide sufficiently elegant or tractable accounts that capture the very flexibility and open-endedness of cognition that their inventors had originally intended to capture.

Consequently, symbolic approaches at a macrostructural level tend to produce models that are brittle, all or nothing, solutions. Symbolic macrostructural approaches to problem solving (hereafter referred to simply as 'symbolic approaches') have a long history of success in the rigor of logic, mathematics and the physical sciences.

Weizenbaum (1976) argues that this history has meant that many associate rigorous inquiry with symbolic formalisms to the effect that they are applied indiscriminately. Consequently, when the domain of a problem is inherently vague, either the vagueness is supplanted with concreteness or the problem is judged an inappropriate object of study.

Additionally, symbolic systems also suffer from what is known as the symbol grounding problem (i.e. the relationship between a word and the object it refers to is basically arbitrary). Modeling intelligence based on approximate descriptions and microstructure level of cognition has gained momentum to the emergence of computational intelligence (also known as soft computing). Additional aspects of intelligence, e.g. approximate or fuzzy reasoning, learning, prediction are being studied in this field (Khosla and Dillon 1997b; Zurada 1994). Unlike artificial intelligence, this field has largely grounded human intelligence in the human brain. A number of computational models have emerged including neural networks, genetic algorithms and fuzzy logic as outlined in section 3.2.1.

Although, the contributions made by artificial and computational intelligence fields have been significant, they still are fairly divergent contributions. These contributions have fallen short of providing a coherent view of human intelligence. Meanwhile, in the human sciences field there have also been significant efforts to model human intelligence. Well known contributions include the work of Boden (1983), and development of semiotics by Pierce and Morris (Pierce 1960; Morris 1971; Morris 1947).

Semiotics deals with basic ingredients of intelligence and their relationships. These ingredients are signs (representations), object (phenomenon), and interpretants (knowledge) as shown in Figure 3.5. The triple (sign, object, interpretation) represents a signic process, or semiosis (Morris 1971). Performing semiosis is to extract meaning of an object or phenomenon. In the process of extracting meaning it essentially studies the basic aspects of cognition and communication. Cognition means to deal with and to comprehend phenomena that occur in an environment. Communication means how a comprehended phenomenon can be transmitted between intelligent beings.

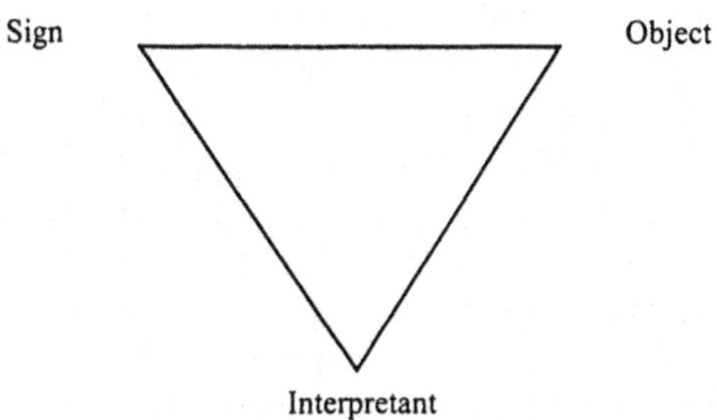

Figure 3.5: Triad of Signs

The basic unit of analysis in semiotics is a sign. Signs are representations, which are either internal or external to a cognitive system, which can evoke – in a cognitive system – internal representations called *interpretants* that represent presumable objects in the world. Interpretants too can act as signs evoking other interpretants. Interpretants only represent presumable objects because any cognitive system will not have absolute privileged access to knowledge about the world.

When we refer to 'dog' we are referring to a sign and not its referent, so 'dog' is a sign for dogs. Incidentally, this kind of sign is known as a *symbol* because it has a conventional and arbitrary relationship with its meaning.

While in some sense the meaning of something can be represented by a word, it is only through a process of decoding this representation that we understand it. To do this requires background knowledge about the meaning associated with it. Obviously, 'dog' does not mean anything unless one has had some kind of experience with dogs and is aware of the conventional relationship between the word and this experience.[2]

The experience of dogs is represented internally by a cognitive system and is distinct from a definition of 'dog' in terms of other high-level symbols. This is a point which seems to be only vaguely recognized by some, working on natural language or other symbolic systems who represent the decoded meaning (or interpretant) of a symbol exclusively in terms of other symbols (for example, a predicate calculus style representation). This practice only serves to specify the *relationships* between symbols in a symbol system without grounding their meaning in experience. One cannot recognize an object in the world without having some knowledge of what the sensory experience of that object (or the objects that comprise it) is like. This problem, called the 'symbol grounding problem' (Gudwin & Gomide 1997 a,b,c), is a recurrent problem in approaches to natural language and symbolic systems in general.

In his work Pierce (1960) developed three trichotomies of signs based on the original triad shown in Figure 3.5. In the first trichotomy a sign, according to Pierce (1960), is one of three kinds: Qualisign (a "mere quality or feeling"), Sinsign (an "actual existent or sensation") or Legisign (a "general law or rational thought"). The second trichotomy relates each sign to its object in one of three ways; as an Icon, Index or Symbol. Icon is "some character in itself", and can be classified as an image, diagram or metaphor. Index represents "some existential relation to an object," like a symptom is causally related to a disease. The symbol represents "some relation to the interpretant". Finally, in the third trichotomy each sign has an interpretant that represents the sign as a sign of possibility (Rheme), fact (Dicent) or reason (Argument). All the three trichotomies are shown in Table 3.1.

Pierce (1960) combined these three trichotomies to develop taxonomy of signs. A description of this taxonomy can be found in Sheriff (1989). Using the correspondence between signs and interpretants, Gudwin and Gomide (1997a) have adapted the taxonomy of signs drawn from semiotics to a taxonomy of associated knowledge types (see Figure 3.6). Figure 3.6 shows a modified version of the taxonomy as outlined by Gudwin and Gomicide (1997a). It includes the fusion, combination and transformation argumentative knowledge types described in section 3.2.1. We now provide a description of each of the knowledge types shown in Figure 3.6.

---

[2] Note that there are actually three distinct kinds of knowledge involved here. One relates to the experience of dogs, one to the experience of 'dogs' (the sign for dogs) and the other to the experience of the mapping between these two concepts.

Table 3.1: Three Trichotomies of Signs (adopted from Sheriff (1989))

A sign is:	a "mere quality" QUALISIGN	an "actual existent" SINSIGN	a "general law" LEGISIGN
A sign relates to its object in having:	"some character in itself" (e.g. metaphor) ICON	"some existential relation to that object" (e.g. symptom to a disease) INDEX	"some relation to the interpretant" SYMBOL
A sign's interpretant represents it (sign) as a sign of:	"possibility" RHEME	"fact" DICENT	"reason" ARGUMENT

3.3.1.1. Rhematic Knowledge

Rhematic knowledge as shown in Figure 3.6 has three types: symbolic, indexical and iconic. Symbolic rhematic knowledge is knowledge of arbitrary names like 'dog' or any symbol which has a conventional and arbitrary relationship to its referent. Indexical rhematic knowledge is knowledge of indices - signs that are not conventionalized in the way symbols are but are indicative of something in the way that smoke is indicative of fire. Indices as mentioned earlier are signs by virtue of their relationships with other phenomena. Such relationships can be causal, spatial, temporal or anything else that has the effect of associating a sign with a phenomenon.

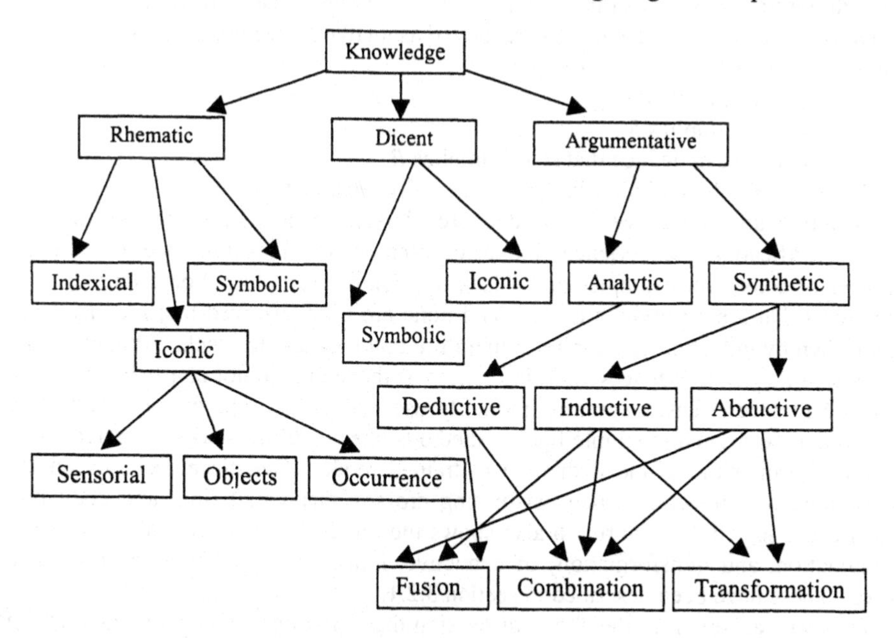

Figure 3.6: Knowledge Types (modified and adapted from Gudwin and Gomide 1997a)

Iconic rhematic knowledge is knowledge of signs that resemble their referents or provide direct models of phenomena. As such, icons unlike symbols are not arbitrarily related to their referents. There is a further subdivision of iconic rhematic knowledge into *sensorial*, *object* and *occurrence* knowledge types. Sensorial knowledge is knowledge from the senses or information to be sent to actuators. It involves interaction with the environment of the cognitive system. Object knowledge is an abstraction of sensory patterns representing an object in the world. Occurrence knowledge is knowledge of events, sequences of events, and involves actions.

3.3.1.2. Dicent Knowledge

Dicent knowledge employs truth-values (or degree of membership) to link sensorial, object or occurrence knowledge to world entities. It has two types: iconic and symbolic. Iconic propositions are propositions whose truth-values are derived directly from iconic rhematic knowledge (i.e., from experience). Symbolic propositions are names for other propositions, which may be either iconic or symbolic. Their truth-values match those of their associated propositions.

Symbolic propositions have been used in classical symbolic logic where we assume that the truth-value of the symbolic propositions are given. They have 0 or 1 truth-value, where truth-value of 1 means the proposition is a fact. They do not involve the semantic complexity (e.g. fuzzy membership) of iconic propositions where the truth-value lies in the interval 0 to 1.

3.3.1.3. Argumentative Knowledge

Finally, argumentative knowledge is knowledge used to generate new knowledge through inference or reasoning. There are three types: deductive, inductive and abductive. Deductive inference is categorized as analytic, meaning it does not require knowledge of the world. For example:

if P is equivalent to Q
and if Q implies R
then we can deduce that P also implies R.

This is true regardless of what P, Q and R actually represent.

Inductive and abductive knowledge are classed as synthetic because they do require verification in experience. Inductive inference involves inference from a large number of consistent examples and a lack of counter examples. For example, if all the crows that are observed are black and none are ever observed that are any other color, then by induction one can infer therefore that they are all black. Abduction is a method of inference that sees valid inferences as those that do not contradict previous facts. Knowledge based systems implicitly use deductive, inductive and abductive knowledge. Computational intelligence methods like neural networks are inductive in training/learning mode and deductive in trained mode. Genetic algorithms on the other hand use induction when performing crossover and mutation, and abduction when using selection. Figure 3.6 also shows the three hybrid configurations (fusion, combination, and transformation) of deductive, inductive and abductive knowledge types. These have been described in section 3.2.1.

Thus we can see from the above discussion that semiotics develops models which are deeper than those developed in artificial intelligence and computational intelligence. Unlike artificial and computational intelligence it considers multiple facets of intelligence through various knowledge types. It supports Weizenbaum's (1976) intuition that intelligent systems are better served by a variety of both symbolic and non-symbolic representational methodologies than by purely symbolic accounts of meaning. The suggestion is not that some aspects of cognition aren't best thought of as the manipulation of symbols, but rather that symbolic approaches cannot account for all aspects of human intelligence.

Gudwin and Gomide (1997a,b, c) have used the three knowledge types (rhematic, dicent and argumentative) in the interpretant space to develop a computational semiotics framework for building intelligent systems. They demonstrate the applicability of their semiotic framework through a control application. Although, the framework seems to be useful for small applications, for large-scale complex applications problems like combinatorial explosion, scalability and maintainability problems are likely to be encountered. Further, the use of primitive knowledge types does not throw any light on how to deal with complexity of large-scale problems. One can get lost in the details of interaction of various knowledge types.

3.3.2. Cognitive Science Theories

In the last section we have looked at human cognition based on semiotics which has its underpinnings in philosophy. In this section we discuss the developments in cognitive science and how they contribute to human-centeredness. We compare and contrast four approaches in cognitive science, namely, traditional approach, radical approach, situated cognition, and distributed cognition. The comparison primarily centers around factors related to human problem solving like external and internal representations, perceptual and cognitive processes, and task context (discussed in chapter 1). External representations are defined in terms of knowledge and structure of the external environment (Zhang and Norman 1994). They can be objects, physical symbols, or dimensions, and external rules, constraints or relations embedded in physical configurations (e.g., visual and spatial layouts of objects in an image, spatial relations between objects in an image). In comparison, internal representations are the knowledge and structure in memory (Zhang and Norman 1994). They can be propositions, productions/rules, schemas, neural networks, etc. Perceptual processes are used to analyze and process external representations, whereas cognitive processes are used to retrieve the information from the internal memory. The four approaches will now be discussed based on these factors.

3.3.2.1. Traditional Approach

The traditional approach (see Figure 3.7) developed by Newell (1990) primarily focuses on internal representations. According to the traditional view, external representations are merely inputs and stimuli to the internal mind. Thus when an intelligent agent has to accomplish a task which involves interaction with the environment, it creates an internal model of the environment through an encoding process, performs mental computations on the contents (symbolic or subsymbolic) in this internal model, and externalizes the output through a decoding process.

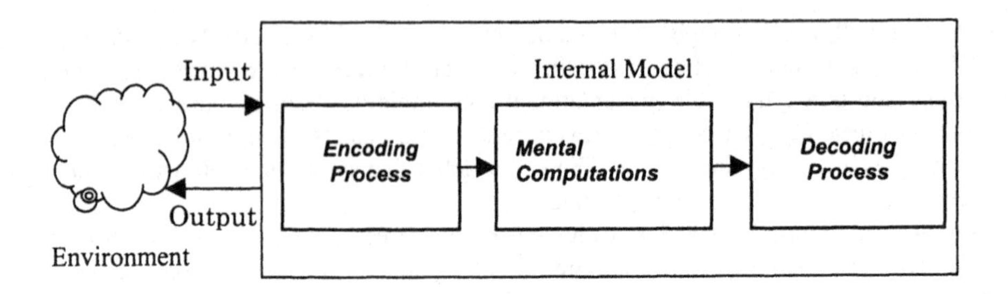

Figure 3.7: Traditional Approach to Cognitive Science

As noted by Zhang and Norman (1994), Kirlik et al. (1993) and Suchman (1987) most studies in traditional cognitive science do not separate external representations from internal representations or equate representations having both internal and external components to internal representations. This confusion often leads one to postulate unnecessary complex internal mechanisms to explain the complex structure of wrongly identified internal representation, much of which is merely a reflection of the structure of the external representation. More so, computer systems developed based on this traditional view often are cognitively rich and perceptually poor leading to a higher cognitive load on its users.

In general, traditional cognitive science based on the physical symbol hypothesis has other problems like the symbol grounding problem discussed in the previous section, frame problem, problem of 'situatedness' (idea that agent's actions are determined through the interaction of the agent with the current situation), lack of robustness (fault and noise tolerance, generalization capacity, adaptability to new situations), failure to perform in real time, and not sufficiently brain-like (Brooks 1991; Clancey 1989, 97 Dorffner 1996; Dreyfus 1992; McClelland and Rumelhart 1986; Harnad 1990; Suchman 1987).

3.3.2.2. Radical Approach

A radically different view (also known as the ecological view) proposed by Gibson (1966, 79) argues that perception is a direct process in which information is simply detected rather than constructed. It is based on the premise that brain is all there is and it is not a computer. According to Gibson, the environment is highly structured and full of invariant information. This invariant information in the environment is adequate to specify the objects and events in the environment, and thus it is sufficient for perception and action. Further, the invariant information can be directly perceived without the mediation of memory, inference, deliberation or other mental computations.

One of the central concepts of this approach is the notion of affordances. That is, what we see as the behavior of a system, object or event is that which is afforded or permitted by the system. For example, in Figure 3.8a the door handle affords grasping or pulling action whereas in the configuration shown in Figure 3.8b it affords a pushing action as against a grasping action. In other words, when the affordances are ambiguous, it is easy for us to make mistakes when trying to interact with an object.

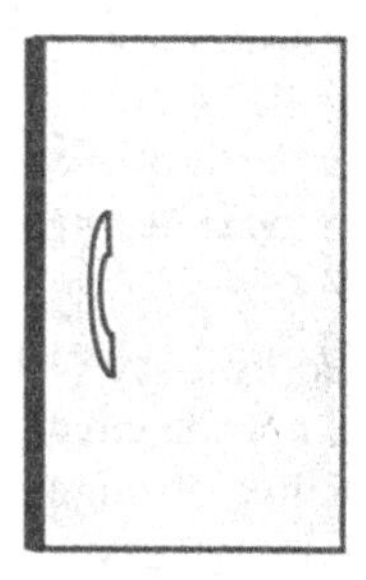

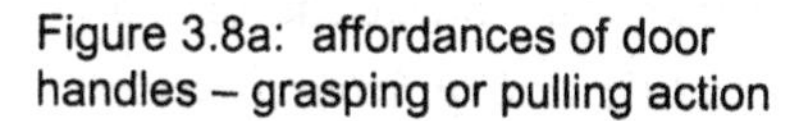

Figure 3.8a: affordances of door handles – grasping or pulling action

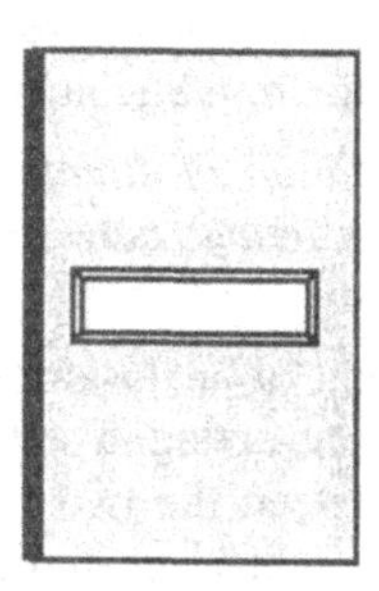

Figure 3.8b: affordances of door handles – pushing action

The strength of the radical approach, from a human-centered perspective, is that it emphasizes the need to consider perceptual aspects of a domain which have been largely ignored by traditional cognitive science. The perceptual perspective can assist in developing systems that minimize cognitive load on its users through direct manipulation. Other strengths include their ability to address fundamental problems like symbol grounding and emergence, and performance problems like real time response and robustness.

The main weakness of this approach is its over emphasis on perceptual aspects of a domain. Systems built with perception and action approaches do not scale-up well to large and complex domains. In complex systems, in order to deal with complexity of the domain (e.g., through abstraction) and guide or prevent the perceptual processes from going astray, it becomes necessary to assist this approach with cognitive processes. The next section describes the situated cognition approach, an emerging field in cognitive science.

3.3.2.3. Situated Cognition

Situated cognition approach has been developed and advocated by a number of researchers (Clancey 1993 1997; Suchman 1987; Brooks 1990, 91; Nehmzow et al. 1989; Winograd and Flores 1986; Pfeifer and Redamakers 1991; Pfeifer and Verschure 1992a, 92b, 95). According to this approach, people directly access situational information and act upon it in an improvisatory, adaptive and emergent manner, rather than in a routine and predictable manner.

Situated action refers to the idea that an agent's actions are determined through the interaction of the agent with the current situation. Situated action directly grows out of the particularities of a situation. Thus the focus of study is the situated activity or practice rather than the formal or cognitive properties of artifacts (Nardi 1996). Situated action is not purely reactive as it depends on the agent's experience, which is not based on a prior knowledge but is acquired through interaction with a situation.

Situated agents, unlike traditional AI systems, are adaptive and can act in real time. Situated models or systems do not passively receive and process input but are inextricably embedded in their environment and in a constant sensori-motor loop with it via the system's own actions in the environment. Situated systems are developed at a very fine-grained level of minutely observed activities that are embedded in a particular situation. This is reflected in the work of Clancey (1997) and Suchman (1987). To quote from Suchman (1987):

The organization of situated action is an emergent property of moment-by-moment interactions between actors, and between actors and the environment of their action.

The basic unit of analysis for situated action as identified by Lave (1988) is "the activity of persons-acting in setting." It is not the individual nor the environment, but the relation between the two. By paying attention to the flux of ongoing activity, situated action emphasizes improvizatory nature of human activity (Lave 1988). Lave (1988) illustrates such improvisation through the well known "cottage cheese" story.

A participant in a Weight Watchers program had the task of fixing a serving of cottage cheese that was to be three quarters of the two-thirds cup of cottage cheese.

The participant after puzzling over the problem a bit, filled a measuring cup two-thirds full of cheese dumped it out on a cutting board, patted it into a circle. Marked a cross on it, scooped away one quadrant, and served the rest of it.

Thus, by emphasizing particularities of a situation in minute detail, situated action de-emphasizes generalization and regularities which span across situations and are essential for dealing with large and complex systems.

Situated action analysis relies on recordable, observable behavior. In this analysis, since ongoing action directs the flow of human action, goals and plans are considered "verbal interpretations" (Lave 1988) and plans as "retrospective reconstructions" (Suchman 1987). As illustrated by Nardi (1996), a meteorologist and a bird watcher can both be looking at the sky with different goals in mind. A meteorologist may be looking skyward to determine the weather whereas, a bird watcher may be looking for birds. The situation is the same in both cases, however the goals are different. A video recording has no way of determining what is in the mind of two individuals.

The development of 'New AI' based on completely autonomous systems (Brooks 1990, 91a, 91b; Nehmzow et al. 1989; Pfeifer and Verschure 1992) and radical connectionism based on self-organization (automatic adaptation via feedback through the environment) subscribe to the purist view of ongoing interaction modeled by situated models. Radical connectionism and other bottom-up approaches (e.g., Edelman 1987; Reeke and Edelman 1988; Edelman 1989; Dorffner 1996) deny the need of internal models. The goal of these approaches is biased towards understanding the nature of natural behaving systems, like animals, than towards developing complex real world applications in design, diagnosis, scheduling, etc.

A more moderate or 'brain-like' version called connectionism has been advocated by Smolensky (1988). It advocates integration of the symbolic and subsymbolic (or microfeature) levels in a connectionist framework using artificial neural networks. The microfeatures, at the subsymbolic level, are distributed and do not have conceptual semantics individually. When considered together as patterns of activity, microfeatures are capable of producing emergent symbolic behavior. Using the massive parallelism and other properties of neural networks, connectionist models also help to alleviate performance-related problems of a traditional approach like real-time response, noise tolerance and generalization. A number of applications of the connectionist model can be found in natural language (Sejnowski and Rosenberg 1987) and other areas (Sun 1991,94). In general the connectionist approach helps to bridge the gap between existing computer models (based on traditional approach) and the brain. The moderate view however, is not without its criticisms. The encoded microfeatures, although subsymbolic, represent a thinner slice of the symbolic world and hence do not adequately address the symbol grounding problem. Further, connectionist models are seen to lack situatedness in the sense that ontology is given by the designer rather than developed independently by the network.

Overall, situated cognition theory has generated a lot of interest. However, it is still not fully developed and there is no set methodology for building real world applications. The research and application stances vary from a purist view of completely autonomous systems to recent attempts by situated cognition researchers to include "routine practices," "routine competencies," and abstraction theories (Sunchman and Trigg 1991; Suchman 1993; Clancey 1997) to account for the observed regularities in the work settings studied. Although, the purist view seems pretty much grounded in human-centeredness, focusing on moment-by-moment interaction, one can get involved into a myriad of details which can make building of large scale systems cumbersome.

Next, we look at another emerging field of cognitive science, namely, distributed cognition.

3.3.2.4. Distributed Cognition

Distributed cognition is an emerging framework which describes cognition as it is distributed across individuals and the setting in which it takes place (Hutchins and Norman 1988; Norman 1988, 91, 93; Zhang and Norman 1994). To quote from Flor and Hutchins (1991):

> *Distributed cognition is a branch of cognitive science devoted to the study of: the representation of knowledge both inside the heads of individuals and in the world ...; the propagation of knowledge between different individuals and artifacts....; and the transformations which external structures undergo when operated on by individuals and artifacts...; By studying cognitive phenomena in this fashion it is hoped that an understanding of how intelligence is manifested at the systems level, as opposed to the individual cognitive level, will be obtained.*

There are two parts in the above definition. The first part relates to the consideration of external (outside the head) and internal (inside the head) representations of artifacts. Problem solving then involves combination of perceptual and cognitive processes that involve external (perceptual) and internal (cognitive) representations (Hutchins 1991; Norman 1992; Zhang and Norman 1994) as shown in Figure 3.9. Zhang and Norman (1994) and Zhang (1997) have illustrated the effect of external and internal representations on problem solving and the cognitive work involved using the tic-tac-toe problem.

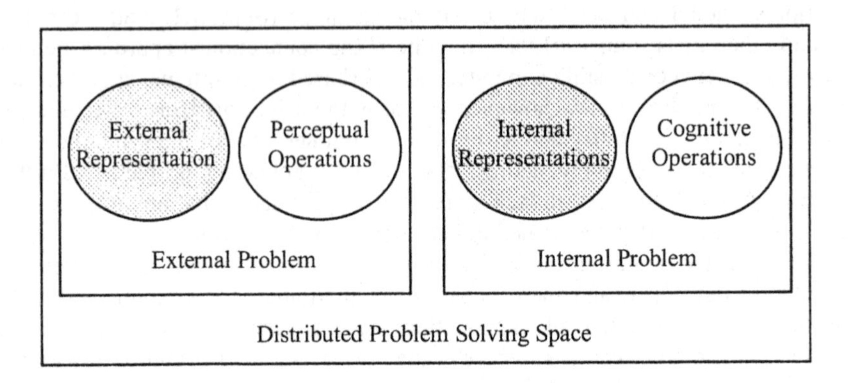

Figure 3.9: Distributed Representations and Problems Solving

Figure 3.10 shows two representations of the tic-tac-toe problem. In the first representation (on the right) a player has to color three squares in a straight line in order to win the game. In the second isomorphic representation (on the left) a player has to color three squares which add up to 15 in order to win the game. As also explained in chapter 1, problem solving in the first representation is accomplished using external representations or perceptual processes. The second isomorph involves cognitive operations (addition of three numbers). Thus depending on the representation chosen (also known as representational effect) the cognitive work involved will be different.

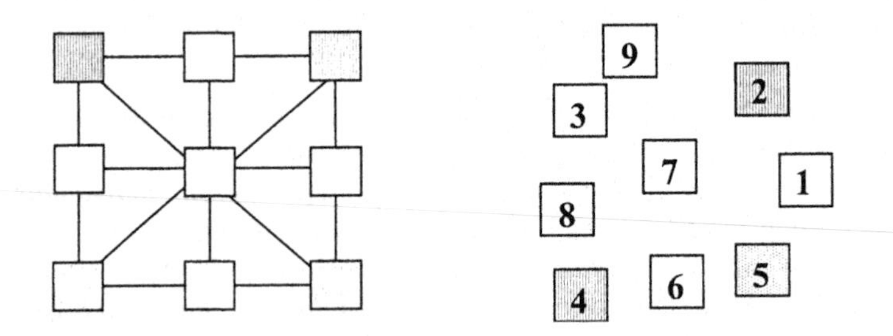

Figure 3.10: Tic-Tac-Toe

The second part of the distributed cognition definition relates to conceptualization of cognitive activities as embodied and situated within the work context in which they occur (Hutchins 1990; Hutchins and Klusen 1992). In other words, it involves describing cognition as it is distributed across individuals (as against its embodiment within an individual) and setting in which it takes place. It involves development of functional systems that determine the relations between a collection of actors or people, computer systems and other artifacts as situated in an environmental setting. Thus distributed cognition shifts the unit of analysis from an individual to the system and its components. At the systems level its primary goal is to analyze how different components of the functional system are coordinated. A number of functional systems like software programming teams (Flor and Hutchins 1991), ship navigation (Hutchins 1990), air-traffic control have been studied in this regard.

A common aspect of situated cognition and distributed cognition is the shift towards real activity in real situations. The difference is lack of goals in situated cognition as against system goals and motives. That is, in distributed cognition system goal or goals is/are the beginning point of analysis. In situated cognition the reference point in moving from one situation into another is not a goal or motive. In other words, condition for situated action does not have to be a goal but a response to dynamically changing conditions in the environment.

This completes the four main cognitive science theories. In the next section we delve into activity theory, an enabling theory from psychology.

3.3.3. Activity Theory

Activity theory has recently gained importance in human-computer interaction for analyzing user interfaces (Bodker 1991), Computer Supported Work Systems (Knutti 1991). Activity is generated by various needs for which people want to achieve a certain purpose or goal. The origins of activity theory can be found in the work done by psychologist Leont'ev (1974), Vygotsky (1978), and more recently by Nardi (1993, 96) and Kuutti (1996). The basic components of activity theory as outlined by Knutti (1996) are shown in Figure 3.11. Knutti (1996) defines an activity as a form of human doing whereby a subject works on an object (as in objective) in order to attain the desired outcome. An object can be a material thing (produce a new car), but it can also be less tangible (satisfying customer need). Activities are distinguished from each other according to their objects. Transforming the object into an outcome motivates the existence of an activity. The subject can be a person or a group of persons involved in an activity. An object (in the sense of objective) is held by the subject and motivates activity, giving it a specific direction (Leont'ev 1974). Behind the object there always stands a need or desire/motive, to which (the activity) always answers (Leont'ev *1*974). Tools or artifacts usually mediate activity. The tool in Figure 3.11 represents a transformation process employed by the subject (or subjects) to transform the object into a desired outcome. The tools can be material tools (e.g., axe, computer systems, procedures) and/or tools of thinking (e.g., plan). For example, in an automated supply system, the motive of a subject (e.g. database designer) may be, improved supply management, career advancement or gaining control over a vital organizational power source). In the supply system, a database designer who is modifying a database schema (object) so that all year columns are four digits (outcome) using a schema editor (tool).

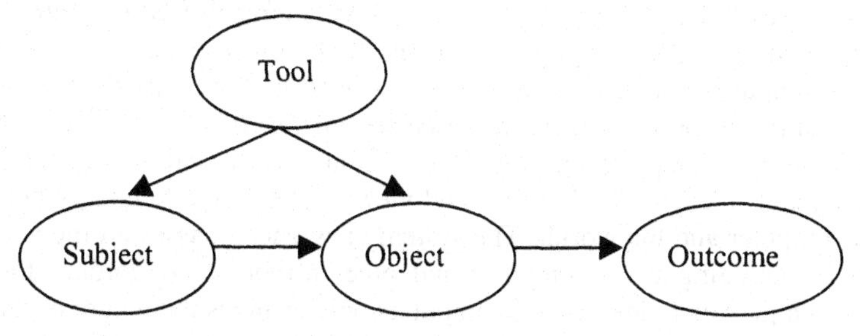

Figure 3.11: Components of Activity

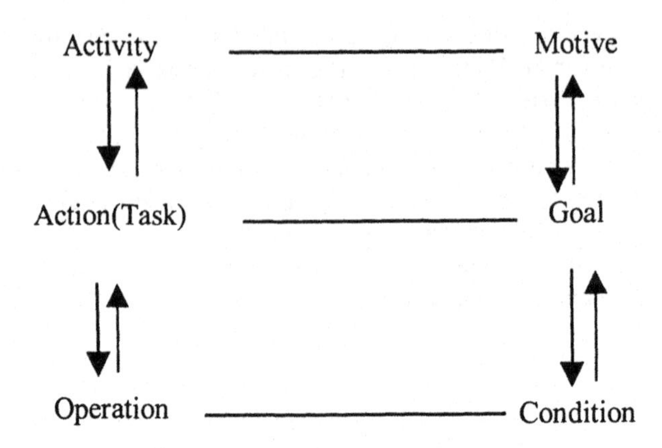

Figure 3.12: Levels of Activity

Activities as described by Leont'ev, (1974), Nardi (1993,96), Knutti (1996) and Kaptelinin (1996) are composed of actions and operations. The actions are goal-directed processes that are undertaken to fulfill an object. The actions are similar to tasks referred to in AI literature (Chandrasekaran et. al. 1992) and human-computer interaction literature (Norman 1991). Operations are low level refinements of actions. They can also be seen to represent unconscious, routinized or automated aspects of actions. For example, when learning to drive a car, changing the gears is a conscious action. However, with practice this conscious action becomes a routinized, unconscious operation.

The constituents of an activity can dynamically change as the conditions change. That is, all levels of activity can move up or down. For example, moving from a country with left-hand drive to a country with right hand drive can result in relearning certain operationalized aspects of driving (e.g. rules for turning left or right).

Like distributed cognition, the unit of analysis in activity theory is an activity or work system. However, unlike distributed cognition where the focus is on system goal or goals, the focus in activity theory is on individual goal or goals. One can say there is an overlap between individual goals and system goals (Nardi 1996), the focus in distributed cognition is to direct analysis at the systems level rather than the individual level.

One of the important contributions of activity theory is the tool mediation perspective. Besides activity theory, the cognitive approach has also introduced the mediation concept called "cognitive artifacts." Norman (1991) defines them as follows: "A cognitive artifact is an artifact designed to maintain, display, or operate upon information in order to serve representational function." Norman (1991) uses this definition to distinguish between the personal view and the system view to human-computer interaction. The personal view relates to the boundary between the user, and computer and the world. The system view encompasses all the components of a system including user, computer and other artifacts. According to Norman (1991), from a personal view, use of computers only changes the nature of the task (to the extent it can make a task easier for the user). It (i.e., use of computers) does not necessarily, empower a user (say, with a skill), from a personal view. The empowerment is true only from a systems view, where certain performance improvements can be realized through use of computers with people and other artifacts. On the other hand, activity theory subscribes to only one view and that is personal view. Further, it states that tools (like computers) not only change the nature of the task but also empower the user or individual even if the external tool is no longer used. Kaptelinin (1996) outlines three stages which lead to empowerment: a) the initial phase, when performance is the same, with or without the tool, because the tool is not mastered well enough to provide the benefits; b) the intermediate phase, when aided performance is superior to unaided performance, and c) the final phase, when performance is the same, with and without the tool, but now because the tool mediated activity is internalized and the external tool is no longer needed. For example, a fresh sales recruit may use a computer based training tool for enhancing their salesperson-customer interaction skills on a day-to-day basis. After a few years the salesperson may not need the training tool because of internalization of the training skills imparted by the tool.

The major limitation of activity theory is that it is not operationalized enough (Kaptelinin 1996). That is, methods and techniques are not rich enough to be directly utilized for developing workplace systems.

3.3.4. Workplace Theory

Till this point in the chapter we have looked at pragmatic considerations which establish a need for human-centered systems, and theories which can contribute towards development of human-centered systems. Another important aspect that has recently gained momentum is the need to situate computer system development in a work environment. That is, rather than engaging in an objective isolated rationalistic approach which may involve use of certain systems methodology (e.g. object-oriented, logic, etc.) there is a need to introduce certain subjectivity through a work-oriented design approach (Ehn et al 1989). This subjectivity entails involvement of stakeholders and end users in the system design process. This perspective has grown out of dissatisfaction experienced by practitioners and researchers (Ehn et al 1989; Greenbaum and Kyng 1991; and others) in workplace environments with traditional (formal) theories and methods of systems design. Workers or users see these approaches as politically motivated to deskill them. More so, human creativity and intelligence is restricted to the limited vocabulary of these methods.

Ehn et al (1989) has suggested a rethink in the existing design processes to include structures which ordinary people can use to incorporate their own interests and goals. Besides, descriptions in design should be flexible enough to enable users to express all their practical competence. Here again, two stances are emerging for developing such structures. One is largely based on ethnographic techniques and the other is based on Work-Centered Analysis (WCA) with emphasis on user and stakeholder expectations, and organizational culture. Both techniques have merit. The ethnographic techniques (which involve use of video and audio techniques for recording every minute aspect of work activity) seem to be grounded in human-centeredness, although the whole process can really become cumbersome for development of large and complex systems. The work-centered analysis and socio-technical framework developed in the information systems area by Alter (1996,99) and Laudon and Laudon (1995), respectively, have been derived from workplace settings and are based on a marriage between social and systemic aspects of computer system development. In this book, the latter approach has been adopted. These structures facilitate development of work-oriented systems by situating system development in organizational and stakeholder contexts.

The WCA framework defined by Alter (1996) is a comprehensive framework for analyzing a business process and use of information technology from a business as well as human perspective. However, for the purpose of building intelligent multimedia systems it does not provide an adequate framework for integrating the business and human perspective with the technical perspective of developing complex intelligent systems.

3.4. Discussion

In the preceding sections of this chapter, we have attempted to describe how pragmatic issues rather than purely technological innovation (as has largely been done in the past) are driving the evolution of various areas in computer science and information systems. The enabling theories from philosophy, cognitive science, psychology, and workplace, which can help to address the human *vs.* technology mismatch and other pragmatic aspects, have been described. In this section, we think it is a good idea to recapitulate the pragmatic and theoretical aspects that will form part of the human-centered framework developed in the next chapter.

In this book the fact that we want to develop intelligent multimedia multi-agent systems, the obvious starting point is intelligent systems described in section 3.2.1. Given the range of intelligent technologies, their strengths and weaknesses, their cognitive, neuronal and evolutionary underpinnings, we showed how task orientation can prove to be an important concept in harnessing the deliberative, fuzzy, reactive, self-organizing, parallel processing and optimizing properties of these intelligent technologies. Task orientation is also an essential component of human-centered software development. We intend to integrate this concept into the human-centered framework.

In a sense, like the intelligent technologies, various software engineering technologies have functional and structural underpinnings. The functional and structural underpinnings of these software engineering technologies impose restrictions on which aspects of a domain need to be captured, and how they are to be modeled. That is, as indicated in section 3.2.3 agents are intuitively suited for modeling of tasks and task based behavior and not for structural modeling, whereas, objects are intuitively suited for structural modeling of a domain. Thus, their inability to capture all aspects of a domain on their own impacts on the syntactic and semantic quality of a computer system from a human-centered standpoint. In our work, we intend to incorporate concepts from object-oriented and agent technologies in the human-centered framework to enhance the syntactic and semantic quality of the framework.

The development of multimedia databases has demonstrated that the conventional DBMS (DataBase Management Systems) are not adequate to handle queries associated with multimedia databases. This aspect is covered in more detail in chapter 9. These queries require semantic correlation between various media artifacts. This has resulted in a need for defining an ontological level above the metadata level associated with each media. This ontological level has more to do with the problem solving ontology of a domain rather the metadata aspects of the media or the media structure. Thus it is media independent and can be domain dependent or independent.

Unlike traditional commerce, most of the electronic commerce systems on WWW are supply centered. As a result, unlike the traditional customer-centered markets, it has become difficult for the customers to obtain the best product at the best price in the electronic marketplace using the general internet search engines. There is a need for developing user centered models or problem solving ontologies that can assist in moving supplier-centered models closer to the traditional customer-centered commerce models.

Data mining is an emerging field and is in high demand in the industry. However, the technology-centered methods used for data mining are beset with problems like meaningfulness and relevance of extracted rules. Here again, there is a need to integrate user-centered problem solving models with the data mining process in order to enhance the meaningfulness of its results.

The technology based vs. task based intelligent associative systems, human cognition based ontological level vs. metadata level in multimedia databases, supplier *vs.* customer-centered models in electronic commerce systems, meaningfulness problem associated with data mining, task *vs.* feature oriented interfaces and underlying user tasks *vs.* underlying system tasks problem of existing user interfaces all suggest a need for task orientation and technology independent problem solving ontologies. Thus we intend to integrate concepts like task orientation, technology independent human-centered problem solving ontology and task based human-computer interaction as building blocks of the human-centered framework. These concepts answer the question, "What needs to be done?" In order to answer the question "How can it be done," we have looked into enabling theories in philosophy, cognitive science, psychology and workplace.

The semiotic, cognitive science, activity and workplace system theories described in this chapter contribute significantly to the content of human-centered systems framework. These theories represent diverse aspects related to human-centeredness. The discussion and comparison of these theories leads us in the direction of amalgamation of concepts from these theories rather than committing ourselves to one particular theory. We feel amalgamation of concepts from various theories will help us to satisfy the pragmatic needs and develop a more comprehensive human-centered framework. In the rest of this section, we highlight various concepts developed by these theories which will form part of the human-centered systems framework.

Semiotic theory covered in this chapter describes a human cognitive system based on taxonomy of linguistic and non-linguistic signs. Any intelligent human-centered system should facilitate a symbiosis between the linguistic and non-linguistic nature of human communication and human-computer interaction. The four cognitive science theories discussed in this chapter model the linguistic and non-linguistic nature of human communication separately and collectively. The traditional approach is heavily biased towards the linguistic nature of human communication and relies on cognitive processes of humans. On the other hand, the radical and situated cognition approaches are more biased towards the dynamic and non-linguistic nature of human interaction. We subscribe to the dynamic aspects of situated cognition that involve, among other aspects, ability to adapt to novel situations and learn incrementally from them. The linguistic and non-linguistic nature of human communication is also linked to the cognitive and perceptual processes of humans. The distributed cognition approach models linguistic and non-linguistic aspects collectively through cognitive and perceptual models. Problem solving in distributed cognition takes place in a distributed problem space involving external and internal representations. Further, distributed cognition, like activity theory, is concerned with finding stable design principles across design problems (Norman 1988, 91; Nardi 1996; Nardi and Zarmer 1993). Like the workplace system theory, the unit of analysis in distribution cognition is a system. The system consists of people and artifacts working in a cooperative and coordinative manner to accomplish system goals. We also subscribe to these aspects of distributed cognition, activity theory and workplace, namely, system as the unit of analysis, system goals, distributed problem solving space, external and internal representations developing stable design principles across problems. However, unlike distributed cognition, we do not equate the human with the machine. We are more comfortable with the tool view as advocated by the activity theory. Computers are tools like other artifacts that enable people to accomplish their tasks. They can be programmed to exhibit certain intelligent properties like humans (and to that extent they can be considered as cognizing tools) but to a large extent are still not close to the range of intelligent behavior exhibited by humans, which includes emotions and other aspects.

Finally, the organizational and stakeholder context of workplace system theory described in section 3.3.4 provides us with a basis for putting people at the forefront of the system development process. In the next two chapters we develop various components of the human-centered framework based on the pragmatic and theoretical aspects described and discussed in this chapter.

3.5. Summary

This chapter looks into various pragmatic issues and the enabling theories for development of intelligent human-centered systems in particular and human-centered systems in general. The pragmatic issues primarily center around the human *vs.* technology mismatch and the epistemological limitations (and strengths) which humans and computers have. It is shown how these pragmatic issues have become a pivotal point in the evolution in a number of areas including intelligent systems, software engineering, multimedia databases, electronic commerce, enterprise modeling, data mining and human-computer interaction. The outcome of the discussion on pragmatic issues and their impact on various areas helps us to identify a set of critical properties that need to form a part of a human-centered system development framework. Whereas, the pragmatic issues help us identify some of the critical properties of a human-centered system development framework, they do not provide us with a theoretical basis for underpinning the framework. For that matter, the chapter then moves on to describe enabling theories in philosophy, cognitive science, psychology, and workplace for development of human-centered system development frameworks. The outcome of discussion on these theories is a set of theoretical concepts on which human-centered system development framework is founded in the next chapter.

References

Albus, J.S. (1991). "Outline for a Theory of Intelligence," in *IEEE Transactions on Systems, Man and Cybernetics* 21(3), May/June.

Anderson, J. and Stonebraker, M. (1994), "Sequoia 2000 Metadata Schema for Satellite Images," in *SIGMOD Record, special issue on Metadata for Digital Media*, W. Klaus, A. Sheth, eds., 23 (4), December, http://www.cs.uga.edu/LSDIS/pub.html.

Aristotle. (1938). *De Interpretaione* (H. P. Cook, Trans.), London: Loeb Classical Library.

Bezdek, J.C. 1994, 'What is Computational Intelligence?' *Computational Intelligence: Imitating Life*, Eds. Robert Marks-II et al., IEEE Press, New York.

Boden, M.A. (1983). "As Ideias de Piaget," in Traducao de Alvaro Cabral – Editora Cultrix – Editora da Universidade de Sao Paulo.

Bodker, S. (1991). *Through the Interface: A Human Activity Approach to User Interface Design*, Hillsdale, NJ:Lawerence Erlbaum

Bohm, K., and Rakow, T. (1994). "Metadata for Multimedia Documents," in SIGMOD Record, special issue on Metadata for Digital Media, W. Klaus, A. Sheth, eds. 23 (4), December, http://www.cs.uga.edu/LSDIS/pub.html

Brooks, R.A. (1990). Elephants Don't Play Chess, in P. Maes, ed., *Designing Autonomous Agents*, Cambridge. MA: MIT Press, Bradford Books, pp. 3-16.

Brooks, R.A. (1991a). Intelligence Without Representation, *Artificial Intelligence* 47 (1-3), Special Volume: Foundations of Artificial Intelligence.

Brooks, R.A. (1991b). Comparitive Task Analysis: An Alternative Direction for Human-Computer Interaction Science. In J. Caroll, ed., *Designing Interaction: Psychology at the Human-Computer Interface*, J. Caroll, ed., Cambridge: Cambridge University Press.

Chandrasekaran, B., Johnson, T.R., and Smith, J.W. 1992, 'Task Structure Analysis for Knowledge Modeling,' *Communication of the ACM*, vol. 35, no. 9., pp. 124-137.

Chen, F., Hearst, M., Kupiec, J., Pederson, J., and Wilcox, L. (1994), "Metadata for Mixed-media Access," in SIGMOD Record, special issue on Metadata for Digital Media, W. Klaus, A. Sheth, eds., 23 (4), December, http://www.cs.uga.edu/LSDIS/pub.html.

Chiaberage, M., Bene. G.D., Pascoli, S.D., Lazzerini, B., and Maggiore, A. 1995, "Mixing fuzzy, neural & genetic algorithms in integrated design environment for intelligent controllers," *1995 IEEE Int Conf on SMC*,. Vol. 4, pp. 2988-93.

Clancey W.J. (1989). "The Knowledge Level Reconsidered: Modeling How Systems Interact," in Machine Learning 4, pp.285-92.

Clancey, W.J. (1993). "Situated Action: A Neuropsychological Interpretation (Response to Vera and Simon)," *Cognitive Science*, 17, 87-116.

Clancey, W.J. (1997). *Situated Cognition*, Cambridge, MA: MIT Press.

Clancey, W.J. (1999). "Human-Centered Computing – Implications for AI Research," http://ic.arc.nasa.gov/ic/HTMLfolder/sld001.htm.

Coad, P. and Yourdon, E. (1990) *Object-oriented Analysis,* Yourdon Press, Prentice-Hall, Englewood Cliffs, NJ.

Coad, P. and Yourdon, E. (1991) *Object-oriented Design,* Yourdon Press, Prentice-Hall, Englewood Cliffs, NJ.

Dorffner G. (1996). "Radical Connectionism - A Neural Bottom-Up Approach to AI," in *Neural Networks and New Artificial Intelligence*, G. Dorrfner ed., International Thomson Press.

Dreyfus H.L. (1992). *What Computers Still Can't Do*, Cambridge, MA: MIT Press.

Edelman, G. 1992, *Bright Air, Brilliant Fire: On the Matter of the Mind.* New York, USA, Raven Press.

Edelman, G.M. (1987). *Neural Darwinism: The Theory of Neuronal Group Selection*, New York: Basic Books.

Edelman, G.M. (1989). *The Remembered Present: A Biological Theory of Consciousness*, New York: Basic Books.

Ehn, P. & Kyng, M. (1989) *Computers and Democracy: a Scandinavian Challenge*, edited by Bjerknes, G., Ehn, P. and Kyng, M. Aldershot [Hanks, England]; Aveburg, pp 17-57.

Flor, N., and Hutchins, E. (1991). "Analyzing Distributed Cognition in Software Teams: A Case Study of Team Programming During Perfective Software Maintenance," in J. Koenemann-Belliveau et al., eds., *Proceedings of the Fourth Annual Workshop on Empirical Studies of Programmers*, Norwood, N.J.: Ablex Publishing.

Fu,L.M.& Fu,L.C. 1990, "Mapping Rule based Systems into Neural Architecture".*Knowledge-Based Systems*. 3(1): 48-56.

Fukuda, T., Hasegawa, Y. and Shimojima, K. 1995, 'Structure Organization of Hierarchical Fuzzy Model using Genetic Algorithm'. *1995 IEEE International Conference on Fuzzy Systems*. vol. 1, pp. 295-9.

Gallant, S. 1988, "Connectionist Expert Systems". *Communications of the ACM.* February 152-169.

Gamma, E et. al., (1995) "Design Elements of Object-oriented Software," Massachusetts: Adisson-Wesley.

Gibson, J.J. (1966). *The Senses Considered as Perceptual Systems*, New York: Houghton Mifflin Company.

Gibson, J.J. (1979). *The Ecological Approach to Visual Perception*, Boston: Houghton Mifflin.

Glavitsch, U., Schauble, P., and Wechsler, M., (1994) "Metadata for Integrating Speech Documents in a Text Retrieval System," in SIGMOD Record, special issue on Metadata for Digital Media, W. Klaus, A. Sheth, eds., 23 (4), December, http://www.cs.uga.edu/LSDIS/pub.html.

Glushko R., Tenenbaum J. and Meltzer B. (1999), "An XML-framework for Agent-Based E Commerce", Communications of the ACM, vol. 42 no. 3.

Goldberg, D.E. (1989), *Genetic Algorithms in Search, Optimization and Machine Learning*, Addison-Wesley, Reading, MA, pp. 217-307.

Greffenstette, J.J. (1990), "Genetic Algorithms and their Applications," *Encyclopedia of Computer Science and Technology*, vol. 2, eds. A. Kent and J. G. William, AIC-90-006, Naval Research Laboratory, Washington DC, pp. 139-52.

Grosky, B. (1994), "A Primer on Multimedia Systems, " IEEE Multimedia pp. 12-24.

Gudwin R. and Gomide, F. (1997a). "Computational Semiotics: An Approach for the Study of Intelligent Systems - Part I: Foundations," *Technical report RT-DCA 09 - DCA-FEEC-UNICAMP.*

Gudwin R. and Gomide, F. (1997c). "An Approach to Computational Semiotics," In *Proceedings of the ISA'97 - Intelligent Systems and Semiotics: A Learning Perspective, International Conference*, Goithersburg, MD, USA, 22-25, September, 1997.

Gudwin R.and Gomide, F. (1997b). "Computational Semiotics: An Approach for the Study of Intelligent Systems - Part II: Theory and Application," *Technical report RT-DCA 09 - DCA-FEEC-UNICAMP.*

Hamada, K., Baba, T., Sato, K. and Yufu, M. 1995, "Hybridizing a Genetic Algorithm with Rule based Reasoning for Production Planning,". *IEEE Expert.* 60-67.

Hamilton S., "Electronic Commerce for the 21st Century", IEEE Computer, vol. 30, no. 5, pp.37-41

Hands, J., Patel A., Bessonov M. and Smith R. (1998), "An Inclusive and Hands Extensible Architecture for Electronic Brokerage", Proc. of the Hawai Intl. Conf. on System Sciences, Minitrack on Electronic Commerce, pp.332-339.

Harnad, S. (1990). "The Symbol Grounding Problem," in *Physica D*, 42 (1-3), pp. 335-46.

Hinton, G.E. 1990, "Mapping Part-Whole Hierarchies into Connectionist Networks," *Artificial Intelligence*. 46(1-2): 47-76.

Hutchins, E. (1990). "The Technology of Team Navigation," in J. Galegher, ed., *Intellectual Teamwork*, Hillsdale, NJ: Lawerence Erlbaum..

Hutchins, E. (1991). *How a Cockpit Remembers its Speeds*. Ms La Jolla: University of California, Department of Cognitive Science.

Hutchins, E. (1995). *Cognition in the Wild*, Cambridge, MA: MIT Press.

Ishibuchi, H., Tanaka, H. and Okada, H. 1994, "Interpolation of Fuzzy If-Then Rules by Neural Networks,". *International Journal of Approximate Reasoning*. January, 10(1): 3-27.

Jacobson, I., (1995). *Object-oriented Software Engineering*, Addison-Wesley.

Jain, R. and Hampapuram, A. (1994). "Representations of Video Databases," in SIGMOD Record, special issue on Metadata for Digital Media, W. Klaus, A. Sheth, eds., 23 (4), December, http://www.cs.uga.edu/LSDIS/pub.html

Jennings, N. R. et al. (1996). "Using Archon to Develop Real-World DAI Applications, Part 1." *IEEE Expert* December, 64-70.

Kaptelinin, V. (1996). "Computer-Mediated Activity," in *Context and Consciousness*, B. Nardi, ed., MIT Press, pp. 17-44.

Kashyap, V., Shah, K., and Sheth, A. (1995). "Metadata for Building the Multimedia Patch Quilt," in *Multimedia Database Systems: Issues and Research Directions*, S. Jajodia and V. S. Subrahmanium, Eds., Springer-Verlag, p. 297-323

Kashyap, V., and Sheth, A.(1994) "Semantics-based Information Brokering," in Proceedings of the Third International Conference on Information and Knowledge Management (CIKM), November, http://www.cs.uga.edu/LSDIS/infoquilt

Kashyap, V., Shah, K., and Sheth, A. (1996). "Metadata for Building the Multimedia Patch Quilt," in SIGMOD

Khosla, R., 1997a, ***Tutorial Notes on Software Engineering methodology for Intelligent Hybrid Multi-Agent Systems,*** Int. Conf. on Connectionist Information Processing and Information Sys, Dunedin, New Zealand, November

Khosla, R. & Dillon, T.S. 1997b, ***Engineering Intelligent Hybrid Multi-Agent Systems.*** Boston, USA, Kluwer Academic Publishers.

Khosla, R. and Dillon, T.S. 1997c, "Neuro-Expert System Applications in Power Systems,". K. Warwick, A. Ekwue & R.K. Aggarwal, eds. ***Artificial Intelligent Techniques in Power Systems.*** UK, IEE press, pp. 238-58.

Khosla, R and Dillon, T., 1997d, "Task Structure Level Symbolic-Connectionist Architecture," chapter in ***Connectionist-Symbolic Integration: From Unified to Hybrid Approaches,*** edited by Ron Sun and Frederic Alexandre, Lawrence Erlbaum Associates in USA, pp. 37-56, November 1997.

Khosla, R. & Dillon, T. 1997e, "Fusion of Knowledge-Based Systems and Neural Networks and Applications,". Keynote paper. *1st Int. Conf. on Conventional and Knowledge-Based Intelligent Electronic Systems.* Adelaide, Australia, May 21-23, pp. 27-44.

Khosla, R and Dillon, T., 1997f, *"Learning Knowledge and Strategy of a Generic Neuro-Expert System Arch. in Alarm Processing,"* in IEEE Trans. on Power Systems, Vol. 12, No. 12, pp. 1610-18, November.

Khosla, R. and Dillon, T.S. 1995a, "GENUES Architecture and Application,". In J. Liebowitz, ed. ***Hybrid Intelligent System Applications.*** New York, Cognizant Communication Corporation, pp. 174-99.
Khosla, R. and Dillon, T.S. 1995b, "Symbolic-Subsymbolic Agent Architecture for Configuring Power Network Faults,". *International Conference on Multi Agent Systems.* San Francisco, USA, June., pp 451
Khosla, R. and Dillon, T.S. 1995c, "Integration of Task Structure Level Architecture with O-O Technology,". *Software Engineering and Knowledge Engineering.* Maryland, USA, June 22-24, pp. 95-7.
Kirlik et al. (1993). " tain Environment: Laboratory Task and Crew Performance," in IEEE Transactions on Systems, Man, and Cybernetics 11(4), pp.1130-38.
Knutti, K. (1991). "Activity Theory and its applications to Information systems research and Development," in H.-E. Nissen, ed., Information Systems Research, Amsterdam: Elsevier Science Publishers, pp. 529-549.
Knutti, K. (1996). "A Framework for HCI Research," in *Context and Consciousness*, B. Nardi, ed., Mit Press, pp. 45-68.
Koenemann-Belliveau et al., eds., *Proceedings of the Fourth Annual Workshop on Empirical Studies of Programmers*, Norwood, N.J.: Ablex Publishing.
Laird, J., Rosenbloom, P. and Newell, A. (1987), "SOAR: An Architecture for General Intelligence" *Artificial Intelligence*, vol. 33, pp 1-64.
Laudon, K.C. and Laudon, J.P., 1995 *Management Information Systems*, Prentice Hall.
Lave, J. (1988). *Cognition in Practice*, Cambridge: Cambridge University Press.
Leont'ev, A. (1974). "The Problem of Activity in Psychology," in *Soviet Psychology*, 13(2):4-33.
Maes, P. (1994) "Agent That Reduce Work and Information Overload," Communications of the ACM, July, pp. 31-40.
Main, J., Dillon, T.. and Khosla, R. 1995, "Use of Neural Networks for Case-retrieval in a System for Fashion Shoe Design,".*Eighth Int. Conf. on Industrial and Engg Apps of AI & Expert Systems*.Melbourne, June, pp. 151-8.
McClelland, J. L., Rumelhart, D. E. and Hinton, G.E. (1986), "The Appeal of Parallel Distributed Processing," Parallel Distributed Processing, vol. 1, Cambridge, MA: The MIT Press, pp. 3-40.
McClelland J., Rumelhart D., et al. (1986) *Parallel Distributed Processing: Explorations in the Microstructure of Cognition*, Cambridge, MA: MIT Press.
Morris, C.W. (1971). "Foundations for a Theory of Signs," *in Writings on the General Theory of Signs*, The Hague: Mouton.
Morris, C.W. (1947). *Signs, Language and Behavior*, New York: Prentice Hall.
Nardi, B. (1993). *A Small Matter of Programming: Perspectives on End User Computing*, Cambridge: MIT Press.
Nardi, B. and Zarmer, C. (1993). "Beyond models and metaphors: Visual formalisms in user interface design," *Journal of Visual Languages and Computing*, March.
Nardi, B. (1996). "Studying Context: A Comparison of Activity Theory. Situated Action Models, and Distributed Cognition," in *Context and Consciousness*, B. Nardi, ed., MIT Press, pp. 69-103.
Nehmzow, U., Hallam, J., Smithers, T. (1989) "Really Useful Robots," *Proceedings of Intelligent Autonomous Systems* 2, Amsterdam.
Newell, A., (1980), "Physical Symbol Systems" *Cognitive Science*, vol. 4, pp. 135-183.
Newell A. (1990*). Unified Theories of Cognition*, Cambridge, MA: Harvard University Press.
Norman, D. (1988). *The Psychology of Everyday Things*, New York: Basic Books.
Norman, D. (1991). "Cognitive Artifacts," in *Designing Interaction: Psychology at the Human-Computer Interface*, J. Caroll, ed., Cambridge: Cambridge University Press.
Norman, D. (1993). *Things That Make Us Smart*, Reading, MA: Addision-Wesley.

Norman, D. and Hutchins, E. (1988). "Computation via direct manipulation," Final Report to Office of Naval Research, Contract No. N00014-85-C-0133. La Jolla: University of California, San Diego.

Nwana, H. S. and Ndumu, D. T. (1997) "An Introduction to Agent Technology," In Nwana, H. S. and Azarmi, N. (eds) *Software Agents and Soft Computing: Towards enhancing machine intelligence; concepts and applications*, Spinger- Verlag, pp. 3-26

Orfali, R. and Harkey, D., *Client/Server Programming with Java and COBRA*, John Wiley Computer Publishing

Perrow, C.1984, *Normal Accidents: Living with High-Risk Technologies*, Basic Books, New York.

Pfeifer, R. and Rademakers, P. (1991). "Situated Adaptive Design," in W. Brauer and D. Hernandez eds., *Kunstliche Intelligence und Kooperatives Arbeitein. Proceedings of the International GI Conference*, Berlin: Springer, pp. 53-64.

Pfeifer, R. and Verschure, P.F.M.J. (1992a). "Distributed Adaptive Control: A Paradigm for Designing Autonomous Agents," in F.J. Varela, P. Bourgine eds.: *Toward a Practice of Autonomous Systems, Proceedings of First European Artificial Life Conference*, Cambridge, MA: MIT Press, Bradford Books, pp. 21-30.

Pfeifer, R. and Verschure, P.F.M.J. (1992b). "Beyond Rationalism: Symbols, Patterns, and Behavior," in *Connection Science* 4, pp. 313-25.

Pfeifer, R. and Verschure, P.F.M.J. (1995). "The Challenge of Autonomous Agents: Pitfalls and How to Avoid Them," in L. Steels., R. Brooks eds.: *The Artificial Life Route to Artificial Intelligence*, Hillsdale, N.J.: Erlbaum, pp. 237-63.

Pierce, C. (1960). "Collected Papers of Charles Sanders Peirce," – vol I – Principles of Philosophy; vol II – Elements of Logic; vol III – Exact Logic; vol IV - The Simplest Mathematics; vol V – Pragmatism and Pragmaticism; vol. VI – Scientific Metaphysics – C. Hartshorne and P. Weiss eds., Cambridge, MA: Belknap Press of Harvard University Press.

Pree, W. (1995), *Design Patterns for Object-oriented Software Development*, Massachusetts: Addison-Wesley.

Preece, J. et al. 1997, *Human-Computer Interaction*, Addison-Wesley.

Reeke, G.N., and Edelman, G.M. (1988). "Real Brains and Artificial Intelligence," in *Daedalus*, Winter, pp. 143-78.

Rumbaugh, J. (1991), *Object-oriented Modeling and Design*, New Jersey: Prentice Hall.

Sejnowski, T.J. and Rosenberg, C.R. (1987). "Parallel Networks that Learn to Pronounce English Text," *Complex Systems* 1, pp. 145-68.

Sethi, I.K. 1990, "Entropy Nets: from Decision Trees to Neural Networks,". Proc. of IEEE, vol. 78(10), pp.1605-13

Sheriff, J.K. (1989). *The Fate of Meaning*, Princeton, NJ: Princeton University Press.

Sheth, A. (1996) "Data Semantics: What, Where and How?," *in Proceedings of the 6th IFIP Working Conference on Data Semantics (DS-6)*, R. Meersman and L. Mark (Eds.), Chapman and Hall, London, UK

Smolensky, P. (1988). "On the Proper Treatment of Connectionism," *Behavioral and Brain Sciences* 11, pp. 1-73.

Srinivasan, D., Liew, A.C. and Chang, C.S. 1994, "Forecasting Daily Load Curves using a Hybrid Fuzzy-neural Approach,". IEE Proceedings on Generation, Transmission, and Distribution. 141(6): 561-567.

Suchman, L. (1987). *Plans and Situated Actions*, Cambridge: Cambridge University Press.

Suchman, L. (1993). "Response to Vera and Simon's Situated Action: A Symbolic Interpretation," in *Cognitive Science* 1:71-76.

Suchman, L., and Trigg. R. (1991). "Understanding Practice: Video as a Medium for Reflection and Design," in J. Greenbaum and M. Kyng, eds., *Design at Work: Cooperative Design of Computer Systems,* Hillsdale, NJ: Lawrence Erlbaum.

Sun, R. (1989), "Rules and Connectionism,", in *Technical Report No. CS-89-136,* Waltham, MA: Brandeis University, Dept. of Computer Science.

Sun, R. (1991), "Integrating Rules and Connectionism for Robust Reasoning," in *Technical Report No. CS-90-154,* Waltham, MA: Brandeis University, Dept. of Computer Science.

Sun, R. (1994), "CONSYDERR: A Two Level Hybrid Architecture for Structuring Knowledge for Commonsense Reasoning,". *Proc. of the1st Int. Symp. on Integrating Knowledge and Neural Heuristics.* Florida, USA, pp.32-9.

Tang, S.K., Dillon, T. and Khosla, R. 1995, "Fuzzy Logic and Knowledge Representation in a Symbolic-subsymbolic Architecture,". *IEEE International Conference on Neural Networks.* Perth,Australia, pp. 349-53.

Tang,S.K.,Dillon,T..and Khosla, R. 1996, "Application of an Integrated Fuzzy, Knowledge-based, Connectionistic Arch. for Fault Diagnosis in Power Systems,". *Int Conf on Intell. Sys App to Power Sys* Florida, Jan. pp. 188-93.

Tenenbaum J., Chowdhry T. and Hughes K. (1998), "eCo System: CommerceNet's Architectural Framework for Internet Commerce", *http://www.commercenet.org*

Vygotsky, L.S. (1978). *Mind in Society,* Cambridge, MA: Harvard University Press.

Weizenbaum, J. (1976) *Computer Power and Human Reason: From Judgment to Calculation.* W. H. Freeman, San Francisco.

Weilinga, J., Schreiber, and Breuker, J. A. (1993). "KADS: a Modeling Approach to Knowledge Engineering," in *Readings in Knowledge Engineering,* Academic Press, pp. 93-116.

Winograd, T., and Flores, F. (1986). *Understanding Computers and Cognition: a New foundation for Design,* Norwood, NJ: Ablex.

Wooldridge, M. and Jennings, N. R. (1994) "Agent Theories, Architectures, and Languages: A Survey," In *ECAI-94 Workshop on Agent Theories, Architectures, and* Languages, Amsterdam, Netherlands

Yourdon, E. Constantine, L.,(1978). *Structured Design: fundamentals of a discipline of computer program and Systems* Design, New York: Yourdon Press.

Zhang, J. (1997) "Nature of External Representations in Problem solving," in Cognitive Science, 21(2), 179-217.

Zhang, J., and Norman, D. A. (1994). "Representations in Distributed Cognitive Tasks," in *Cognitive Science,* 18, 87-122

4 HUMAN-CENTERED SYSTEM DEVELOPMENT FRAMEWORK

4.1. Introduction

This chapter builds on the foundations laid down in the previous chapter. It describes the human-centered system development framework for intelligent multimedia multi-agent systems based on human-centered criteria outlined in the first chapter and the pragmatic considerations and theories discussed in chapter 3, which contribute towards realization of those criteria. The human-centered framework is described in terms of four components, namely, activity-centered analysis, problem solving ontology, transformation agents, and multimedia interpretation, respectively. The three human-centered criteria are used as guidelines for development of the human-centered framework. The pragmatic considerations and contributing theories are used to develop the structure and content, or knowledge base, of the four components. The structure and content are described at the conceptual and computational (transformation agents) level. We start this chapter by describing the external and internal planes of action which underpin the development of the human-centered framework. We follow it with the description of two components human-centered system development framework, namely, activity-centered analysis and problem solving ontology. In the next chapter we continue with the description of the problem solving ontology component and describe the transformation agent and multimedia interpretation component. These four components have been used to define the external and internal planes of human interaction with the environment.

4.2. Overview

The human-centered framework developed and applied in this book involves a conceptual level and a computational level. The conceptual level determines how the three human-centered criteria discussed in the first chapter are going to be modeled, based on the contributing theories described in chapter 3. The computational level, on the other hand, looks at how various conceptual levels can be realized using technology-based artifacts. The technology-based artifacts are chosen based on their semantic and syntactic quality, and the pragmatic considerations discussed in the previous chapter. In order to set the scenario, we firstly describe the external and internal planes which among other aspects underpin the development of the framework. Then we describe the various components of the framework.

4.3 External and Internal Planes of Human-Centered Framework

In this book we intend to design and develop computer-based artifacts or software systems based on a human-centered approach. In this section we wish to outline the philosophy behind this approach.

First and foremost, as indicated in chapter 3, we employ *system* as the unit of analysis in our human-centered system development framework. Rather than looking at the system from a purely technical perspective, we look at a system as a unit of analysis from a social and stakeholder perspective. The social perspective, unlike the technical perspective (which is primarily based on principles of rationality and objectivity), emphasizes the role played by people and organizations (particularly organizational culture and policies) in a system and the impact of change (especially technology-based change) on people and incentives to people for accepting the change. In order to look at these aspects more closely, a system is seen as consisting of five components, namely, an activity (more specifically work activity), stakeholder, product, data and tool. An activity consists of goals, tasks and actions or operations. Stakeholders are people or groups who have a stake in the outcome of work activity. Stakeholders can be managers or regulatory bodies who sponsor a work activity (as shown in Figure 4.1). They can be the participants who enter, process and use information in a work activity, and they can be customers (internal or external to an organization) who use the product produced by the work activity. A product, as outlined by Alter (1996), represents an outcome of the activity and can be defined in terms of its physical content, information content and service content. Data used in a work activity can exist in various forms, namely, sensor readings (numerical), text, graphics, audio, video, etc.

Tool is an artifact, which is used to perform a task directly or is used to assist people in performing a task. A tool can be an external tool (e.g., hardware, software packages, etc.) or an internal tool (e.g., problem solving strategy or plan). A customer, as defined by Alter (1996), can be an internal customer or an external customer. An internal customer is within the same organization whereas an external customer is outside the organization.

From a social perspective, any changes in a system either through computerization or otherwise, are made as a result of optimization of all the five components rather than any one component (e.g., technology). In order to understand the meaning of this statement we need to look at the traditional system design approach. Traditionally, system design has been largely targeted towards meeting business goals. These product or business goals are management sponsored and are largely driven by quantitative improvements (e.g., reduced cost, reduced defect rate, reduced cycle time, increased efficiency, etc.). Technology is invariably used as a means for satisfying the business goals and a system's success is determined in a business context rather than a human context. Although technology today enables organizations to suitably respond to external (e.g., competition) and internal (e.g., efficiency) pressures, its underlying principles of rationality and objectivity are not adequate tools for dealing with social and organizational reality in which the technology & other system components operate. Further, it does not have adequate tools to deal with the subjective reality of the stakeholders. As a result, one is likely to end up with a successful technology rather than a successful system. A successful system, unlike a successful technology, not only considers management sponsored business goals but also attempts to marry these goals with the goals and incentives of its direct stakeholders and requirements based on organizational culture. The incentives can involve computer-modeled tasks, which would enhance direct stakeholder competence, increase the degree of involvement of the stakeholders in a work activity, or help them in breakdown situations in a work activity. Assisting the stakeholders in breakdown situations can provide the motivation to the stakeholders to engage with the computer-based artifact as an integral part of their work activity. As mentioned earlier, these breakdown situations relate to those tasks which stakeholders are unable to accomplish in a non-computerized work activity or find it difficult to accomplish. The direct stakeholder incentives and organization culture tend to emphasize important qualitative system improvements.

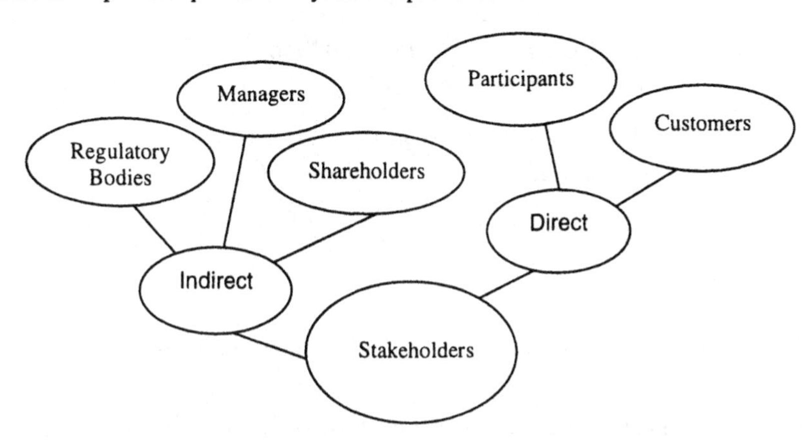

Figure 4.1. Direct and Indirect Stakeholders

In other words, business goals cannot be considered independently from direct stakeholder incentives and culture and vice versa[3]. That is, people cannot survive in organizations without satisfying business goals and organizations cannot survive for long without addressing stakeholder goals and incentives. Thus there is a need to integrate the two perspectives to realize successful systems (rather than just successful technologies).

In order to account for the business and social and stakeholder perspectives a system needs to be analyzed in an external context or plane of action and an internal context or plane of action. The external context defines the problem setting or context in which a system exists. The problem setting or the external environment can be defined in terms of objective aspects of the physical, social, and organizational reality in which a system exists. The physical reality primarily identifies various system

[3] For example a system goal might be reduce the cost of a particular work activity. However, a stakeholder may not participate in the work activity for realizing system goals unless some of their personal or professional goals are also satisfied.

components involved in the system. The social and organizational reality on the external plane involves the social, competitive, technical, and regulatory environment in which the system operates. It includes the division of labor between the stakeholders and tools (e.g., computers), overall business or product goals, competitive forces, and organizational policies. The workplace and, to an extent, situated cognition theories which emphasize the inclusion of physical, social and organizational realities, can be considered as enablers in modeling the external context or plane of action.

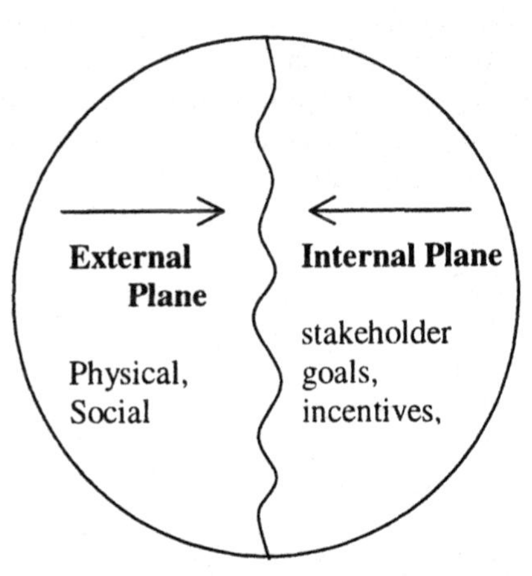

Figure 4.2. External and Internal Plane of Human-Centered Systems

The internal context, unlike the external context, involves subjective reality. This subjective reality can be studied at the individual or group level in terms of stakeholder goals, incentives, internal representations and external representations (as advocated by the distributed cognition theory) of system components (particularly the data component), and the problem solving strategy. Since we are dealing with systems that include computer-based artifacts in the external plane, the problem solving strategy can also be seen as a means of integrating and transforming the human solution or model to a software solution. The human solution can involve both perceptual tasks (based on external representations of data) and cognitive tasks (based on internal representations) involving deliberate structure and deliberate or automated reasoning. Thus, distributed aspects (representational and problem solving) of distributed cognition theory, stakeholder goal-oriented nature of the activity theory, task and problem driven ontology (discussed in section 4.5) can be considered as enablers for modeling the internal plane.

The use of computer-based artifact also brings into focus the problem of human-machine communication and interpretation of computer generated artifacts (e.g., software system results) by the stakeholders in the external plane. Multimedia artifacts like text, graphics, video and audio, and perceptual aspects of the distributed cognition theory can be considered as enablers for modeling the human-computer (and machine-machine) interface.

It may be noted that the external and internal planes represent two ends of the system development spectrum. These two planes also satisfy human-centered criteria for system development. Firstly, the external plane situates the use of computer-based artifacts among other system components in a work activity. This broadening of the scope of analysis of a human-centered system is more conducive to a problem or work driven design rather than a purely technology driven design. Secondly, the emphasis on stakeholder goals and incentives and problem solving strategy on the internal plane broadens the role of stakeholders from human factors to human actors. Thirdly, the consideration of internal (cognitive) and external (perceptive) representations and the role multimedia artifacts can play in modeling external representations assists in accounting for the representational context in which humans operate. From a human-centered system development perspective these two contexts or planes need to be bridged in a seamless manner for building successful systems.

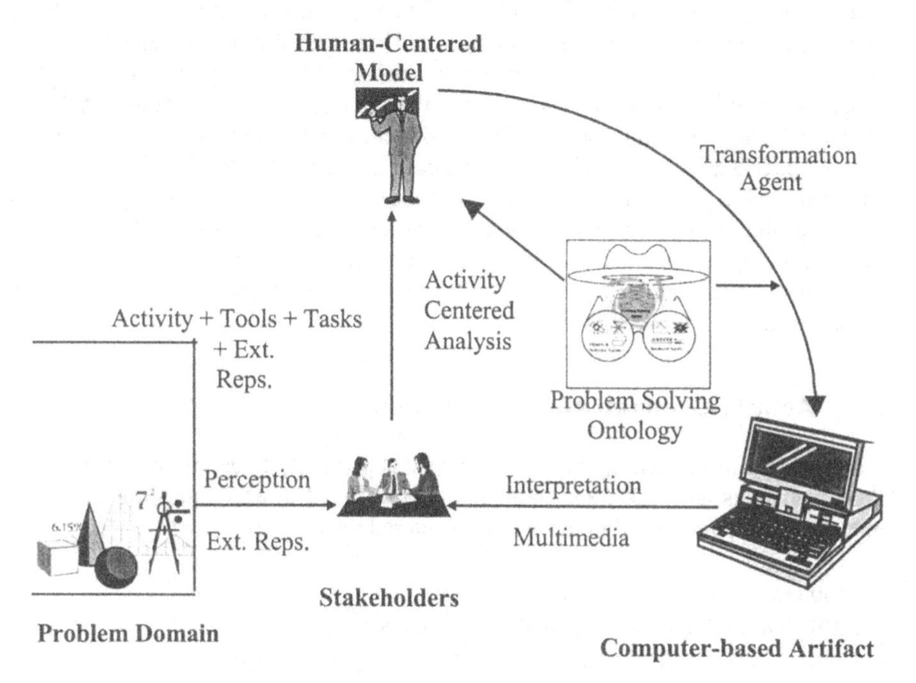

Figure 4.3. Components of Human-Centered System Development Framework

4.4. Components of the Human-Centered System Development Framework

The discussion on the external and internal planes in the preceding section has set out the broad framework for development of human-centered systems as shown in Figure 4.3. In order to focus our attention on various aspects of this broad framework we have conceptualized it into four components. These are the activity-centered analysis component, problem solving ontology component, transformation agents and the multimedia based interpretation component. The purpose of the activity-centered analysis is to account for the physical, social and organizational reality on the external plane and the stakeholder goals, tasks, incentives and organizational culture on the internal plane. We have chosen to separately account for the problem solving strategy in terms of the problem solving ontology component for two reasons. Firstly, we think the role of problem solving generalizations and routines grounded in experience play an important role in systematizing and structuring complex computer-based systems. Thus by accounting for it separately, we can more effectively employ these generalizations on the outcomes of the activity-centered analysis component that primarily focuses on the existing problem setting or situation. The problem solving generalizations employed by us are also used as a means for transforming a human or stakeholder solution to a software solution. This means that the problem solving ontology will interface with conceptual or task aspects of the activity-centered analysis component as well as computational or transformation aspects of the computer-based artifacts. These transformational aspects are modeled by the third component transformation agents and involve use of various technology-based artifacts. Finally, the multimedia interaction component focuses on the human-computer interface in terms of how multimedia artifacts can be effectively used to model external representations, reduce the cognitive load of the computer-based artifacts on the stakeholder and enhance the perceptive aspects of problem solving. These four components are shown in Figure 4.3 as part of the human-centered system development framework. The ontology of each of these components is described in this chapter and the next chapter.

4.5. Activity-Centered Analysis Component

The purpose of the activity-centered analysis is primarily to determine product and stakeholder goals and tasks, and tools and data required in accomplishing the tasks. There are six steps involved in the activity-centered analysis component.

- Problem Definition and Scope
- Performance Analysis of System Components
- Context Analysis of System Components
- Alternative System Goals and Tasks
- Human-Task-Tool Diagram
- Task-Product Transition Network.

4.5.1. Problem Definition and Scope

This step primarily involves discussions with the sponsors (e.g., management, regulatory bodies, etc.) of the work activity. The discussions include overview of the problem and the background and motivation behind the need for reengineering an existing work activity. The background and motivation are discussed in context of the business challenges (external and internal) which face the organization. The outcome of this step is a set of desired business goals that the sponsors will like to be realized through reengineering of the work activity.

Based on the overview of the problem in the first step, the content and scope of various system components like product, activity, customer, stakeholders, is then determined. That is, what steps/tasks are undertaken in the activity to produce the product, what is the physical, information and service content of the product, who are the internal and external customers of the product, who are the direct stakeholders (i.e. day-to-day participants directly responsible for the outcomes/products of the activity) and indirect stakeholders (e.g., sponsors), what data and information are being used in the activity, and what tools are being employed to realize the outcomes.

The bi-directional arrows in Figure 4.5 indicate that all system components influence the work activity and are influenced by it. Further, (although not shown) besides the work activity, the other system components can also be influenced by each another. For example, the outcomes of an activity and its tasks may be influenced by the stakeholder component in terms of satisfaction/dissatisfaction of stakeholder goals. The type of data (e.g. multimedia, noisy and incomplete) may influence the tasks in the activity and type of technological tools used to process the data. The type of tool (e.g., computers) may influence the participation of the stakeholders in the activity based on their perceptions and knowledge of the tool.

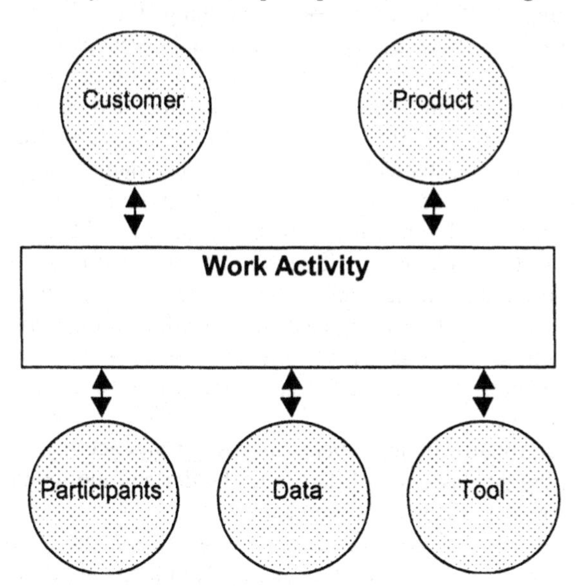

Figure 4.4: Activity-Centered Analysis Components

The content of various system components provides the framework for their analysis. In the next step we do the analysis in terms of performance of various system components and the context (social, organizational, technical and competitive) in which these components operate.

4.5.2. Performance Analysis of System Components

The performance of various system components is analyzed with a view to identify the role and the goals of the computer-based artifacts in an alternative system. The performance parameters of various system components provide an objective basis for determining a comprehensive set of goals, leading to improvement in not one or two but all system components.

The performance of a system can be analyzed in terms of its effectiveness and efficiency. Effectiveness is related to the product really being what the customer wants. In other words, is it the right product? It measures the performance of the product component in terms of cost, quality, responsiveness, reliability, and conformance to standards. Cost is measured in terms of money, time, and effort required using the product. Quality encompasses the customer's perception of a product's quality and measures such as defect rate. For, physical products, customer's perception of the quality relates to the function and aesthetics (e.g., climate control, and computerized directions through global positioning satellites in cars). For information based products, quality is perceived in terms of accessibility and presentation of data and information. The quality of service based products is perceived in terms of the level of customer satisfaction.

On the other hand, efficiency involves doing things the right way and is related to the optimal use of resources of the tool, data and participant components by the activity component for producing the product. The performance variables related to an activity are rate of output, productivity, cycle time, and flexibility. The rate of output involves an estimation of the number of units (e.g., cars) produced per hour or per week. Productivity is typically measured by evaluating output per labor hour, ratio of outputs to inputs, and cost of rework. Cycle time is measured in terms of the turnaround or start to finish time for producing the product. Flexibility, on other hand tests the rigidity of the work activity in terms of the number of product variations the work activity can handle. It determines the extent to which the product of a work activity can be customized to varying customer specifications. It may be noted that in different work activities one set of performance variables may be considered more relevant than others and hence different work activities may consider different performance variables.

Although the performance improvements in product and activity components are the most important, the performance of data, stakeholder and tool components also need to be analyzed for total system improvement.

The performance of data component is measured for its quality, accessibility, presentation and security. The quality is measured in terms of accuracy, precision and completeness of data. Accessibility is determined in terms of ease of data manipulation. Presentation is determined by how effectively various media are used to communicate data/information content. Finally, security involves the extent to which information is controlled and protected from inappropriate, unauthorized access and use.

The stakeholder component performance is determined in terms of the skills and involvement of the participants in the activity. Skills relate to the extent of experience of the participants in the activity.

The involvement relates to the extent to which participants have been involved in determining the tasks and tools to be used in activity. The involvement can range from no involvement to a very high involvement, where all the participants have been consulted in identifying tasks, tools, and data to be used in a work activity.

Finally, the performance of the tool component is determined in terms of its functional capabilities, and their use compatibility and ease of use, and maintainability.

4.5.3. Context Analysis of System Components

The performance analysis identifies the desired goals of the system. The context analysis determines the context in which these goals need to be realized. It determines the nature of tasks which need to be modeled in a computer-based artifact for realization of the goals and acceptance of the computer-based artifact, incentives and goals, technical, competitive and security realms in which these components exist. Unlike the performance analysis where quantitative measures are used, this analytical step involves largely those qualitative constraints that impact upon the successful operation and use of the system. It is important to consider these constraints as they can make a difference between a successful or unsuccessful system. They help to reengineer the tasks and tools in a work activity that lead to the development of a successful system. In the rest of this section the qualitative constraints are determined through analysis of each component.

4.5.3.1. Work Activity Context

The activity context is studied in terms of organizational culture and policies. Organizational culture represents the fundamental set of unwritten assumptions, values, and ways of doing things that have been accepted by most of its members (Laudon and Laudon 1998). For example, in universities it is assumed and accepted that teachers have more knowledge than the students, and that self-learning computer-based artifacts on the internet are not easily accepted by the traditional academics. On the other hand, because of deregulation of many service based industries like electrical utilities and domestic and international couriers, putting service to the customer first is an aspect of organizational culture which can be found in many customer based computerized systems like paying your bills through phone and internet and hour to hour details to customers about the progress of their documents from one destination to another. By studying organizational culture, one can determine not only what tasks can be computerized but also how task modeling needs to be sensitized to various assumptions, values and policies of the organization. These sensitivities may introduce additional tasks and constraints that need to be modeled in the computer-based artifact. The analysis is done in terms of their impact on the tasks being performed in the activity. The outcome of this analysis forms constraints on how various tasks are to be accomplished.

4.5.3.2. Direct Stakeholder Context

The direct stakeholders are participants and customers who enter, process and use information in a work activity and are directly effected by its outcomes. Since our motivation for doing activity-centered analysis is to determine the applicability and use of computer-based artifact in a work activity, we analyze the stakeholder context in terms of the incentives the computer-based artifact has to offer the direct stakeholders to facilitate its acceptance and use. The incentives are determined based on the goals and incentives of the direct stakeholders and how the computer-based artifact satisfies these goals. Unlike the business perspective where traditionally the primary motive for use of a computer-based artifact is automation, the direct stakeholder incentives from a social perspective are analyzed among other aspects, in terms of the breakdowns encountered by the direct stakeholders in accomplishing their tasks in a work activity. These breakdowns can involve those decision-making points in a task where the work activity participants and customers need assistance, and computer-based artifact can be effectively used to model/complete that task. For example, in a sales force recruitment activity, a sales manager may find it difficult to distinguish between two equally good candidates or in fact determine their goodness w.r.t. the existing successful salespersons during an interview. A sales recruitment software can be used (as will be shown in chapter 8) to benchmark existing successful salespersons or compare the profiles of two equally good candidates. In this way computer-based artifacts are likely to be used as partners by the direct stakeholders rather than as technologies which are imposed on them through user manuals and principles of rationality. Further, the accomplishment of goals is analyzed in terms of the stakeholder's perspective. This may result in incorporating flexibility in the computer-based artifact to facilitate its acceptance and use. The outcome of context analysis is a set of direct stakeholder-centered tasks to realize the goals identified.

4.5.3.3. Product Context

The product context is studied in a competitive realm. It is determined whether the product or products produced by the activity can be done away with altogether by the stakeholders or substituted by other products produced outside the present activity. The outcome of this analysis is whether to go ahead with the activity and/or the tasks and constraints which need to be incorporated in the existing activity to make it worthwhile.

4.5.3.4. Data Context

The data is analyzed in terms of the structure of the data used, and policies and practices for information sharing and privacy in an organization. For example, in the medical profession the privacy constraints on patient's data and medical decisions are far more stringent than student's data in a university. These privacy and information sharing constraints need to be properly respected by a computer-based artifact for its acceptability and use. The outcome of this analysis also results in tasks and constraints on processing and use of data.

4.5.3.5. Tool Context

The tool context is studied in the technical realm. That is, whether the existing technology is good enough or new technological artifacts need to be considered for more intuitive modeling of tasks in a work activity. For example, multimedia artifacts are being used today to enhance the perceptual design of tasks accomplished by a computer-based artifact.

4.5.4. Alternative System Goals and Tasks

This step builds upon the outcomes of the performance and context analysis step in terms of the goals and corresponding tasks for an alternative computer-based system. These goals and tasks form the basis for developing a human-centered domain model shown in Figure 4.3.

In order to develop such a model we firstly need to determine the division of tasks between the participants/customers and the computer-based artifact. Further, we need to determine the underlying assumptions or preconditions for accomplishment of these human-centered tasks. This is done in the next two steps.

4.5.5. Human-Task-Tool Diagram

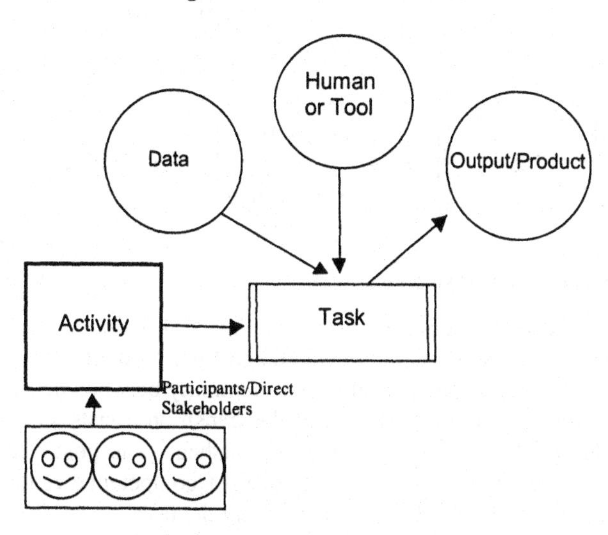

Figure 4.5: Human-Task-Tool Diagram

The purpose of the human-task-tool diagram is to determine the division of tasks between the participants/customers and the computer-based artifact. It assists in identifying the human interaction points with the computer-based artifact and data involved in the interaction. This data is later on used by the multimedia interpretation component for selecting suitable media artifacts. The notations used in the human-task-tool diagram are shown in Figure 4.5. It shows the data used by each task and the intermediate/final product produced after completion of the task. This information is useful for organizing the task-product transition network and in determining the correspondence between task and data to be used later on by the problem solving ontology component.

4.5.6. Task Product Transition Network

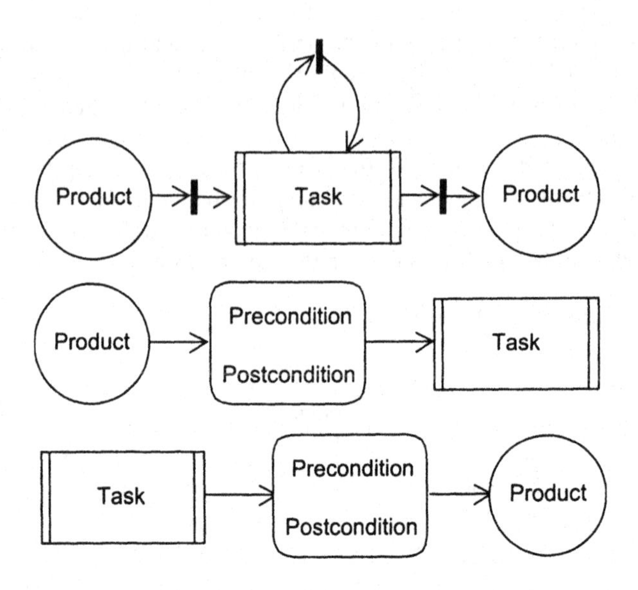

Figure 4.6: Task-Product Transition Network

The task-product transition network shown in Figure 4.6 defines the relationship between the tasks and elementary, intermediate and final products of a work activity. It can also help us in identifying parallelism, sequentiality between tasks and cyclic or repetitive tasks. Further, the precondition of the transition assists us in defining the assumptions under which the task will be accomplished. The postcondition reflects not only the new product state but also the level of competence required from the technological artifact or tool used for accomplishing the task.

4.6. Problem Solving Ontology Component

The problem solving ontology component shown in Figure 4.3 is used to transform a human solution (obtained through activity-centered analysis) to a software solution (in form of a computer-based artifact). In this section, we firstly review some of the work done on problem solving ontologies in the literature. We follow it with the description of problem solving ontology employed in this book.

An ontology is a representation vocabulary, typically specialized to some technology, domain or subject matter. However, here we are dealing with upper ontology, i.e., ontology that describes generic knowledge that holds across many domains. Further, we are dealing with problem solving knowledge or generic (e.g., tasks) about problem solving. In this section we start by covering problem solving ontologies and determining their strengths and weaknesses. We then describe the problem solving ontology used in this book.

4.6.1. *Strengths and Weaknesses of Existing Problem Solving Ontologies*

The research on problem solving ontologies or knowledge-use level architectures has largely been done in artificial intelligence. The research at the other end of the spectrum (e.g., radical connectionism) is based more on understanding the nature of human or animal behavior rather than developing ontologies for dealing with complex real world problems in control, diagnosis, design, etc. It is well acknowledged that to deal with these complex problems one cannot completely rely on the particularities of a real world problem (as suggested by the situated cognition approach). One also has to be in a position to benefit from the generalizations or persistent problem solving structures that hold across domains. It is with this motivation we look into some of the work done in artificial intelligence on the knowledge-use level, as against application level.

The work on knowledge-use level in artificial intelligence can be traced back to the work done by Clancey (1985) on heuristic classification. Clancey analyzed the inference structure underlying a series of expert systems and cognitive models in general. He found that the reasoning pattern of these programs involved a) abstracting descriptions of situations in the world (also called data abstraction), b) heuristically (by cause, frequency, or preference) relating these abstractions to a second classification (also called solution abstraction), and then c) refining the second description to a level suggesting particular actions. He called this reasoning or inferencing pattern heuristic classification. Heuristic classification represents a significant empirical generalization of expert system development. These three stages (data abstraction, solution abstraction through heuristic match and refinement) of heuristic classification have been found useful in focussing expert system development at the knowledge-use level. However, the highly abstract generalizations do not provide enough application vocabulary for the problem solver.

Besides, there are certain omissions like lack of contextual validation of data (its absence can result in nonproductive abstractions) before abstracting it, and lack of consideration of epistemological limitations that humans and computers have and pragmatic constraints associated with complex real world problems. The epistemological limitations relate to the need for making decisions in finite time, finite memory and storage structure of computers, imprecision associated with human observations and the need for inductively derived models (to improve model accuracy and prediction) based on real world data and interactions with humans.

Around this time, another approach, namely, model based approach towards system modeling was being developed (Hart 1984, Steels 1985; Steels and Van de Velde 1985; Simmon 1988). The model based approach focussed on the domain models underlying expertise rather than the inferencing pattern used in heuristic classification. The model-based systems emphasized the need for deep or complete knowledge of the domain of study rather than surface or shallow knowledge that focussed only on portions of deep knowledge. Based on deep knowledge, part-whole, geometric and functional models of a domain were developed. Although, by definition the model-based approach sounds comprehensive, it suffers from certain weaknesses. First of all, it assumes that exact domain theory is known. This is not the case especially, in complex domains where it becomes difficult to develop complete domain models. Secondly, it relies on the correctness of observed data, whereas in complex real world problems, the data is often noisy and incomplete. Finally, the main strength of model based systems i.e., deep or complete knowledge, can work against itself especially in real time situations where exploration of large spaces or combinatorial explosion of rules can lead to unacceptable response times.

In 1988, McDermott developed the problem solving method approach which is somewhere in between model based approach and data base approach. A problem solving method is a knowledge-use-level characterization of how a problem might be solved. For example, a medical diagnostic problem can be solved using a cover-and-differentiate method in which, firstly, explanations covering the observed symptoms are determined, and then the cause is determined by differentiating among various explanations (Eshelman 1988).

A problem solving method specifies the domain knowledge required from the expert to solve a particular problem. For example, in a medical diagnostic system the domain knowledge is represented as a infection model that explicitly represents relations between symptoms (e.g., fever, cough) and infections (e.g., bronchitis). The specified domain knowledge may only form a small portion of the complete domain knowledge as defined in model based systems. Thus the problem solving method approach accounts for some of the problems (e.g., combinatorial explosion of search space) associated with model based systems. However, its strength can also become a drawback in establishing the completeness of the system. Further, problem solving methods also suffer from what Steels (1990) calls the grain size problem. In other words, because a problem solving method intends to solve the complete problem, it may use other problem solving methods for handling various subtasks in the problem domain that may be somewhat different in structure than itself. For example, propose-and-revise method (Chandrasekaran 1990) can involve use of a classification method for proposing different designs in the propose phase. This leads to a proliferation of problem solving methods for a solving a particular problem. More recently, Fensel and Groenboom (1996), Fensel (1997) and Chandrasekaran, Josephson and Benjamins (1998) have suggested use of adapters as a means of mapping a problem solving method on to a task domain ontology. These adapters are different than those used by the Gamma et al. (1995) in software engineering. In software engineering design patterns and adapters are defined as low level primitives that link two software design artifacts. On their own these adapters are not sufficient to solve the complete problem. Besides, they are designed from the perspective of software design rather than problem solving. The adapters defined by Fensel (1997), and Chandrasekaran, Josephson and Benjamins (1998) are used for modeling complete solutions for complex real world problems. These complex problems are solved using intelligent methods which (unlike adapters define by Gamma et. al. (1995)) require assumptions to be made in terms of type domain knowledge needed. However, the adapter based approach of Fensel, and Chandrasekaran, Josephson and Benjamins apparently presupposes use of one or the other problem solving method, domain ontology and domain model for solving a problem besides being only suited for knowledge based systems. In many complex problems more than one domain ontology and domain model may be used (Steels 1990). Thus the adapters should facilitate use of multiple domain-models and domain ontologies.

Another line of research, namely, task structure analysis developed by Chandrasekaran (1983), Chandrasekaran and Mittal (1983), Chandrasekaran, Johnson and Smith (1992) focuses on modeling domain knowledge using generic tasks and methods as mediating concepts. Typical generic tasks are classification, diagnosis, interpretation, and construction. For each generic task (say diagnosis) a task structure analysis is done. The task structure analysis represents the interplay between methods (e.g., abduction) and subtasks for a given generic task. The task structure analysis as outlined by Chandrasekaran, Johnson and Smith (1992) does alleviate, to some extent, problem solving method granularity problems with the problem solving method approach. However, it only employs methods as mediating concepts for task accomplishment and not representations. The distributed cognition approach described in the previous chapter clearly establishes the role of external and internal representations in problem solving. The task structure analysis approach implicitly assumes internal representations and does not take into account external representations.

Table 4.1 : Problem Solving Ontologies – Strengths and Weaknesses

Approach	*Strengths*	*Weaknesses*
Heuristic Classification (inference pattern) Clancey 1985	Good empirical generalization	No distinction between different classification methods. (e.g. weighted evidence combination). Not enough vocabulary. Pragmatic constraints not considered.
Model Based Systems (part-whole, causal, geometric, functional) Steels 1985; Simmons 1988	Principled domain models, complete knowledge	Combinatorial explosion, assumes all observations are correct & exact domain theory is known.
Problem Solving Ontologies (between model based and data based) based on Problem Solving Methods - (cover-and-differentiate, propose-test-refine, etc) McDermott 1988	Helps to determine type of domain knowledge required for problem solving, eases the knowledge acquisition bottleneck.	When to stop the knowledge acquisition, when is the system complete. Do not consider the role of representations or tasks Problem Solving Ontologies.
- Generic Task Based (classification, interpretation, diagnosis & construction/design) Chandrasekaran 1983, Chandrasekaran & Josephson 1997, Chandrasekaran & Johnson 1993, Steels 1990.	Reuse, basis for interpreting acquired data, can build generic software environments.	Generic task categorization is not generic enough because they (e.g. diagnosis) can be accomplished using many different domain models, and different methods (depending on problem granularity), pragmatic constraints not considered. Tasks only mediated by methods.
Generic Ontology – KADS methodology (domain layer, inference layer, task layer, strategy layer) Breuker and Weilinga 1989,91 Weilinga et al 1993.	Segregates knowledge modeling into four layers.	System modeling done with low level primitives. Suitable for knowledge based problems only. Does not consider pragmatic constraints.

In the preceding paragraphs we have outlined four distinct knowledge level approaches to problem solving. The first one developed by Clancey (1985) employs inference structure or pattern as an empirical approach to problem solving. Hart (1984), Steels and Van de Velde 1985 on the other hand use causal, structural and functional domain models to solve problems. McDermott (1988) and Simmon, (1988) adopt a problem solving method approach towards problem solving. Finally, Chandrasekaran, Johnson and Smith (1992). In Europe, Breuker and Weilinga (1989, 1991) have developed Knowledge Acquisition and Design System (KADS) methodology, which includes pertinent aspects of the four approaches. The expertise model of the KADS approach defines three layers, namely, domain layer, inference layer, and task layer to model knowledge based systems. It defines a set of primitives for each of these three layers and employs them in a bottom up fashion in a problem domain to develop higher level of analysis. It has been used successfully to develop a number of expert systems in the field. However, it is not clear as to how the KADS approach accounts for external representations in problem solving, epistemological limitations of humans and computers, and pragmatic constraints associated with real world problems. Further, none of the approaches provides insight into how to deal with the complexity of large-scale real world problems. That is, to what extent applications developed by using these problem solving methods or ontologies will be scalable, evolvable and maintainable. Table 4.1 shows a summary of various problem solving approaches discussed in this section, along with some of their weaknesses.

Besides the above limitations, the existing approaches do not lend themselves towards human-centered research and design and more specifically towards satisfying criteria 1, 2 and 3 of human-centeredness outlined in the first chapter. That is, most of these approaches facilitate an objective way to model solutions to real world problems. They are motivated by answering the following question: what is (or are) the most appropriate approach (s) for solving a particular problem or task? They do not necessarily answer the question: what are the underlying user's goals and tasks, or what problem solving strategy does the user. Further, because these ontologies are defined at a high level of abstraction, they do not provide adequate vocabulary (e.g. Activity Centered Analysis component in this book) or assistance for a non-specialist to solve a particular problem. Additionally, most of the above approaches in Table 4.1 are embedded in the knowledge based system technology. They (e.g., some Generic Task Based approaches), tend to subscribe to best practice approach, an approach which has been recently criticized in the software development community (emergence of patterns is a consequence of the best practice myth). Further, they do not adequately address the pragmatic task constraints modeled or satisfied by other technologies like neural networks, fuzzy logic, and genetic algorithms. A lack of this consideration has resulted in unsatisfactory results (in terms of satisfaction of constraints and quality of solution) from implementation of these problem solving methods in the field.

4.7. Summary

This chapter builds on the foundations laid down in the previous chapter. It describes a human-centered system development framework for developing intelligent multimedia multi-agent systems. The human-centered approach involves a seamless integration of external and internal planes or contexts of action.

The external context defines the problem setting or context in which a system exists. The problem setting or the external environment can be defined in terms of objective aspects of the physical, social and organizational reality in which a system exists. The internal context, unlike the external context, involves subjective reality. This subjective reality can be studied at the individual or group level in terms of stakeholder goals, incentives, organizational culture, internal representations and external representations of data in a work activity, and generic problem solving strategy that is adopted by stakeholders in a work activity.

The external and internal planes represent two ends of the system development spectrum. These two planes are conceptually captured with the help of four system development components, namely, activity-centered analysis component, problem solving ontology component, transformation agent component, and multimedia interpretation component. The activity-centered analysis component described in this chapter defines the scope of six system components, namely, product, data, customer, work activity and tool. It conducts a performance and context analysis of the existing situation as defined by the six components. The outcome of the performance and context analysis is a set of goals and tasks for a computer-based artifact that forms a part of an alternative system. These goals and tasks form the basis for a human-centered domain model. The terminology and notations for a human-task-tool diagram are outlined in order to determine among other aspects, the division of labor between the direct stakeholders and computer-based artifact. It also helps to define the human interaction points in a computer-based system which are used later on by the multimedia interpretation component. A task-product transition network is also drawn to define among other aspects, the preconditions and postconditions for each task.

The results of the activity-centered analysis and the task-product transition network are used by the problem solving ontology component to develop a human-centered domain model. Another role of the problem solving ontology component is to systematize and structure the tasks outlined in the task-product transition network. This chapter covers some existing work done by researchers in the evolution and development of problem solving ontologies. It outlines the strengths and weaknesses of some of the problem solving ontologies. The next chapter describes the problem solving ontology used in this book and transformation agent and multimedia interpretation components.

References:

Alter, S., 1996, Information systems – A Management Perspective, second edition, Benjamin/Cummings Publishing Company.

Breuker, J.A. and Weilinga, B.J., 1989 "Model Driven Knowledge Acquisition" in B. Guida & G. Tasso eds. *Topics in the Design of Expert Systems*, Springer-Verlag, pp 239 – 280.

Breuker, J.A. and Weilinga, B.J., 1991 *Intelligent Multimedia Interfaces*, Edited by Mark Maybury, AAAI Press, Menlo Park, CA.

Chandrasekaran B. and Josephson, J.R., 1997 "Ontology of Tasks and Methods", *AAAI 97 Spring Symposium on Ontological Engineering,* March 24-26, Stanford University, CA California, USA.

Chandrasekaran, B. 1983, "Towards Taxonomy of Problem Solving Types", *AI Magazine* Vol 4 No. 1, Winter/Spring pp 9-17.

Chandrasekaran, B., and Johnson, T.R. 1993, "Generic Tasks and ask Structues: History Critique and New Directions", *Second Generation Expert Systems*, G.M. Davies, J.P. Krivine and R. Simmons.

Clancey, W.J., "Heuristic classification", Artificial Intelligence, 27, 3 (1985), 289-350.

Eshelman, L. 1988. "Mole: A Knowledge Acquisition Tool for Cover-and-Differentiate systems" in *Automating Knowledge Acquisition for Expert Systems*, Ed. S. Marcus, 37-79. Boston: Kluwer

Fensel, D (1997), "The Tower-of-Adapter Method for Developing and Reusing Problem-Solving Methods," *EKAW*, pp. 97-112

Fensel, D. and Groenboom, R., (1996), "MLPM: Defining a Semantics and Axiomatization for Specifying the Reasoning Process of Knowledge-based Systems,".*ECAI*, pp 423-427

Gamma, E et. al., (1995) "Design Elements of Object-Oriented Software," Massachusetts: Adisson-Wesley.

Hart, P. 1984. "Artificial Intelligence in Transition" in *Knowledge-Based Problem Solving*, Ed. J. Kowalik, 296-311. Engelwood Cliffs, N.J. Prentice-Hall.

Laudon. K.C. and Laudon, J.P., 1998, Management Information Systems, Prentice Hall International.

McDermott, J., Preliminary steps toward a taxonomy of problem solving methods. In *Automated Knowledge Acquisition for Expert Systems,* S. Marcus, Ed., Kluwer Academic 1988, pp. 225-256.

Simmons, R. 1988. "Generate, Test, and Debug: A Paradigm for Solving Interpretation and Planning Problems". Ph. D diss., AI Lab, Massachusetts Institute of Technology.

Steels, L. (1990) "Components of Expertise", AI Magazine, 11, 28-49., 11, 28-49

Steels, L. 1984. "Second-Generation Expert Systems" presented at the *Conference on Future Generation Computer Systems,* Rotterdam. Also in *Journal of Future Generation Computer Systems* (1)4: 213-237.

Steels, L., and Van de Velde, W. 1985. Earning in Second-Generation Expert Systems" in *Knowledge-Based Problem Solving*, Ed. J. Kowalik. Englewood Cliffs, N.J.: Prentice-Hall.

Steven, S.S. (1995). "On the Psychological Law", *Psychological Review*, 64(3), 153-181

Weilinga, B.J., Ath. Schreiber and J.A. Breuker, 1993 "KADS: A Modelling Approach to Knowledge Engineering", Readings in *Knowledge Acquisition and Learning*, eds. Buchanan, B.. & Wilkins, D., San Mates California, Morgan Kaufmann pp 92-116.

5 HUMAN-CENTERED VIRTUAL MACHINE

5.1. Introduction

The objective of this chapter is to outline the computational framework of intelligent multimedia multi-agent systems based on the human-centered approach. The title human-centered virtual machine encapsulates the integration of conceptual components of the human-centered system development framework and the technology based artifacts used to realize the conceptual components at the computational level.
In chapter 4 we described some of the existing problem-solving ontologies, their strengths and weaknesses. In this chapter we start with the description of the problem-solving ontology component. We follow it with description of the transformation agent component and multimedia interpretation component. The transformation agent component is constructed through integration of the problem solving ontology with various technological artifacts like intelligent technologies, agent and object-oriented technologies, multimedia presentation, XML and distributed processing technologies. The multimedia interpretation component, on the other hand, deals with interpretation of data content by the direct stakeholders/users. It does that by mapping the data characteristics to media characteristics and media expressions of different media artifacts. The chapter concludes by outlining the emergent characteristics of the human-centered virtual machine.

5.2. Problem Solving Ontology Component

As mentioned in the previous chapter the main aim of the problem solving ontology component is to develop a human-centered domain model based on the stakeholder goals and tasks model (outcome of activity-centered analysis), stakeholder representational model, and stakeholder domain model for various tasks. As shown in Figure 5.1. It does that by systematizing and structure these aspects using five information processing phases, namely, preprocessing, decomposition, control, decision, and postprocessing.

The information processing phases and their generic tasks have been derived from actual experience of building complex systems in engineering, medicine, bioinformatics, management, internet and e-commerce. Further, they have been based on number of perspectives including neurobiology, cognitive science, learning, forms of knowledge, user intelligibility, and others (Khosla and Dillon 1997).

Each information processing phase in turn is defined in terms of generic goals, generic tasks, constructs for analyzing external representations and sensor data, underlying assumptions on the domain data, knowledge engineering strategy (top down or bottom up), soft (e.g., non-symbolic methods like neural networks and genetic algorithms) and/or hard (e.g., symbolic rule based systems) computing techniques used for accomplishing various tasks. Although, the five information processing phases represent domain independent tasks, domain dependent tasks can also be integrated into these phases.

Each phase is encapsulated using a problem solving adapter construct. A problem solving adapter construct besides distinguishing between different information processing phases is used to establish a signature mapping between user's or practitioner's goals and task (as determined in the by the activity-centered analysis component), external (perceptual) and internal (interpreted/linguistic and non-linguistic) representation ontology and the domain model. The problem solving adapter definitions do not constrain the user or practitioner in terms of domain model or models employed, problem solving technique employed.

In the remainder of this section, we outline the problem solving vocabulary of the five problem solving adapters. Before we do that, it is useful to define the terms used in the vocabulary.

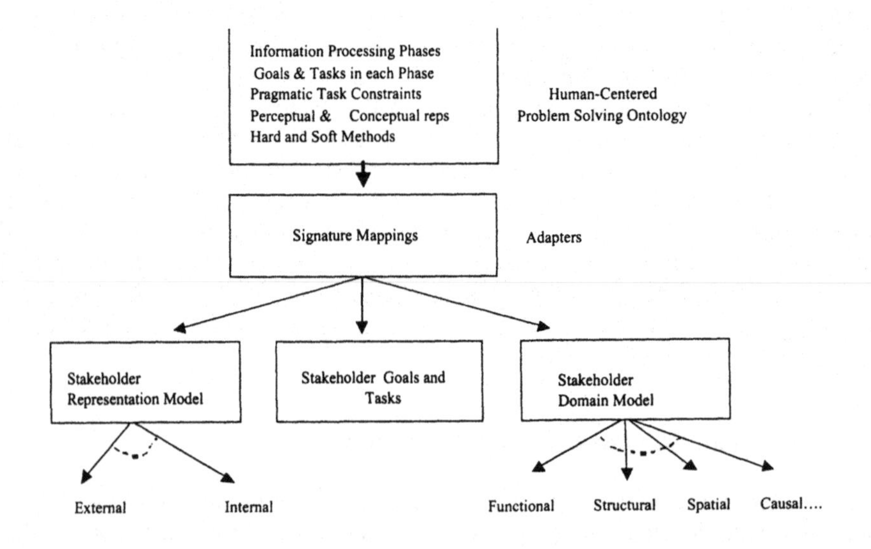

Figure 5.1. Human-centered Problem Solving Ontology

5.2.1. Definition of Terms Used

Information Processing Phase:

- a distinct step or event in problem solving

Goal:

- a desire or desired outcome or state

Task:
- Tasks are goal directed processes in which people consciously or unconsciously engage.

Task Constraints:
- are pragmatic constraints imposed by the stakeholders and the environment for successful accomplishment of a task. The task constraints primarily determine the selection knowledge required for selecting a technological artifact (e.g., a computing technique) for accomplishing a task. The task constraints are a byproduct of epistemological limitations that humans and computers have and the environment in which a computer based software artifact will operate (Steels 1990). Human limitations relate to the need to make decisions in finite time. Thus those models or techniques which lead to deep solution hierarchies (e.g., symbolic rule based systems) and large response times cannot be used in software systems supporting humans in real world tasks requiring fast response times. Similarly, computers have finite space and memory. Therefore, models and techniques (e.g. breadth first search) requiring large search spaces cannot be used. Other human limitations include lack of domain knowledge in certain tasks which means techniques like self-organizing neural networks need to be used to inductively learn the domain model and concepts used to accomplish such tasks. Finally, human or sensor observations may be imprecise. Therefore hard computing artifacts which rely on precision cannot be used. That is, the epistemological limitations lead to a number pragmatic considerations or constraints for selection of appropriate techniques. These include dealing with information or data explosion, noisy and incomplete data, need to avoid search or use search techniques which are not constrained by necessary and sufficient conditions. Besides the above constraints imposed by epistemological limitations, human ability to adapt also constrains use of those technological artifacts, which can adapt like humans do in new or similar situations. Therefore, techniques that do not have adaptive behavior cannot be used to model tasks that require adaptation. In summary human and computer related task constraints can be knowledge and data related (e.g. imprecise/incomplete data, learning), conceptual and software design related (e.g., scalability, maintainability), and domain performance related (e.g., response time and adaptation).

Precondition:
- helps us to define underlying assumptions for task accomplishment.

Postcondition:
- defines the level of competence required from the technique or algorithm used for accomplishing the task.

Represented Features:
- are linguistic (e.g. symbolic, fuzzy) and non-linguistic (e.g. numeric) features in a domain.

Representing Dimension:
- is the physical or abstract dimension used to represent a feature. It can be seen as capturing the perceptual representation or category of a feature. The perceptual representation is a stable signature (e.g. oval shape of a face or

pattern in a raw sensory signal). These representing dimensions can be shape, color, distance, location, orientation, density, texture, etc

Psychological Scale:

is the abstract measurement property of the physical or abstract dimension of a represented feature. There are four types of scales, namely, nominal, ordinal, interval and ratio. The four psychological scales devised by Steven (1957) are based on one or more properties, such as category, magnitude, equal interval and absolute zero. The category refers to the property by which the instances on a scale can be distinguished from one another. The magnitude denotes the property that one instance on a scale can be judged greater than, less than, or equal to another instance on the same scale. The equal interval refers to the property that the magnitude of an instance represented by a unit on the scale is the same, regardless of where on the scale the unit falls. Finally, absolute zero: is a value that indicates the nonexistence of the property being represented. The nominal scale is based on the category property only. The ordinal scale includes the category as well as magnitude properties. The interval scale includes category, magnitude and equal interval properties. Finally, the ratio scale includes all the four properties (i.e., category, magnitude, equal interval and absolute zero).The purpose of using the representing dimension and scale information is twofold. Firstly, from a human-centered perspective the representing dimension and scale information provide insight into distributed representations (external and internal) used in problem solving (Zhang & Norman 1994; Stevens 1957). The distributed representations account for the representational context of human-centered criteria no. 2. Through the representing dimension and scale information, we can determine what part or parts of a task can be accomplished perceptually. For example, in a energy forecasting domain, the represented features of a energy consumption profile are the hourly energy consumption data points. The representing dimension of the energy consumption profile is its shape (on the nominal scale). The shape can be seen to represent the external representation whereas numeric data values of the data points are internal representations of the profiles. Thus, certain tasks like eliminating noisy consumption profiles involving valley or straight line shapes can be done perceptually using the representing dimension of shape. Secondly, the representing dimension and scale information can assist in developing more efficient and effective means of communicating the data content perceptually to the user/direct stakeholders of a computer-based artifact. This will also help in reducing the cognitive load on the users

Technological Artifact:

- can be a software artifact like an agent, object and/or a hard or soft computing technique to accomplish a given task. Objects and classes can be used to structure the represented features/data and/or the devices/components/objects used by different problem solving adapters to accomplish various tasks. Thus the technological artifacts and relations are defined in the task context as outlined by users. Agents can be used to model various tasks associated with the adapter. Soft or hard computing techniques can be used for accomplishing various tasks. The selection of a soft or hard computing technique will depend upon the knowledge engineering strategy and the task constraints. For upon the knowledge engineering strategy and the task constraints. For example, in case domain knowledge is not available then soft computing techniques have to be used. On the other hand, if domain knowledge is available (top-down knowledge engineering strategy) then hard computing techniques can suffice.

Knowledge Engineering Strategy:

- Top down or bottom up knowledge engineering strategy is simply indicative of the availability or non-availability of domain knowledge, respectively, for accomplishing a task. A bottom-up strategy is contingent upon use of soft computing techniques for accomplishing a given task, whereas a top-down strategy can use hard computing technique like a symbolic rule based system for accomplishing a task. Further, in a bottom-up strategy, learning and adaptation are a necessity, whereas, in a top-down strategy learning and adaptation may be used for enhancing performance of a task. A number of complex problems employ a mixture of top-down and bottom-up strategy for accomplishing different tasks.

5.2.2. Problem Solving Adapters

In the next few sections we define the problem solving adapters based on the terms defined in the preceding section. These are preprocessing, decomposition, control,, decision, and postprocessing adapters. These adapters are built on five information processing phases developed by Khosla and Dillon (1997).

5.2.2.1. Preprocessing Phase Adapter:

- The preprocessing adapter shown in Figure 5.2 can be used by all the phase adapters except the postprocessing phase adapter.

Goal: As shown in Figure 5.2, the goal of the preprocessing adapter is to improve data quality.

Task: Noise Filtering

- employs heuristics or other algorithmic/non-algorithmic techniques for removing noise from a domain at a global or a local level of problem solving. The non-algorithmic techniques can involve perceptual or visual reasoning (e.g. distinguishing a irregular shape from a regular one). The noise represents peculiarities that are specific to a problem domain and need to be removed in order to improve the quality of data. It can involve removing irrelevant parts of a natural language query, eliminating skin look-alike regions from actual skin regions in a face recognition problem, eliminating highly irregular shaped energy consumption profiles from standard profiles in a energy prediction problem, eliminating nuisance alarms, faulty alarms in a alarm processing and diagnosis problem, etc.

Task: .Input Conditioning

- this task may require simple formatting of the input data and/or transforming the data from one format to another (e.g., transforming different image formats, etc.), dimensionality reduction (e.g., combining and/or removing ineffective data points, using existing domain knowledge to aggregate/partition data, etc.).

Task: Problem Formulation

- involves sequencing of various actions required to accomplish the above tasks.

Tasks: Other Domain Dependent Tasks

- This includes those tasks which are peculiar to a domain

Task Constraints:

- Domain and application dependent

Represented Features:

- Since the task like noise filtering is heuristic in nature, the represented features in the domain can be qualitative/linguistic (binary, structured and fuzzy) or continuous in nature. For example, in a alarm processing problem, a alarm may be filtered based on its existence (binary), based on multiple occurrences of it or based on the topology of the network (structured). Further, fuzzy variables (e.g., adjectives in a natural language query) may be used to eliminate particular type of queries. In domains like signal processing fast forward transforms are applied on continuous numeric data.

Psychological scale:

- The represented features can be analyzed based on the nominal, ordinal, interval or ratio scales. These psychological scales which have been developed by Steven (1957) are used by humans to derive perceptual and conceptual semantics of real world objects.

Perceptual Representing Dimensions:

- The perceptual dimensions on which the psychological scales are applied could be shape (e.g., eliminating noisy energy consumption profiles based on shape), distance (e.g., suppressing sympathetic alarms emerging from parts of network beyond certain threshold distance from the faulty component), color, etc.

Knowledge Engineering Strategy

- Top-down or bottom-up.

Technological Artifacts: hard or soft computing techniques.

- In our problem solving ontology we consider computational or algorithmic techniques as well as perceptual or non-algorithmic techniques. Besides, we also consider software-engineering techniques like object-oriented methodology as a means of accomplishing a task. For example, a dimension reduction task can be accomplished using an object-oriented technique by aggregating or partitioning the data. The computational techniques can be hard or soft, depending upon the task constraints and the represented features. On the other hand, perceptual techniques exploit the perceptual representing dimensions of the represented features.

The preprocessing phase adapter definition based on the goal, task, precondition/postcondition and other definitions, as outlined in this section, is shown in Figure 5.2. Figure 5.3 shows the representation and task signature mapping for preprocessing adapter. The signature mappings represent those aspects (e.g. goals and tasks) of the preprocessing adapter definition in Figure 5.2, which are invariably used by computer-based applications. The tasks indicated as optional or not shown in Figure 5.3 and shown in Figure 5.2 are optional and may or may not be used in a particular application.

Phase: Preprocessing
Goal: Improve data quality
Task: problem Solving Context – Global; Input Context – raw symbolic or continuous data`
Task: Noise filtering – form – time based noise filtering, content and task context based noise filtering
Task: Input conditioning – form – Dimensionality reduction, Data transformation (e.g. color transformation), input formatting
Task: Problem formulation – form – conceptual ordering of actions
Task Constraints: Domain/application dependent
Precondition: Raw or processed data
Postcondition: Conditioned data
Represented Features: *Qualitative/linguistic* – binary, structured
Non-Linguistic – continuous features
Psychological Scale: Nominal; Ordinal, Interval, Ratio
Representing Dimension (Perceptual): Shape, Location, Position, Color, etc.
Knowledge Engineering Strategy: Top-down or bottom up
Technological Artifacts: Hard (e.g. symbolic rule based), soft (e.g. neural networks), etc.

Figure 5.2: Preprocessing Phase Adapter Definition

Domain Representation Signature

Phase: Preprocessing

Goal: Improve data quality

Represented Features:

Qualitative/linguistic- binary, structured,

Non-Linguistic- continuous features

Psychological Scale: Nominal; Ordinal, Interval, Ratio

Representing Dimension (Perceptual): Shape, Location, Position, Color, etc.

Knowledge Engineering Strategy top-down or bottom-up

Domain Task Signature

Phase: Preprocessing

Goal: Improve data quality

Precondition:raw or processed data

Task : Noise filtering (optional)

Task: Data conditioning - form - Dimensionality reduction, Data transformation (e.g. color transformation), data formatting

Task Constraints domain/application dependent

Domain Model:functional, structural, spatial, Causal, spatial, etc.

Postcondition conditioned data

Knowledge Engineering Strategy: top-down or bottom-up

Figure 5.3. Signature Mapping for Preprocessing Adapter

5.2.2.2. Decomposition Phase Adapter

Goal:

- The primary goal of the decomposition phase adapter is to restrict the context of the input from the environment at the global level. The secondary goals are to reduce the complexity and enhance overall reliability of the computer-based artifact.

Task: Restrict input context

- The input context at the global level is restricted in terms of user's or stakeholder's perception of the task context. The user's task context can be used to restrict the input in terms of different types of users (e.g., medical researcher, and evolutionary biologist in a human genome application), different player configurations in a computer game application, different control models in an optimum control system modeling application, different categories of alarms in a real-time alarm processing application or different subsystems in a sales management application. Thus user's task context is captured with the help of concepts that are generally orthogonal in nature. This also enables a reduction in the complexity of the problem as well as enhancement of the reliability of the computer-based artifact. Further, these concepts are abstract and do not provide a direct solution to the task in hand.

Task: Concept Validation

- In a number of multimedia applications (e.g., image retrieval applications) the search is guided by feedback from the user during run time. That is, the nature of the user query or input data in general may not be adequate and feedback from the user in terms of pursuing the search in one of many directions may help to reduce the search time as well as enhance the quality of the results. For example, in an electronic commerce application, an initial user query may only specify buying a shirt. It may not specify what type of shirt and/or collar. This information can be ascertained by prompting the user to select from a range of shirts with different types of collars.

Task Constraints: The generic task constraints associated with the decomposition phase adapter are scalability and reliability. The concepts used to restrict the input context in the decomposition phase should be scalable vertically as well as horizontally. One way of satisfying this task constraint is to ensure that the concepts defined in this phase are orthogonal or un-correlated. This will also enhance the reliability and quality of results produced by other phase adapters (e.g., control), which depend on the competency of the decomposition phase adapter. It may be noted that these task constraints also serve a useful purpose in terms of future evolution, maintenance and management of the computer based artifact.

Represented Features:

- The qualitative or linguistic features employed in this phase by the user are coarse-grain features. These coarse-grain features may have binary and/or structured values. For example, coarse-grain binary features to partition a global concept like an animal (into mammal, bird) in the animal kingdom domain, may be has_feathers, gives_milk, has_hair, etc. On the other hand, structured features like player_configuration (with values like 1, 2, 3,4) may be used in a computer game application.
- The features representing concepts in this phase can also be numeric or continuous in nature. For example in an image processing application like face recognition, orthogonal concepts, like skin regions and non-skin regions, can be distinguished based on the skin color pixel data.

Domain Models:

- As shown in Figure 5.4 the domain models used for restricting the context and identifying the represented features can be structural, functional, causal, geometric, heuristic, spatial, shape, color, etc.

Psychological Scale:

- The psychological scale used by the decomposition phase adapter is the nominal scale. The nominal scale is the lowest psychological scale with formal property category. It is suitable for determining orthogonal concepts represented by binary and structured qualitative features.

Representing Dimension:

- The representing dimension of the represented features can be shape, position, color etc. measured on the nominal scale. For example, in a face recognition application, the representing dimension for distinguishing between orthogonal concepts like skin-region and non-skin-region is the skin-tone color.

Name Decomposition

Goal: Restrict Data Context, Reduce complexity, enhance reliability

Precondition : Conditioned or transformed/ filtered data

Task : Determine abstract orthogonal concepts - form - subsystems, categories, regions, control models, game configurations, system user-based configurations, etc.

Task: concept validation (for relevance feedback systems, e.g. multimedia product search)

Domain Model structural, functional, causal, geometric, heuristic, spatial, shape, color, etc

Task: Problem formulation

Task Constraints orthogonality reliability, scalability,

Represented Features: *Qualitative/linguistic-* binary, structured
Non-Linguistic- continuous features

Psychological Scale: Nominal

Representing Dimension (Perceptual): Shape, Location, Position, etc. on the nominal scale

Knowledge Engineering Strategy top-down or bottom-up

Problem Solving Methods hard (e.g. symbolic rule based), soft (e.g. neural networks)

Postcondition Domain decomposed into orthogonal concepts.

Figure 5.4. Decomposition Phase Adapter Definition

Knowledge Engineering Strategy:

- Top-Down or Bottom-up

Technological Artifacts:

- Objects and classes can be used to structure the represented features/data and/or the devices/components/objects used by the decomposition problem solving adapter to accomplish a task. Agents can be used as software artifacts to model various tasks associated with the adapter. Soft or hard computing mechanisms can be used for accomplishing various tasks. As explained earlier, the selection of a soft or hard computing technique will depend upon the knowledge engineering strategy and the task constraints.

Precondition:

- Conditioned or transformed/noise filtered data

Postcondition:

- Domain decomposed into orthogonal concepts

Domain Representation Signature	Domain Task Signature
Name : Decomposition **Represented Features:** *Qualitative/linguistic* - binary, structured *Non-Linguistic* - continuous **Psychological Scale:** Nominal; *Formal Property*: category **Representing Dimension (Perceptual):** Shape, Location, Position, etc. on the nominal scale	**Goal:** Restrict Data Context, Reduce complexity, enhance reliability **Precondition:** Conditioned or transformed/ filtered data **Task :** Determine abstract orthogonal concepts - form - subsystems, categories, regions, control models, game configurations, system user-based configurations, etc.. **Task:** concept validation (Optional - for relevance feedback systems, multimedia product search) **Domain Model**: structural, functional, causal, geometric, heuristic, spatial, shape, color, etc. **Task Constraints**: orthogonality, scalability, reliability **Postcondition**: Domain decomposed into orthogonal concepts **Knowledge Engineering Strategy**: top-down or bottom-up

Figure 5.5. Signature Mapping for Decomposition Adapter

5.2.2.3. Control Phase Adapter

The control phase adapter definition is shown in Figure 5.6.

Goal: Establish the decision control constructs for the domain based decision classes as identified by stakeholders/users.

As explained in the decomposition phase adapter definition, the goal of the decomposition phase adapter is to reduce the domain complexity by restricting the input context. However, the decomposition phase adapter does not account for the specific problem being solved in terms of decisions/outcomes required from the computer-based artifact. The primary goal of the control phase adapter is to establish the decision control constructs for the domain based decision classes as identified by stakeholders/users. The decision classes are defined for each abstract concept defined in the decomposition phase.

Task: Noise filtering and input conditioning

- The preprocessing phase adapter, as mentioned earlier, accomplishes these tasks. Whereas, the preprocessing phase adapter is used in the global context prior to the decomposition phase, in the control phase it is used in the local context to filter out noise and condition the data within each abstract concept defined in the decomposition phase.

Task: Determine decision level classes

- Decision level classes are those classes inference on which is of importance to a stakeholder/user. These classes or concepts represent the control structure of the problem. These decision-level classes generally exist explicitly in the problem being addressed. These decision level classes could represent a set or group of network components in a telecommunication network, possible faulty section/s in a electric power network, possible set of control actions in a control application, potential set of diagnosis in a medical diagnostic application, possible face regions in a face recognition application, possible behavioral categories in a sales recruitment application, etc. These concepts can be determined using functional, structural, causal, or other domain model/s used by the stakeholder/user.

- The granularity of a decision level class can vary between coarse and fine. The coarseness and the fineness of a decision level class depend on the context in which the problem is being solved and the decision level class priority in a given context. In one context, a decision level class may be less important to a problem solver, and thus a coarse solution may be acceptable, whereas, in another context the same decision level class may assume higher importance and thus a finer solution may be required. That is, if the decision level class priority is low, then its granularity is coarse, and the problem solver is satisfied with a coarse decision on that class. Otherwise, if the decision level class priority is high then the decision-level class has fine granularity and the problem solver wants a fine set of decisions to be made on the decision-level class, which would involve a number of microfeatures in the domain space. In case of coarse granularity distinct control and decision phase adapters (described in the next section) may not be required and can be merged into one.

Task: Concept validation

- Like in the decomposition phase adapter, this task is required in applications where problem solving is largely guided by relevance feedback from the stakeholder/user. This is especially true in a number of image retrieval applications on the internet.

Task: Conflict Resolution

- It is possible that the decisions made by a decision level class may conflict with the decisions by another decision level class. For example, in a telecommunication network diagnostic problem, two decision level classes may represent two sections of a telecommunication network. If these sections predict fault in two different network components (given that only one of them can be actually faulty), then there is a conflict. Similarly conflict may arise if two gene classifications result in a human genome application or two different control actions result in a control application.

  The conflicts can also occur with respect to previous knowledge or in situations involving temporal reasoning. In the case of temporal reasoning the previous result may become invalid or conflict with the result based on new data. The conflict may be resolved by looking at the structural, functional, spatial indisposition of the decision level classes or their components or even through concept/decision validation (which would involve validation/feedback from the stakeholder/user on the conflicting set decisions).

Task: Problem Formulation

- involves sequencing of various actions required to accomplish the above tasks.

Task Constraints:

- Learning and adaptability are the additional domain independent task constraints in this phase, besides scalability and reliability.

Represented Features:

- *Qualitative/linguistic* - binary, structured, fuzzy

  The qualitative or linguistic features employed by the control phase adapter include semi-coarse grain binary, structured and fuzzy features. The granularity of the binary and structured features used by the control phase adapter is finer than those used in the decomposition phase. In the decomposition phase binary and structured features are used for determination of abstract independent orthogonal concepts at the global level. In the control phase adapter the binary and structured features are used at the local level within each abstract concept. More so, the binary and structured features are used many times with fuzzy features in order to identify the decision level concepts in a domain. The fuzzy features are used in the control phase instead of the decomposition because fuzzy features cannot be used to distinguish between abstract orthogonal concepts. For example, let us assume mammal and bird are two abstract concepts in an animal classification domain. Then the interpretation of a large mammal is not the same as a large bird. That is, the fuzzy variable *large* qualifying a mammal and bird carry different perceptual as well as conceptual meanings and thus cannot be used universally at the global level for discriminating between abstract concepts.

- *Non-Linguistic - continuous features*

  Continuous valued features used by the control phase adapter are limited to a abstract concept determined in the decomposition phase. For example, in a face recognition application pixel data related to the skin region concept is analyzed.

Domain Models:

- The domain model used for accomplishing various tasks and identifying the represented features can be structural, functional, causal, geometric, heuristic, spatial, shape, color, etc. For example, in the face recognition application, shape and area models are used to determine the decision classes. On the other hand, in a genome classification application a functional model may be used to determine gene decision (classification) classes based on their functionality or in an alarm processing application, structural configuration of various components in the network may be used for determining the faulty sections.

Psychological Scale:

- Besides the nominal scale, ordinal, interval and ratio scales can used by the control phase adapter. The fuzzy features used by the control phase adapter can be seen to represent information on the ordinal, interval or ratio scales.

Representing Dimension:

- The representing dimension of the represented features can be shape, position, color etc. measured on the nominal and/or ordinal, interval and ratio scales. For example, in a face recognition application, area and shape of the skin-regions are the representing dimensions of the various face-recognition decision classes.

Name: Control

Precondition : orthogonal concept defined, concept data/expertise available

Goal: Establish domain decision control constructs for orthogonal concepts based on desired outcomes from the system

Task: Local noise filtering (done by Preprocessing Adapter) – form – time based noise filtering, content and context based noise filtering

Task: Determine decision level concepts – form - secondary codes, potential fault sections/regions, potential explanation sets/ cause sets/diagnosis sets, decision categories based on structural, functional shape, color, location, spatial and heuristic domain models

Task: Decision level concept validation (optional – for relevance feedback systems)

Task: Conflict resolution (optional) – form – decision conflicts between decision categories,

Task: Problem formulation

Task Constraints: scalability, reliability, maintainability, learning, adaptability,

Domain Models : structural, functional, causal, geometric, heuristic, spatial, shape, color, etc.

Represented Features: *Qualitative/Linguistic* - binary, structured, fuzzy data
Non-Linguistic - continuous data related to an orthogonal concept

Psychological Scales: Nominal, Ordinal, Interval, Ratio

Representing Dimensions (perceptual): shape, size, length, distance, density, location, position, orientation, color, texture

Knowledge Engineering Strategy : top-down or bottom-up

Technological Artifacts: hard (symbolic), soft (e.g. neural networks, fuzzy logic, genetic algorithms) or their hybrid configurations

Postcondition: decision level concepts defined, decision control constructs/actions defined.

Figure 5.6. Control Phase Adapter Definition

Knowledge Engineering Strategy:

- top-down or bottom-up knowledge engineering strategy can be used.

Technological Artifacts:

- The computing technique can be hard (e.g., symbolic) or soft (e.g. neural networks, fuzzy logic, genetic algorithms) or a hybrid configuration of hard and soft computing techniques (Khosla & Dillon 1997) depending upon the task constraints and the knowledge engineering strategy. We have also shown structural relationships in Figure 5.7 which can be used for identifying the relationships between data entities. It can also be used in other problem solving adapters.

Precondition:

- The control phase adapter assumes that orthogonal concepts in the domain have been defined. Further, if top-down strategy is employed, it is assumed that qualitative data is available. However, if bottom-up strategy is used it is assumed raw case data is available. Based on the above description, the signature mappings of the control phase adapter are shown in Figure 5.7.

Postcondition:

- Defines the competence of the control adapter in terms of defining the decision control constructs for the decision level concepts.

Domain Representation Signature

Name: Control
Represented Features: *Qualitative/Linguistic*- binary, structured, fuzzy data
Non-Linguistic -continuous data related to an orthogonal concept
Psychological Scales: Nominal, Ordinal, Interval, Ratio
Representing Dimensions (perceptual): shape, size, length, distance, density, location, position, orientation, color, texture
Knowledge Engineering Strategy top-down or bottom-up

Structural Relationships (optional):
Inheritance, composition, association

Domain Task Signature

Name: Control
Goal: Establish domain decision control constructs for orthogonal concepts based on desired outcomes from the system
Precondition: orthogonal concept defined, concept/ case data
Task: Determine decision level concepts – potential fault sections/regions, potential explanation sets/ cause sets/diagnosis sets, decision categories based on structural,functional shape, color, location, spatial and heuristic domain models
Task: Decision level concept validation (optional – for relevance feedback systems)
Task: Conflict resolution (optional)– decision conflicts between decision instances
Task Constraints:scalability, reliability, maintainability, learning, adaptability,
Domain Models: structural, functional, causal, geometric, heuristic, spatial, shape, color, etc.
Postcondition: decision level concepts defined, decision control constructs/axioms defined

Figure 5.7. Signature Mapping for Control Phase Adapter

5.2.2.4. Decision Phase Adapter

Goal:

- Provide decision instance results in a user/stakeholder defined decision concept.. Whereas, the control phase adapter primarily controls the invocation of various decision level classes and conflicts between them, the decision phase adapter is responsible for providing specific outcomes required by the user/s or stakeholder/s in each decision class. These outcomes can include specific faulty component/s in a telecommunication network, specific infection and treatment in a clinical support system, actual faces in a face recognition problem, legal move in a computer game, product with desired features in a electronic commerce application, and so on.

Task: noise filtering and input conditioning

- The preprocessing phase adapter, as mentioned earlier, accomplishes these tasks. In the decision phase adapter, the preprocessing phase adapter is used to filter out noise and condition data in a decision class.

Task: Determine decision instance

- This task entails determination of specific decisions or decision instances required by the user/stakeholder(s). Decision instance or instances represent partly or wholly user defined outcomes from a computer-based artifact. These outcomes are realized within each decision class invoked by the control phase adapter. For example, in a face recognition problem, this task may involve identification of a face (or faces) among various face candidates (which represent the decision classes) or in telecommunication this may involve determination of a faulty component or components in the candidate section or sections (decision classes) of the telecommunication network. Similarly, in an alarm processing and fault diagnosis application in a power system control center, this task may involve determination of various fault instances like single line fault, multiple line fault, etc. in candidate sections (e.g., 220kv, 66kv, etc.) of the power network. On the other hand, in a control system application, this task may involve selecting a control action among various candidate control actions.

Task: Viability/Utility of Decision (optional)

- In some real time systems it may become necessary to compute the computational resources and the time required by different decision level classes to determine the solution. Thus, certain decision level classes may not be considered viable under these constraints and thus may not be activated. For accomplishing this task symbolic or fuzzy techniques are used.

Task Constraints:

- scalability, reliability, maintainability, learning, generalization, adaptability, domain dependent

Domain Model:

- Here again, one or more domain models like structural, functional, causal, geometric, heuristic, spatial or location, shape, color, etc. can be used for determining the decision instances. For example, in the face recognition problem, shape and location domain models of the face and facial features like eyes, nose and mouth are used. In the alarm processing and fault diagnosis problem structural and spatial models are used for determining the fault instances in different sections of the network. The structural domain model is used in terms of the connectivity of different components in a given section of the power network. The spatial model is used in terms of spatial proximity of the alarms emanating from different parts of a network from the faulty component. That is, the further away an alarm is from the location of the faulty section or component, the lesser is its importance.

Represented Features:

- *Qualitative/Linguistic*- binary, fine grain fuzzy data,

  The qualitative or linguistic features employed can be fine grain fuzzy or even binary. For example, in the alarm processing problem two properties of the alarm data are used. Firstly, existence or absence (i.e. binary property) of a circuit breaker alarm in a decision class (candidate faulty section) is determined. Secondly, fine grain fuzzy contribution value of a circuit breaker alarm and associated relay towards a fault in a particular network component is modeled in terms of their protection proximity to a possible faulty component (Khosla and Dillon 1997, pp. 319-22). This contribution value is determined in terms of activity level of a path (consisting of alarm and relay).

  Similarly, in a animal classification (more specifically, tiger classification) domain, whereas *large*, *medium* and *small* semi-coarse grain fuzzy features are used by the control phase adapter to distinguish between lion, tiger, puma and a domestic cat, fine grain fuzzy feature like *heavy cheek hair* is used to distinguish between different types of tigers in the decision phase.

- *Non-Linguistic - continuous decision data*
  For example, in the face recognition problem, color pixel data related to a face candidate and spatial coordinates of facial features (like eyes, mouth and nose) are used to identify actual faces and track eye movements in the decision phase.

Psychological Scales: Nominal, Ordinal, Interval, Ratio or none

- The nominal scale can be used to measure binary features (like existence or non-existence of an alarm) whereas fine grain fuzzy features can be measured on the ordinal, interval or ratio scales, depending on the scale properties, by the fuzzy features. For example some of the scale properties of fuzzy feature heavy cheek hair are category (cheek hair), magnitude (heavy > light) and absolute zero (no cheek hair). These properties represent the ratio scale.
- Representing Dimensions (perceptual): shape, size, length, distance, density, location, position
- As mentioned earlier representing dimension is useful for determining the perceptual aspects of data and reasoning in a problem domain. For example, in the animal classification domain, the representing dimension of the fuzzy feature is density.

Knowledge Engineering Strategy:

- The decision to use top-down, bottom-up or a mix of both will depend upon availability/non-availability of domain knowledge for various tasks.

Technological Artifacts:

- hard (symbolic), soft (e.g. neural networks, fuzzy logic, genetic algorithms), hybrid configurations, or other statistical/mathematical algorithms.

  Broadly hard symbolic computing mechanisms (like rule based systems) can be used for high level tasks (like problem formulation) subject to availability of qualitative domain knowledge for the task. On the other hand soft computing mechanisms can be used for decision instance task which may involve pattern recognition, learning, generalization and adaptability. As a consequence of satisfying task constraints (like learning, generalization and adaptability) optimization may be another constraint that may need to be satisfied. Genetic algorithms are ideal for satisfying the optimizing learning and generalization characteristics of soft computing mechanisms (like neural networks). More details on use of various soft computing mechanisms in isolation and in hybrid configuration can be found in Khosla and Dillon (1997). Similarly, hard symbolic techniques can be used for accomplishing the task.

Precondition: raw and/or qualitative case data, user specified decision instances.

Postcondition: Unvalidated Decision instance results from the computer-based artifact based on user/stakeholder defined decision concepts/classes.

The adapter definition and signature mapping are shown in Figures 5.8 and 5.9, respectively.

Phase: Decision
Goal: Provide decision instance results based on user/stakeholder defined decision concepts/classes from the computer-based artifact
Precondition Decision concepts defined (for top-down KE strategy), decision control constructs defined (optional), decision concept data/expertise available
Task: Context validation – Problem Solving context – Decision level; Input context: Local decision concept data
Task: Decision concept noise filtering (done by preprocessing adapter)
Task: Define specific decision instances for each decision concept
Task: Validate/Utility of decision
Task: Other user/stakeholder defined decision instance related tasks
Task: Problem formulation
Task Constraints: Learning, generalization, adaptability, domain dependent
Precondition: Raw or processed data
Postcondition: Conditioned data
Represented Features: *Qualitative/linguistic* – binary, fine grain fuzzy decision concept data
Non-linguistic – continuous decision concept data
Psychological Scale: Nominal; Ordinal, Interval, Ratio or none
Representing Dimension (Perceptual): Shape, size, length, distance, density, location, position, orientation, color, texture
Knowledge Engineering Strategy: Top-down or bottom up
Technological Artifacts: Hard (symbolic), soft (e.g. neural networks, fuzzy logic, genetic algorithms), hybrid configuration or other statistical/mathematical techniques Structural relationships based on object-oriented technology can also be used.
Postcondition: Unvalidated decision instance results

Figure 5.8. Decision Phase Adapter Definition

Domain Representation Signature

Name : Decision

Represented Features:
Qualitative/Linguistic - binary, fine grain fuzzy decision concept data,
Non-Linguistic - continuous decision concept data
Psychological Scales: Nominal, Ordinal, Interval, Ratio or none

Representing Dimensions (perceptual): shape, size, length, distance, density, location, position, orientation, color, texture
Structural Relationships (optional)
Inheritance, composition, association

Domain Task Signature

Name : Decision

Goal: Decision instance results based on user/stakeholder defined decision concepts/classes
Task : Determine decision instance

Domain Model: - functional, structural, causal, spatial, color, etc.

Task Constraints: learning, generalization, adaptability

Precondition: Decision concepts defined (for top-down) decision control constructs defined (optional), decision data/expertise available

Postcondition: Unvalidated decision Instance results

Knowledge Engineering Strategy: top-down or bottom-up

Figure 5.9. Signature Mapping for Decision Phase Adapter

5.2.2.5. Postprocessing Phase Adapter

Goal: Establish outcomes as desired outcomes, Satisfy user/stakeholder.

Logic and provability are the hallmarks of our conscious interactions with the external environment. Thus, the goal of the postprocessing phase adapter is to validate outcomes from the decision phase adapter as desired or acceptable outcomes.

The decision phase adapter in the postprocessing phase uses symbolic techniques to validate or evaluate decisions made in decision phase.

Task: Decision instance result validation – form – model based instance result validation.

- For example, in a face recognition application the actual faces and facial movements as determined by the decision phase adapter, need to be validated by the user. In a control system application, feedback from the environment establishes whether the selected/executed control action has produced the desired results. For example, a control action taken by the inverted pendulum control system may result in the pole balancing or falling over. This result is the feedback from the environment validating or invalidating the control action by the inverted pendulum control system. In an alarm processing application, the operator may instruct the system or computer based artifact to explain how certain components in the network have been identified as faulty.
- The validation task can be accomplished by perceptual and/or hard/soft computing mechanisms. For example, in a real time alarm processing application a power system control center operator may validate a decision made in the decision phase by using graphic display of the power network and by querying the system on the fault model of the faulty component. On the other hand a control system application may validate a control by using perceptual mechanisms (e.g. the location/position of the inverted pendulum) task model base.

Name: Postprocessing

Precondition: specific unvalidated decision outcomes available, decision data available. (optional).

Goal: Validate decision outcomes as desired outcomes, Satisfy user/stakeholder

Task: Context validation - Problem solving context - Postprocessing ; Input context: decision instance result data and/or model

Task: Decision instance result validation through domain model, user/stakeholder or environment

Task: Decision instance result explanation

Task: Problem formulation

Task Constraints: provability, reliability

Represented Features : *Qualitative/Linguistic* - binary, fine grain fuzzy, *Non-Linguistic* - continuous

Psychological Scales: Nominal, Ordinal, Interval, Ratio or none

Representing Dimensions (perceptual): shape, size, length, distance, density, location, position, orientation, color, texture

Knowledge Engineering Strategy: top-down or bottom-up

Technological Artifacts : hard (symbolic), soft (e.g. neural networks, fuzzy logic, genetic algorithms)

Postcondition: Decision results validated and explained to the user

Figure 5.10. Postprocessing Phase Adapter

In control systems and multimedia based relevance feedback problems the postprocessing phase adapter can be seen as part of the feedback loop where validation on the outcomes of the decision phase is provided by the environment or the human user. In other domains if validation models are not available, they can be developed for each decision concept or class (defined by the control phase adapter.) by employing learning artifacts like neural networks.

The validation and explanation tasks in the postprocessing phase (shown in Figure 5.10) can also be seen to represent logic and provability which are the hallmarks of our conscious interactions with the external environment.

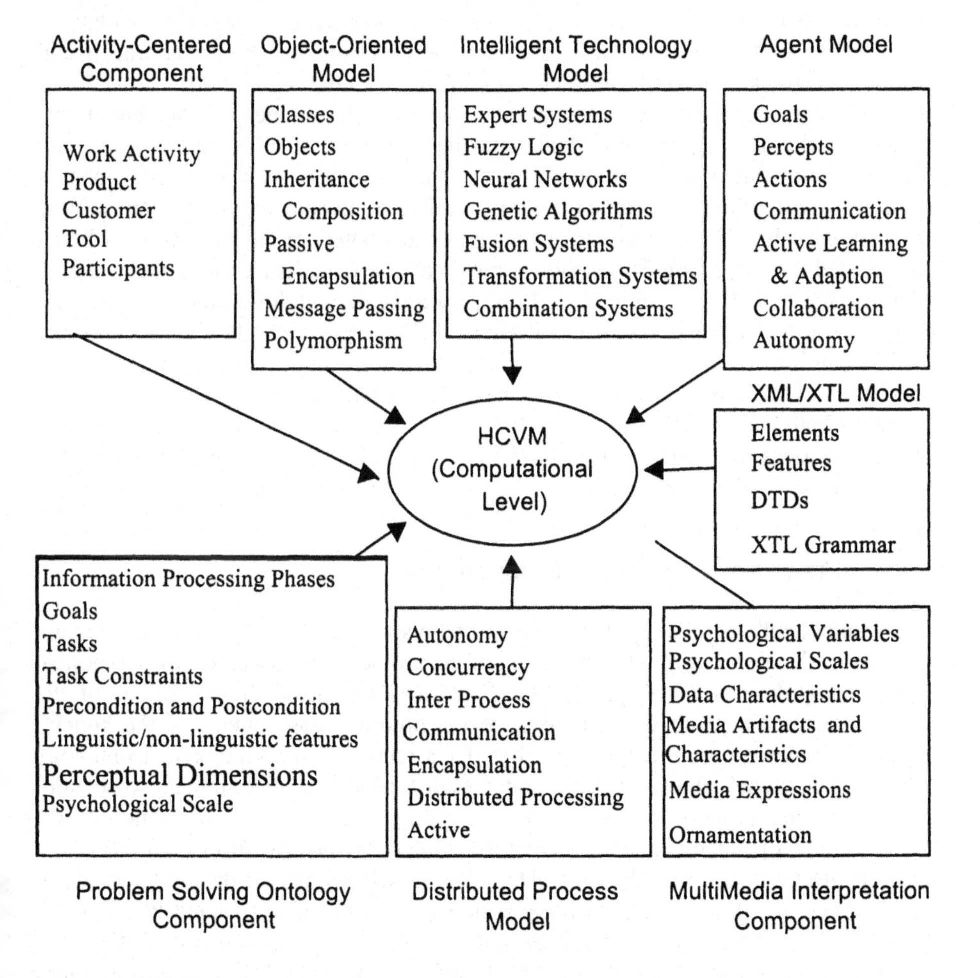

Figure 5.11. Human-Centered Virtual Machine (HCVM)

5.3. Human-Centered Criteria and Problem Solving Ontology

The problem solving ontology component developed by us has been designed to satisfy criteria 1and 3 of human-centeredness outlined in the first chapter. That is, it is derived from the problem solving pattern or consistent problem solving structures/strategies employed by practitioners while designing solutions to complex problems or situations (criteria 1: Human-centered research and design is problem/need driven as against abstraction driven (although there is an overlap)). It facilitates use of perceptual (external) as well as conceptual (internal) representations for problem solving (as also advocated by the distributed cognition approach and criteria 3: Human-centered research and design is context bound). Further, it constrains the perceptual and conceptual representations of the environment in the context of the activity being studied with the help of five information processing phases. From a situated cognition viewpoint, the five information processing phases represent the routines or problem solving structures people employ when solving complex problems. These phases are situated in the context of the work activity being studied and the technological artifacts employed to accomplish various tasks can be adaptive and evolutionary in nature. In other words, the ontology facilitates use of a range of intelligent technologies for satisfying different pragmatic task constraints and thereby minimizing task generation. Finally, the problem solving structures of the problem solving ontology component have been derived from studying complex problems both inside and outside knowledge based systems area and include problems in image processing, data mining, process control, electronic commerce, diagnosis, forecasting, and sales recruitment.

5.4. Transformation Agent Component

This component's purpose is to transform the systematized human-centered domain task model developed through application of the problem solving ontology component into a computer-based software artifact. It does that through integration of the activity-centered analysis and problem solving ontology components of the human-centered framework with technological artifacts related to the intelligent technology model, agent model, object-oriented model, multimedia (described in the next section) and distributed process model as shown in Figure 5.11. The outcome of this integration is a Human-Centered Virtual Machine (HCVM) shown in Figure 5.12. It consists of five layers, namely, the object layer, which defines the data architecture or structural content in the context of the work activity, the software agent layer, which helps to define the distributed processing constructs, the multimedia design and generation constructs and the XML/XTL based constructs used for transforming task and representation constructs of the problem solving adapters into an XML representation for internet based applications. The intelligent agent layer defines the constructs for intelligent technologies (Khosla and Dillon 1997). The hybrid layer defines constructs for intelligent fusion, combination and transformation technologies. Finally, the problem solving agent layer defines the constructs related to the problem solving adapters described in section 5.2. The five layers facilitate a component based approach for agent based software design. The generic agent definition used for defining the transformation agents in the problem solving agent layer, intelligent hybrid agent layer, intelligent agent layer and software agent layer is shown in Figure 5.13. Based on the generic agent definition, a neural network agent is shown in Figure 5.14.

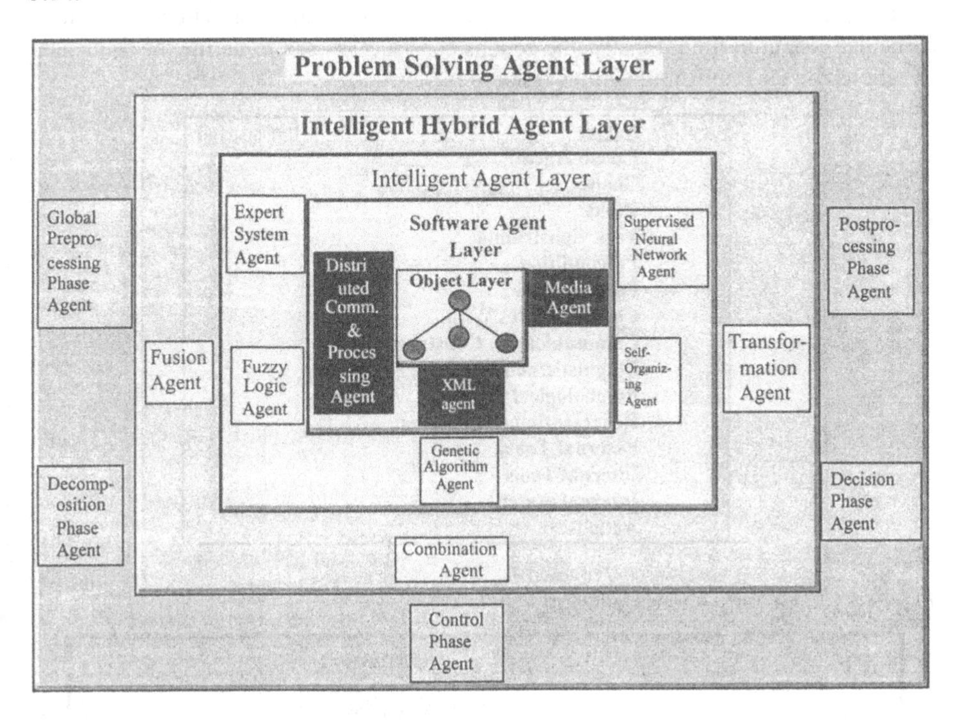

Figure 5.12: Five Layers of HCVM

The generic definition of the transformation agent includes communication constructs employed by the transformation agent. These communication constructs are based on human communicative acts like request, command, inform, broadcast, explain, warn and others (Maybury 1995). The linguistic and non-linguistic features represent the sensed data from the external environment as well as computed data by the agent. The sensed and computed data are used by the multimedia interpretation component (described in the next section) to gather data from the environment (in this case human is the data source) and also assist the direct stakeholders in interpreting the computing data.

The *parent agent* construct identifies the generic agents in the four agent layers, whose constructs and services have been inherited by a particular application or domain based transformation agent. The *communication with* construct in Figure 5.13 identifies all the agents and objects that a transformation agent communicates with in the five layers. The *external tools* construct in Figure 5.13 refers to those computer-based or other tools that are external to the definition of an agent. On the other hand, *internal tools* are those tools that are defined internally by a transformation agent. For example, Figure 5.14 shows the agent definition of a neural network agent. The *external tools* include simulated training data files used by the agent. On the other hand, the sensitivity algorithms and the back propagation rule are *internal tools* defined and used by the neural network agent. Since the neural network agent is a generic agent it does not have any parent agent or communication constructs.

The *internal state* construct refers to the beliefs of a transformation agent at a particular instant in time. Finally, the actions construct is used to define the sequence of actions for accomplishing various tasks.

Name:
Parent Agent:
Goals:
Tasks:
Task Constraints:
Precondition:
Postcondition:
Communicates With:
Communication Constructs:
Linguistic/non-linguistic Features:
Psychological Scale:
Representing Dimensions:
External Tools:
Internal Tools
Internal State:
Actions:

Figure 5.13: Generic Agent Definition

- Name:
  - Neural Network
- Goals:
  - create NN model of the domain
- Tasks:
  - perform backpropagation to learn weights
  - test convergence with test data
- Tasks Constraints
  - computing resources
- Precondition:
  - Training data available (continuous/discrete)
  - Initial network structure available
  - Training data normalized
  - convergence criteria
- Post Condition:
  - Converges on global minimum
- Represented features:
  - Training data
  - Training/ Test set error
  - convergence criteria
- Psychological Scale: Ratio
- Representing Dimension (Perceptual):
  Shape (graph, plot)
- External tools
  - simulated data files
- Internal tools
  - Sensitivity algorithms
  - Backpropagation
  - Parallel distributed learning
- Actions
  - Feed weights parameters to network
  - return model parameters
  - return training set error

Figure 5.14. Neural Network Agent

5.5. Multimedia Interpretation Component

From a social perspective, the role that computer-based artifacts play in mediating work activity is the result of social interaction between users and their environment and between the users and the artifact. The social interaction determines the way the users perceive, use and learn the artifact. This social interaction is determined by the psychological apparatuses or structures employed by the users and computer-based artifacts can be seen as extensions of the psychological apparatuses of their users (However, this is not intended to mean that humans and computer-based artifacts are necessarily cognitively equal). In this chapter and the last one we have studied these psychological apparatuses through the problem solving ontology component and the activity-centered analysis component. The role of the multimedia interpretation component is to make the psychological apparatuses mapped in the computer-based artifact transparent to its users. This transparency will provide an immersive environment for the users and enable uninhibited interaction between the users and the artifact.

Keeping in view that computer-based artifacts are extensions of the psychological apparatuses of their users, their interpretation should be based on among other aspects, the psychological scales and representing dimensions employed by the users on the psychological variables of interest rather than physical variables used and computed by the computer-based artifacts to do the computations (Norman 1988). For example, in a sales recruitment system (used for determining the selling behavior profile of a salesperson and described in chapter 8) the computer-based artifact computes a behavioral category score (a physical variable) based on a number of physical variables (like numerical values of answers to various questions, weights of various questions, etc). However, the psychological variable of interest to a sales manager (a user) is the degree of fit of the salesperson's behavioral profile into a frontline sales job or a customer service role or, for that matter, the training needs of the salesperson for different roles, similarity/dissimilarity with benchmark profiles, etc. This is an example where psychological variables determine the information content of the physical variables, to be presented to the user for interpreting the outcomes or results of a computer-based artifact. Additionally, the physical variables used for computing the results (e.g., questions, answers in the sales recruitment example) also need to be modeled based on the psychological scales and representing dimensions employed by the users/direct stakeholders in the work activity. That is, for effective data gathering, appropriate psychological scales and representing dimensions have to be used. Given this background, the main focus of the multimedia interpretation component is to identify and analyze the psychological scales and representing dimensions employed by direct stakeholders for gathering and providing information (e.g. to a computer-based artifact) as well as psychological variables of interest used for interpretation of results. Based on the analysis, media artifacts like graphics, text video and audio are then used to model the data perceptually in order to reduce the cognitive load on the users. This process is encapsulated in the following three steps. They are also shown in Figure 5.15.

1. Data Content Analysis
2. Media, Media Expression and Ornamentation Selection, and
3. Media Presentation Design and Coordination

5.5.1. Data Content Analysis

Data content analysis involves firstly, the identification of data for various tasks in a work activity context. Secondly, it involves determination of content of data to be communicated to the stakeholders based on its dimensionality, psychological scales, and representing dimensions as perceived by its users in the work activity and task context. Finally, it involves analysis of other data characteristics like granularity, transience, urgency and volume as defined by Arens et. al (1994). Figure 5.15 shows the influence of the task context, user context and data characteristics on the data content analysis stage.

The tasks in a work activity establish the context in which data is to be interpreted and used. The human-task-tool diagram and the task product transition network assist in identifying the human-computer interaction tasks and other computational tasks. The primary aim of the multimedia interpretation component is to model the data content linked to human-computer interaction tasks. These tasks define the human-computer interaction points for data gathering as well as interpretation of computed results.

The five phases of the problem solving ontology systematize the tasks and assist in identifying data at different levels of abstraction. These different levels of abstraction in problem solving can also be mapped to different levels of media expression in order to situate the users or direct stakeholders in the computerized system and make information processing in these transparent to them. This aspect of media representation will be explained in the next section.

The psychological scales and representing dimensions have been defined by the problem solving adapters of the problem solving ontology component. These psychological scales and representing dimensions are associated with data used for different tasks. The primary purpose of using the psychological scales and representing dimensions in the problem solving component was to determine whether perceptual reasoning techniques could be used for accomplishing the tasks. However, these psychological scales and representing dimensions of data that have been identified in the user and task context can also be used for data content analysis by the multimedia interpretation component. An application of the scale information and representing dimensions for data content analysis will be shown in the next chapter.

The other data characteristics used for data content analysis include dimensionality, granularity, transience, urgency, and volume. These are described in the rest of this section.

Dimensionality: of a data item refers to the number of degrees of freedom on which a data item is perceived by the user in a particular task context. For example, a medical symptom like *ear drum red or yellow and bulging* is perceived by the user on **two** dimensions namely, color and shape.

Granularity: determines the granularity of variation in data value that carries meaning for the user. Granularity can be continuous or discrete. Continuous is a class in which small variations along a dimension of interest carry meaning. Information in such a class is best supported by a medium that supports continuous change. Discrete is a class in which there exists a lower limit to variations on the dimension of interest (e.g. types of cars made in Australia).

Transience: refers to whether the information to be presented expresses some current/changing state or not. The changing state can be *live* or *dead*. Live information consists of a single conceptual item of information that varies with time along some linear ordered dimension. On the other hand, dead information does not reflect current state but rather past state.

Urgency: Urgent information requires presentation in such a way that it draws the user's attention. The characteristic takes the values *urgent* or *routine*. For example, in a medical symptom like high blood pressure, patient's blood pressure reading of 210/130 may be represented by a media with high or medium to high default detectability to draw doctors' attention.

Volume: A batch of information may contain various amounts of information to be presented. If it is single fact (e.g., name), it is called *singular*; If more than one fact (e.g., a database record) but still little relative to some task and user-specific threshold - it is called *little*; otherwise (e.g. on-line help) it is called *much*. A batch of information with volume *much* (like on-line help) will require use of medium like written text with a transient property *dead*. Whereas, a single fact can be represented by a medium with transient property *live*.

5.5.2. Media, Media Expression and Ornamentation Selection

The main aim of this step is to map the data to the appropriate media, media substrate and media expression. The data characteristics described in the preceding section primarily influence the media, media substrate and media expression selection. Media, as is obvious, specifies the type of medium used (e.g., text, graphics/image, video, audio, etc.). Media substrate is a background to a simple exhibit. It establishes to the consumer, physical or temporal relation and the semantic context, within which new information is presented to the information consumer or user, e.g., piece of paper or screen (on which information may be drawn or presented); a grid (on which a marker might indicate the position of an entity).

Media expression on the other hand, determines the abstraction level of media. The three abstraction levels, elaborate, representative and abstract for text, graphics/image, sound, and motion are shown in Table 5.2 (Heller and Martin 1995). For example, in order to display a particular medical symptom, elaborate or representative image and text may be used as the media or medium, a sliding scale may be used as a substrate with continuous granularity for indicating the severity of the symptom.

The three levels of media expression shown in Table 5.2 also influence the media characteristics shown in Table 5.1. That is, the abstract level of media expression has discrete granularity and high baggage, whereas, the elaborate level of media expression has continuous granularity and low baggage.

The medium and substrate are then selected based on the correspondence between data characteristics and media characteristics (shown in Table 5.1). In order to do the mapping firstly, the psychological scale information and representing dimension characteristic of the data is matched with the internal semantics of the media artifacts. For example, a representing dimension like location can be represented by a picture or map which has internal semantics of spatial location or animated picture with internal semantics of spatial location and motion. Secondly, we look at other characteristics of data like transience and urgency to enhance or upgrade the existing selected medium. For example, if a location has a transient property live then an animated picture rather than a simple picture would be used. Further, if the object or information carrier in the animated picture has an urgency property as urgent then besides animation (which has high detectability) it may have to be further enhanced with flashing bright color. Arens et al (1994) defines a set of transformation rules for selecting the medium and substrate. Some of these rules are defined below.

Transience: If the transience property is live, as a carrier, use a medium with the temporal endurance characteristic transient if the update rate is comparable to the lifetime of the carrier signal. If the data update rate is much longer, as a carrier, use a medium with the temporal endurance characteristic permanent. As substrate, unless the information is already part of an existing exhibit, use neutral substrate.
If the transient property is dead use a carrier/media with permanent temporal endurance.
Urgency: If the urgency property is urgent then if the information is not yet part of a presentation instance, use a medium whose detectability has the value high either for substrate or carrier. If the information is already displayed as part of a presentation instance, use the present medium but switch one or more of its channels from fixed to the corresponding temporally varying state, e.g. flashing. On the other hand, if the property is routine, choose a medium with low default detectability and a channel with no temporal variance.

Thirdly, we look at how we can complement selected media with one or more media by integrating the selected media with other media at different levels of media expression or abstraction. That is, an elaborate selected media can be integrated with abstract or representative forms of other media to enhance understanding and develop a more immersive environment for the users.

The level of media expression also facilitates mapping the media to different levels of problem solving. At higher levels of problem solving it is likely that abstract or representative levels of media expression will be mapped to data. whereas, at lower levels of abstraction elaborate levels of media expression are likely to be used. Further, as mentioned in the last section, the three levels of media expression can be effectively used to situate the user appropriately among the information processing phases of the computer-based artifact.

Finally, we also look into the ornamental aspects of the overall presentation from an industrial design perspective. For example, laptops and pagers have traditionally come in black or grey colors. These colors have been associated with top level corporate executives and professionals. Similarly the color background of a computer-based presentation should reflect the social characteristics of the direct stakeholders and the environment in which they work. A medical diagnostic system presentation, for example, should use serene colors instead of bright colors (e.g. yellow) to reflect the social characteristics of the medical practitioners and clinical environment they work in.

Thus conceptually, the various aspects of data content analysis, media and media expression selection discussed in this section and the preceding section are meant to reduce cognitive load on the users through perceptual presentations. They are meant to facilitate direct manipulation of data through multimedia artifacts, and situate users in the information processing phases of the computer-based artifact through use of multiple levels of media expression.

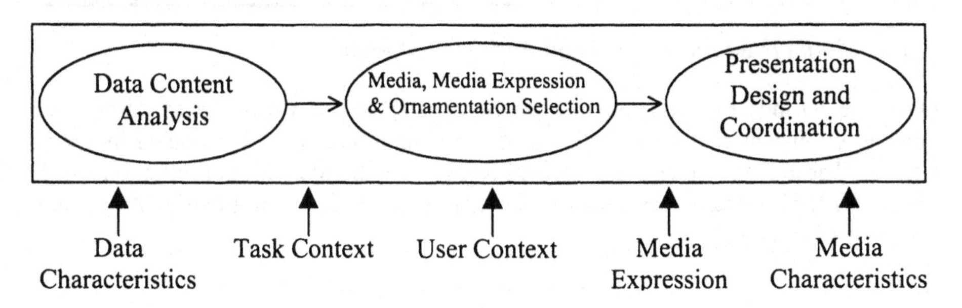

Figure 5.15: Media Analysis, Selection and Design Steps

Table 5.1: Media Characteristics

Medium	Carrier Dimension	Temporal Dimension	Granularity	Medium Type	Default Detectability	Baggage
Map	2D	Perm	Continuous	Visual	Low	High
Picture	2D	Perm	Continuous	Visual	Low	High
Table	2D	Perm	Discrete	Visual	Low	High
Form	2D	Perm	Discrete	Visual	Low	High
Graph	2D	Perm	Continuous	Visual	Low	High
Ordered List	1D	Perm	Discrete	Visual	Low	Low
Sliding Scale	1D	Perm	Continuous	Visual	Low	Low
Written Sentence	1D	Perm	Continuous	Visual	Low	Low
Spoken Sentence	1D	Perm	Continuous	Aural	Mhigh	Low
Animation	2D	Trans	Continuous	Visual	High	High
Music	1D	Trans	Continuous	Aural	Mhigh	Low

Table 5.2: Levels of Media Expression © IEEE

MEDIA TYPE	ELABORATE MEDIA EXPRESSION	REPRESENTATIVE MEDIA EXPRESSION	ABSTRACT MEDIA EXPRESSION
Text	Fully expressed written text	Abbreviated text, titles, bulleted items	Shapes, Icons
Graphics	Fully expressed photograph	Abbreviated blueprint, layout	Graphic icon
Sound	Fully expressed speech	Abbreviated tones	Sound effects
Motion	Fully expressed film footage	Abbreviated animation, news clips, film preview	Animated model/icon

5.5.3. Media Presentation Design and Coordination

Once the various media artifacts have been selected for various data items, their generation, display and coordination at the computational level is modeled by media agents. These media agents are defined using the generic agent definition shown in Figure 5.13. The media agents coordinate their action with the problem solving agents of the problem solving agent layer.

5.6. Emergent Characteristics of HCVM

In order to get an overall picture of the HCVM, it is useful to look at its emergent behavior. We define the emergent behavior of HCVM by outlining its architectural, management and domain application characteristics.

5.6.1. Architectural Characteristics

The architectural characteristics define the significance of HCVM in terms of its emergent design characteristics. Some of the emergent design characteristics are outlined in this section.

5.6.1.1. Human-Centeredness

HCVM has been grounded in the three human-centered criteria outlined in the first chapter. These criteria have been built into the four components, namely, activity-centered analysis, problem solving ontology, transformation agent and multimedia interpretation component. These four components have been used to define the internal and external plane of a system, respectively. The external plane captures the physical, social and organizational reality, whereas the internal plane captures the subjective reality related to stakeholder incentives, organizational culture and other aspects.

5.6.1.2. Task Orientation vs Technology Orientation

The solutions to real world problems are determined by engineers, designers, accountants, sales managers, etc. in a task context (Chandrasekaran 1992; Peerce et al 1997) rather than a techniqueological context. Various intelligent technologies, like knowledge based systems, fuzzy logic, neural networks and their hybrid configurations (fusion, transformation and combination), propose a technology-based solution to real world problems. The problem solving ontology component of HCVM is a task oriented system in which technological artifacts are considered as primitives for accomplishing various tasks. The use of one or more technological primitives is contingent upon satisfaction of task constraints. The task orientation enables HCVM to match a given task to one or more technologies among a suite of technologies rather than match a given technology to tasks in a work activity.

5.6.1.2. Flexibility

Most complex real world problems require satisfaction of a number of task constraints ranging from incomplete and noisy information, learning, fast response time to explanation and validation and one technology is not enough to provide a satisfactory solution. A technology-based solution constrains a problem solver to force-fit particular software design onto a task or problem. The five problem solving adapters of HCVM allow the problem solver to use multiple domain models. The intelligent hybrid agent layer and the intelligent agent layer of the HCVM provide flexibility in terms of multiplicity of intelligent techniques and their hybrid configurations that can be employed to satisfy various task constraints. Further, HCVM also provides flexibility in terms of pursuing different decision paths based on user competence and experience. This will be demonstrated in the medical diagnosis and treatment support application in the next chapter. That is, the user can follow the five phases in different sequences. A decision making sequence can include five or less phases as will be shown in chapter 6.

5.6.1.3. Versatility

Technologies like expert systems and fuzzy logic rely heavily on availability of domain knowledge. In a number of real world problems (e.g. data mining) explicit domain knowledge is not available or may involve a long and cumbersome knowledge acquisition process. HCVM is versatile in that they can model solutions in the presence or absence of domain knowledge.

5.6.1.4. Forms of Knowledge

Real world problems involve use of multiple forms of knowledge (e.g. continuous, discrete symbolic and fuzzy). Unlike a number of intelligent technologies, associative systems are not limited to one or two forms of knowledge but can model any real world problem with continuous, discrete, fuzzy knowledge because of the multiplicity of techniques used by it.

5.6.1.5. Learning and Adaptation

The ability to learn new tasks and adapt to novel situations are essential properties of HCVM. HCVM involve task based learning in which a problem solver employs multiplicity of learning techniques (e.g. supervised, self-organized, evolutionary, their variations and hybrid configurations) to match the needs of various learning tasks.

5.6.1.6. Distributed Problem Solving and Communication - Collaboration and Competition

In order to deal with the complexity of real world problems (e.g. real-time alarm processing in a power system control center, building design) in general and World Wide Web (WWW) based problems (e.g. Web searching, Web mining, internet games) in particular, distributed problem solving has become a necessity. The task oriented approach of HCVM not only enables distribution of tasks among different system components (which may be executed on remote machines) but also facilitates collaborative and competitive problem solving. That is, agents can collaborate with each other by performing different tasks. They can be mobile and perform computations on remote machines. On the other hand, because of availability of multiple techniques, agents can compete with each other on the same task (by performing it using different techniques like neural networks, knowledge-based systems, etc.) thus enhancing overall system reliability.

5.6.1.7. Component Based Software Design

The five layers of the HCVM lead to a component based software design. The generic agents of the problem solving agent layer, intelligent hybrid agent layer, intelligent agent layer and software agent layer facilitate corresponding component definitions in the domain of study.

5.6.2. Management Characteristics

The management characteristics define the significance of HCVM in terms of management considerations that determine the use and maintenance of computer-based artifacts. Some of these characteristics are outlined next.

5.6.2.1. Cost, Development Time and Reuse:

The optimization of human and computing resources is an important management consideration for using information technology today. It has become an essential consideration in the deregulated industrial climate of the late 90's. In associative systems the tasks can be distributed and implemented over various machines to enable optimization of computing resources as well as save valuable human time. The multiplicity of intelligent techniques employed by HCVM enables users (i.e., designers, engineers, etc.) to reduce the system development time. It also helps them to create an optimum system model (i.e., various intelligent techniques and their hybrid configurations can be employed simultaneously to save development time as well as assist in determining optimal system design). The component-based approach of HCVM facilitates reuse in terms of application-related objects (object layer), software agents, intelligent agents and problem solving agents. It reduces the need for building new applications from scratch.

5.6.2.2. Scalability and Maintainability:

The multilevel and component-based properties of HCVM enable the scalability (horizontal and vertical) and maintainability of the agents. The five problem solving adapters of the HCVM assists in systemizing and structuring software artifacts. This results in easier maintainability as well as scalability of the artifacts.

5.6.2.3. Intelligibility:
Humans form an important part of solutions to most real world problems. Thus it is imperative that any software system architecture should enable reduction of cognitive barriers between the user and the computer. This is vital for two reasons, namely, acceptability and effectiveness. That is, systems with low cognitive compatibility lead to low acceptability because the system's behavior may appear surprising and unnatural to the user. Further, systems with low cognitive compatibility will lead to low effectiveness because of lack of user involvement (resulting in unsatisfactory performance and major accidents (Perrow 1984)).

5.6.3. Domain Characteristics

The significance of an HCVM can also be seen in terms of the problems which can be modeled with it. The architectural and management characteristics of HCVM and outlined in this section establish that HCVM can be used for a wide range of complex, data /knowledge intensive, distributed and time critical problems.

5.7. Summary

The objective of this chapter is to define the computational level of the human-centered system development framework outlined in the chapter 4. It does that by developing the Human-Centered Virtual Machine (HCVM) through integration of activity-centered analysis component, problem solving ontology component and multimedia interpretation component with various technological artifacts. These technological artifacts include intelligent technologies, agent and object-oriented technologies, distributed processing and communication technology and XML technology.
The problem solving ontology component is described with the help of five problem solving adapters, namely, preprocessing, decomposition, control, decision and postprocesssing. These adapters are grounded in the experience derived from developing various complex systems. They capture human generalizations and persistent structures used for modeling complex systems. They help in systematizing and structuring the human-centered tasks and representations in a form suitable for transforming a human solution into a scalable, evolvable and maintainable software solution. The transformation is realized by defining a set of transformation agents. These transformation agents are derived through integration of activity-centered analysis component, problem solving ontology component and multimedia interpretation component with various technological artifacts. The outcome of the integration process is HCVM with a problem solving agent layer, intelligent hybrid agent layer, intelligent agent later, software agent layer and an object layer. The transformation agents in the four agent layers are modeled with the help of a transformation agent definition. The agent definition encapsulates characteristics of

these transformation agents like goals, tasks and actions, representation, communication, external and internal tools used, and others.

The primary aim of the multimedia interpretation component is to model the data content of the human-computer interaction tasks using various media artifacts. It does that in three stages or steps. These are data content analysis, media, media expression and ornamentation selection, and presentation design and coordination. Finally, in order to get an overall emerging picture of the HCVM, we outline its emergent behavior. We outline the emergent behavior in terms of its architectural characteristics, management characteristics and domain characteristics.

In chapters 6, 7, 8,10 and 11 we describe applications of the HCVM in medical diagnosis and treatment support, face detection and annotation, internet games and sales recruitment, e-commerce and medical image retrieval.

References:

Arens, Y., Hovy, E.H., and Vosser, M. (1994) "On Knowledge Underlying Multimedia Presentations", *Intelligent Multimedia Interfaces*, Mark T. Maybury, Eds: AAAI Press, pp. 280-306.

Chandrasekaran, B., Johnson, T.R., and Smith, J.W. 1992, 'Task Structure Analysis for Knowledge Modeling,' *Communication of the ACM*, vol. 35, no. 9., pp. 124-137.

Heller , R. & Martin, B., (1995), "A Media Taxonomy, " *IEEE Multimedia*, pp. 36-45.

Khosla, R. & Dillon, T.S. 1997, ***Engineering Intelligent Hybrid Multi-Agent Systems***. Boston, USA, Kluwer Academic Publishers.

Maybury, M. T. (1994) "Planning Multimedia Explanation Using Communicative Acts", *Intelligent Multimedia Interfaces*, Mark T. Maybury, Eds: AAAI Press, pp. 60-74

Norman, D. A. (1988). The Psychology of Everyday Things. Basic Books: New York

Perrow, C.1984, *Normal Accidents: Living with High-Risk Technologies*, Basic Books, New York.

Preece, J., et al (1997), *Human-Computer Interaction*, Massachusetts: Addison-Wesley Pub.

Steven, S.S. (1995). "On the Psychological Law", *Psychological Review*, 64(3), 153-181

Zhang, J., and Norman, D. A. (1994). "Representations in Distributed Cognitive Tasks," in *Cognitive Science*, 18, 87-122

6 INTELLIGENT MULTIMEDIA MULTI-AGENT CLINICAL DIAGNOSIS AND TREATMENT SUPPORT SYSTEM

6.1. Introduction

The objective of this chapter is to show the application of the conceptual and computational facets of the HCVM in a clinical drug prescription monitoring activity. The application is in the area of infectious diseases (Gorbach et al. 1998; Barrows et al. 1991) and addresses problems related to the gathering of patient symptomatic data, providing diagnostic assistance, and finding inconsistencies in practitioner prescribed treatments compared to those recommended by therapeutic guidelines (TG 1998).

The chapter starts with the activity-centered analysis of the drug prescription monitoring activity. A human-centered domain model is then developed for drug prescription monitoring activity. This is achieved among other aspects, by integrating stakeholder tasks from the activity-centered analysis with the problem solving adapters of the HCVM. The problem solving adapter definitions of the human-centered domain model are then transformed into intelligent problem solving agents at the computational level using generic agent definition of the transformation agent component. Media agents which generate, display and coordinate the multimedia artifacts modeled for gathering symptomatic data are also defined. Implementation results of the problem solving agents and media agents are shown along with their definitions.

6.2. Activity-Centered Analysis – Drug Prescription Monitoring Activity

The activity-centered analysis as described in chapter 4 involves the following steps:

- Problem Definition and Scope
- Performance and Context Analysis
- Alternate System Goals and Tasks
- Human-Task-Tool Diagram
- Task-Product Transition Network

6.2.1 Problem Definition and Scope

The current state of clinical medical practice sees great inconsistencies between the reality of therapeutic treatments prescribed by practicing clinicians, and therapeutic treatments recommended by existing clinical guidelines. Such inconsistencies towards guideline recommended treatments in therapeutic prescriptions are of great concern to the regulatory bodies and the medical community in general. This is largely a result of an existing inclination towards over-prescription of antibiotic therapies, coupled with a trend in the general population of increasing resistance to antibiotic therapy (Sarbani 1997). Furthermore, not only do these inconsistencies exist between the reality of practitioner prescription habits and guideline recommendations, but they also exist to a large extent between general practitioners.

This current state of disparity between the prescription habits of general practitioners and guideline recommendations has been empirically monitored and established by the "Inner South Eastern Division of General Practice," located in Alfred Hospital, Melbourne, Victoria. The scope of the various components of this drug prescription monitoring activity is shown in Figure 6.1.

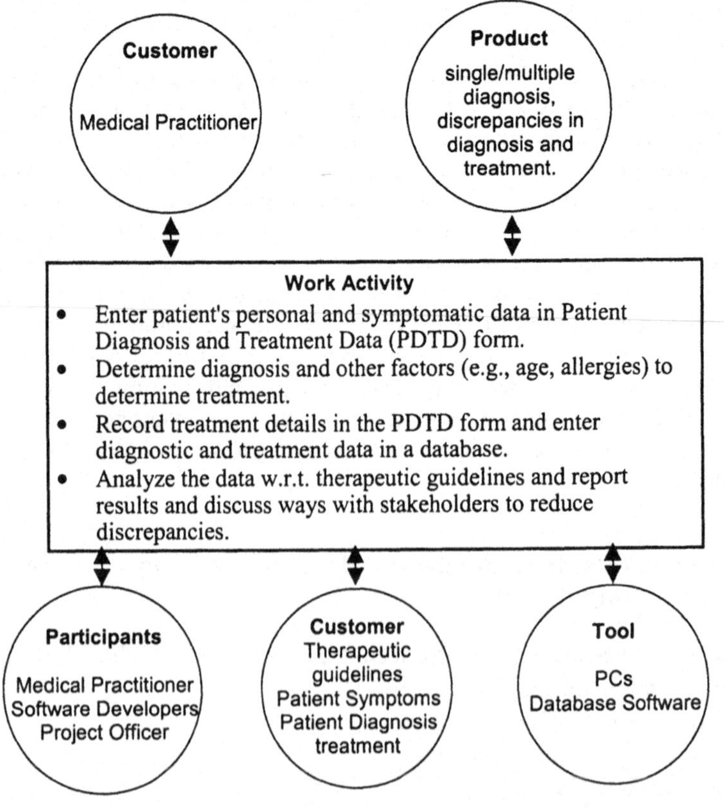

Figure 6.1: Existing Scope of Drug Prescription Monitoring Activity Components

Over the past two years, the Inner South Eastern Division of General Practice has monitored and recorded the prescription habits of 20 participating general practitioners based in the Melbourne metropolitan area, for a duration of 6 consecutive winter weeks in each year. The empirical data gathered by this investigation shows that inconsistencies prevail not only in drug prescription but also to some extent in medical diagnosis.

As the aim of an investigation such as this is to ultimately minimize or remove such inconsistencies, an alternative system is sought. Such an alternative system should:

- provide diagnostic decision and treatment support
- identify the inconsistencies that exist between the practitioner prescription data and the corresponding guideline recommendations, on an ongoing basis.

6.2.2. Performance and Context Analysis

The performance and context analysis of the Drug Prescription Monitoring Activity (DPMA) is undertaken to identify the role, goals and tasks of an alternative computer-based system. The role and the goals are determined through quantitative analysis of performance of the six system components shown in Figure 6.1. It identifies the role in terms of goals of an alternative system. The context analysis, on the other hand, does a qualitative analysis of the six system components in DPMA It analyses the realization of the goals in terms of the direct stakeholder and cultural context, data structure and data security context, and competition and emerging technology context. It determines the nature of tasks to realize the goals. The outcome of the context analysis is a set of tasks for realizing the goals.

6.2.2.1. Performance Analysis of DPMA

In this section we outline the performance analysis of four system components of DPMA, namely, product, activity, data and tool.

Product

The performance of the product is analyzed in terms of its cost, quality, responsiveness, reliability and conformance to standards as defined in section 4.5.2 of chapter 4.

Cost: is determined in terms of cost involved in providing treatment discrepancy related information, and advice for minimizing the discrepancies to the medical practitioners. The cost of administering collation and analysis of patient data for determining treatment discrepancies and distribution of this information runs into hundreds of thousands of dollars at present.

Quality: the drug prescription monitoring activity is measured in terms of difference between the practitioner based treatments and those recommended by therapeutic guidelines. There is approximately 30% difference between practitioner-based treatments and those recommended by therapeutic guidelines. Some of the reasons found for this anomaly are different practitioners prescribing different treatments for the same symptomology and, in some instances, different practitioners arriving at multiple diagnosis.

Responsiveness: is measured in terms of availability and time difference between the treatment discrepancy and advice information and its use by the medical practitioners. At present, this information takes about one year after patient data collection.

Conformance to Standards: relates to adherence of treatment standards set by the medical experts in the form of therapeutic guidelines. As mentioned in the quality analysis of the product component, a 30% discrepancy exists between practitioner-based treatments and those recommended by therapeutic guidelines.

The performance analysis of the product component in this section identifies a need for a computer-based artifact that aims to:

- reduce the cost of patient data collation and analysis (Goal 1, G1),
- improve quality by minimizing discrepancies (G2),
- improve responsiveness by making the treatment discrepancy and advice information available quickly (G3), and
- improve public confidence in medical practitioners by conformance to the therapeutic guidelines (G4).

Now we look at the performance of the data component of the activity-centered analysis.

Data

The performance of data is analyzed in terms of its accessibility, quality, and presentation.

Quality: is determined in terms of accuracy/precision and completeness of symptomatic data, diagnostic data and treatment data. The symptomatic data (e.g., toxic looking, acute sore throat) shown in Figure 6.2, by definition, is fuzzy and imprecise. However, they are shown in Figure 6.2 in discrete form using check boxes. The ticks or crosses entered in the check boxes are an inaccurate representation of a medical practitioner's feedback on a patient's symptom. This inaccuracy in the symptomatic data can lead to inaccurate treatments.

Accessibility: is determined in terms of availability of data desired by the practitioners for diagnosis and treatment. At present, practitioners cannot access the therapeutic guidelines relevant to a particular diagnosis on-line. They have to sift through manuals and books related to the guidelines. Further, they cannot access historical diagnostic data related to difficult diagnostic cases.

Inner South East Melbourne Division of General Practice

Patient ID	Patient Sex	Patient Age
....	....	....
Date of Visit	GP Initials	Office Use Only
....	....	....

Diagnosis, symptoms & signs

[A] ☐ **Undifferentiated upper respiratory tract infection**
[1] ☐ Fever
[2] ☐ Cough
[3] ☐ Runny nose
[4] ☐ Mild sore throat

[B] ☐ **Acute sore throat**
[5] ☐ Hard to swallow
[6] ☐ Toxic looking
[7] ☐ Follicular or exudative tonsillitis
[1] ☐ Fever
[8] ☐ Tender lymph nodes
[9] ☐ Age over 4
[10] ☐ Existing rheumatic heart disease
[11] ☐ Absence of cough
[12] ☐ Aboriginality
[13] ☐ Scarlet fever

[C] ☐ **Acute otitis media**
[14] ☐ High fever
[15] ☐ Sore ear
[16] ☐ Child screaming
[17] ☐ Child tugging ears
[18] ☐ Eardrum mild reddening or dullness
[19] ☐ Eardrum red or yellow and bulging
[20] ☐ Discharging ear
[21] ☐ Past history of perforation
[22] ☐ Has grommets

[D] ☐ **Otitis media with effusion**
[23] ☐ Glue ear < 3 months duration
[24] ☐ Glue ear > 3 months duration
[25] ☐ Poorly moving drum

[E] ☐ **Acute sinusitis**
[26] ☐ Nasal discharge clear
[27] ☐ Nasal discharge purulent
[28] ☐ Prolonged fever
[29] ☐ Facial pain
[30] ☐ Tenderness over the sinuses
[31] ☐ Headache

Treatment

[1] ☐ **None**
[2] ☐ **Paracetamol and rest**
[3] ☐ **Antitussives or decongestants**
[4] ☐ **Bronchodilators**

Antibiotic
[5] ☐ Amoxycillin/ampicillin
[6] ☐ Amoxycillin and clavulanate
[7] ☐ Cephalexin/cephradine/cefaclor
[8] ☐ Dicloxacillin/flucloxacillin
[9] ☐ Erythromycin/roxithromycin
[10] ☐ Phenoxymethyl penicillin/phenethicillin/procaine penicillin
[11] ☐ Sulphamethoxazole and trimethoprim
[12] ☐ Tetracyclines/doxcycline
[13] ☐ Other

[14] ☐ **Antibiotic script if symptoms worsen**
[15] ☐ **Antibiotic samples if symptoms worsen**
[16] ☐ **Other (specify)**
....
....
....

Reasons for choice

[1] ☐ **Symptoms and signs**
[2] ☐ **Intuitive feeling**
[3] ☐ **Coexisting illness/smoker**
[4] ☐ **Age risk**
[5] ☐ **Prolonged illness**
[6] ☐ **Past history of recurrent infections**
[7] ☐ **Patient expectation**
[8] ☐ **Lack of time to explain**
[9] ☐ **Child unexaminable**
[10] ☐ **Just in case / diagnosis unsure**
[11] ☐ **Other cases in community/family**
[12] ☐ **Microbiology**
[13] ☐ **Recent hospitalisation**
[14] ☐ **Not responding to the antibiotic prescribed**
[15] ☐ **Other (specify)**
....
....
....

P.t.o.

Figure 6.2: A Sample PDTD Forms Used in Existing DPMA

Presentation: is measured in terms of whether presented data is based on psychological variables of interest to the medical practitioner for ascertaining the symptoms, diagnosis and treatment, or simply physical variables related to these aspects are presented. For example, besides the inaccuracies in representing the symptomatic data, the PDTD form only employs the text media, whereas psychologically, the practitioner is perceiving the image of the symptom in the patient and interpreting it in discrete or fuzzy terms on different psychological scales.

Thus the performance analysis of the data component identifies a need for an artifact (e.g., computer based) which can:

- enhance the representation accuracy of the symptomatic data (G5),
- improve accessibility to the therapeutic guidelines (G6), and
- enhance presentation based on psychological variables of interest (G7).

The next section analyses the performance of the human participants and customer, work activity, and tool components, respectively.

Participant and Customer

We have merged the participant and customer because medical practitioners are both customers and participants in the drug prescription monitoring activity. The participants in the drug prescription monitoring activity are medical practitioners, project officer and data entry operators. Their performance is measured in terms of skills and involvement in the work activity.

Skills and Involvement: Skill is measured in terms of the experience of the participants in the work activity. Involvement is measured in terms of degree or extent of the participant involvement in determination of the tasks in the work activity. The medical practitioners involved in the work activity have 8 to 25 years of experience, which is considered adequate for the tasks in the work activity. The experience levels of the project officer and programmers are also considered adequate. However, the degree of involvement of medical practitioners in the work activity is low, as they are not comfortable with the process of discrepancy identification and recommendation.

Customer Satisfaction: Some medical practitioners are not satisfied with the design of the PDTD forms shown in Figure 6.2 as they limit completeness and accuracy of information that can be provided on the patient.

Work Activity

Two criteria are measured in this component, namely, cycle time and consistency.

Cycle time: measures the total time elapsed between start and finish of the drug prescription monitoring activity. The cycle time of the drug monitoring activity is one year.

Consistency: measures the extent to which the same treatment is given for the same diagnosis, symptoms and allergies. Although, we did not get a precise quantitative figure on this measure, there is enough discrepancy in prescribed treatments between different practitioners to cause concern. This is partly due to the influence of the profit driven drug industry on the medical practitioners.

Tool

The tools involved in the drug prescription monitoring activity are the PDTD forms, computers and medical instruments. Since our focus here is to identify the role of computer-based artifacts, we will concentrate on the performance analysis of computer-based tools. The computer tools employed are database software for storing patient related data and spreadsheets for analyzing the data. At present, there is under utilization of computer *cpu and power.* Further, computers are only being used to assist the data entry operators, programmers and project officer for storing and analysis of patient data. They are not being used to assist the medical practitioners in patient diagnosis or treatment prescription.

The performance analysis of the work activity and tool components suggests that a computer-based artifact can be used to

- reduce cycle time (G8),
- improve consistency of patient diagnosis and treatment prescription, and (G9)
- assist medical practitioners in diagnosis and treatments (G10).

Further, the involvement parameter of the participants component suggests that medical practitioners need to be directly involved in determining the tasks in the work activity that need to be restructured. The performance analysis in this section has identified the role and goals of the computer-based artifact vis-a-vis various system components. In the next section we undertake the context analysis of the drug prescription monitoring activity.

6.2.2.2. Context Analysis of the Drug Prescription Monitoring Activity

In the preceding section, the quantitative analysis of various system components has provided an objective means of determining the factors for improving different components of the existing drug prescription monitoring activity as well as enabled us to identify the need for an alternative computer-based system

These factors take the form of product goals. In this section, we undertake a qualitative or subjective analysis of the various system components in order to understand the context in which these goals can be effectively achieved. This context is analyzed in terms of social (participants and tools) organizational (culture), data structure and security context, product substitution, and new emerging tools context.

Participants and Work Activity:

Personal incentives and goals of system participants are often the key determinant of whether the proposed use of computer-based artifact for achieving the product goals will succeed. These incentives influence the degree of involvement of the participants in the work activity. In this section, we look at these incentives as well as cultural issues in the medical profession which affect the use of computer-based artifact. We study them in the context of the problems identified in the performance analysis stage for participant and work activity components, respectively. We translate the incentives and cultural issues into human-centered tasks to be modeled by the computer-based artifact.

Improving degree of involvement and consistency in patient diagnosis and treatment: Traditionally, medical practitioners do not have confidence in using computerized diagnostic systems. However, they do realize the importance of consistency in patient diagnosis and treatment. They are prepared to be more actively involved in improving the consistency of patient diagnosis and treatment prescription and the incorporation of computer-based artifact if the following factors or constraints can be modeled in the work activity.

- Flexibility of using or bypassing the diagnostic system.

  This constraint translates into the following tasks:

  i) Flexibility of moving from symptoms to diagnosis,
  ii) Flexibility of moving from practitioner selected diagnosis to symptoms, and
  iii) Flexibility of bypassing diagnostic stage and going from diagnosis to treatment stage.

- Flexibility of determining diagnosis and treatments based on practitioner experience or based on the guidelines.

  This constraint translates into the following tasks:

  i) The proposed diagnostic system is to be developed for upper and lower respiratory infections. 16 medical practitioners have been involved in the drug prescription monitoring activity. Based on patient symptoms, the practitioners determine a potential diagnosis set for further examination. They use the potential diagnosis set to determine the actual diagnosis/infection and treatment. Most of the practitioner experience is operationalized/internalized for tasks involving potential diagnosis and actual diagnosis and thus has to be learnt from previous patient data related to symptoms and diagnosis.
  ii) Although, the regulatory bodies have suggested that therapeutic guidelines developed by medical experts should form an integral part of any system, the practitioners have also suggested that the option of using treatments based on practitioner experience should also be incorporated.

Reducing cycle time:

The medical practitioners have suggested decentralization of the drug prescription monitoring activity by installing computer-based diagnosis and treatment support systems on their desktop personal computers. This will help the medical practitioners take corrective action without delay. Further, culturally, the medical community is a close-knit community in which some practitioners do not prefer prescription inconsistencies to be made available centrally for medical and political reasons. Thus, for therapeutic guidelines to be effective it is felt that they should be made available on the desktop computers or on a local area network for determining and correcting treatment inconsistencies.

Data

The data context is analyzed in terms of data related issues, identified in the performance analysis, being addressed from the participants viewpoint, relationship between various data entities in the work activity, data sharing and privacy concerns of the participants and other stakeholders.

Improving data quality: In the opinion of some participants incorrect representation of symptomatic data may be partly responsible for the treatment discrepancies. An effective mechanism involving multimedia artifacts for gathering patient symptomatic data has to be developed. This translates into a task for modeling patient symptomatic data using multimedia component of the human-centered conceptual framework.

Data structure: The relationship between various data entities in the context of the drug prescription monitoring activity is shown in Figures 6.3 and 6.4, respectively. Object-oriented notations have been used to describe the relationship between various entities in the PDTD forms and those in the therapeutic guidelines. As Figures 6.3 and 6.4 suggest, we are looking into symptoms, diagnosis and treatment of respiratory infections in this application. Figure 6.4 also shows the attributes and methods associated with different objects.

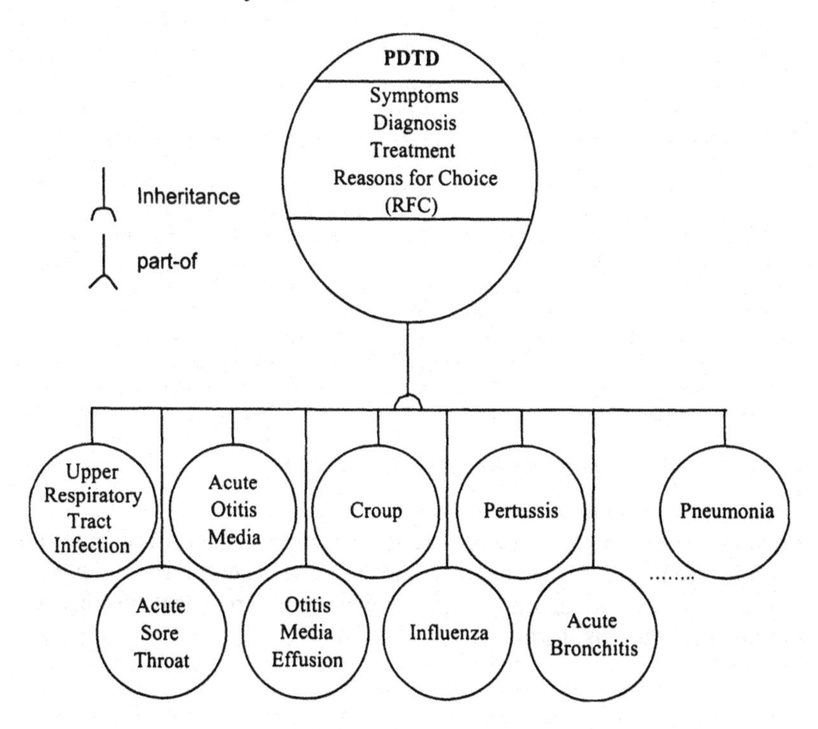

Figure 6.3: Diagnostic Data for Respiratory Infections as Shown In PDTD Form

Data Sharing and Privacy Concerns: The medical practitioners would like to retain information related to their patients on their personal computers. Information related to inconsistencies will be provided on a voluntary basis.

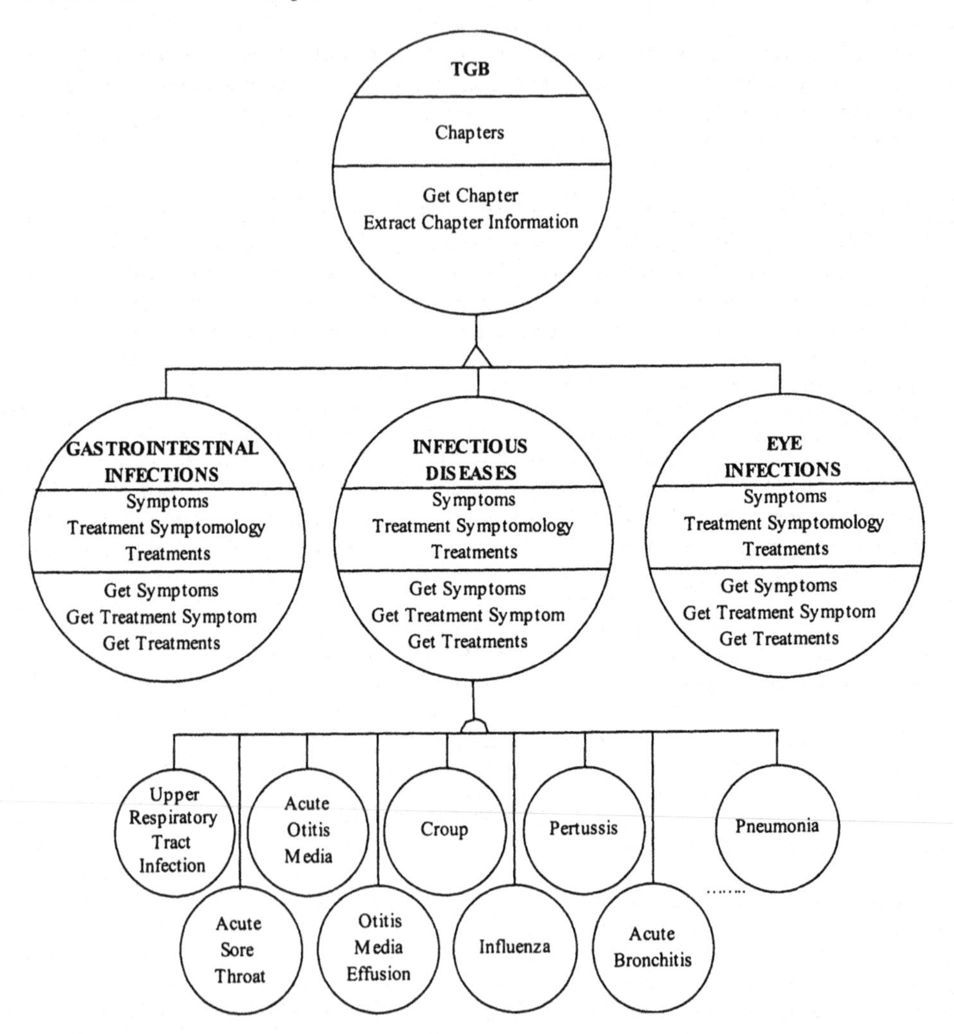

Figure 6.4: Data Relationships in Therapeutic Guidelines

Product Competition/Substitution

The existing sources available to the medical practitioners form reducing treatment discrepancies are books on therapeutic guidelines, and general medicine literature. A computerized system for assisting them in reducing the treatment discrepancies is a better option than the existing one.

6.2.3. *Alternate System Goals and Tasks*

In this step, we consolidate the outcomes of the performance and context analysis in terms of the goals and corresponding tasks for an alternative computer-based system.

The goals and their corresponding tasks are shown in Table 6.1. These goals and tasks form the basis for developing a human-centered domain model. In order to facilitate formulation of such a model, we need to firstly determine the underlying assumptions or preconditions for accomplishment of these human-centered tasks. Further, we also need to determine the division of labor (tasks) between the medical practitioner and computer-based artifact in Table 6.1. These two issues are addressed through a task-product transition network and human-task-tool diagram in the next two sections.

Table 6.1: Grouping goals and tasks for an alternative computer-based system in DPMA

Goals	Corresponding Tasks
G1, G2, G5 and G7	Model and design multimedia artifacts for gathering patient data.
G2, G4, G6, G9 and G10	Flexibility of using or bypassing the diagnostic system, flexibility of determining diagnosis and treatment based on practitioner experience, learning potential diagnosis sets, actual diagnosis, and treatments based on practitioner experience.
G3 and G8	Decentralization of DPMA by automating patient data gathering and making diagnosis and therapeutic guidelines locally available on practitioner's desk top computers.

6.2.4. *Human-Task-Tool Diagram*

As mentioned earlier, the purpose of the human-task-tool diagram is to determine the division of tasks between the medical practitioners and the computer-based artifact. Figures 6.5 and 6.6 show a sample task done by a medical practitioner and the computer-based artifact respectively. The human-task-tool diagram also shows the data used by each task and the intermediate/final product produced after completion of the task. This information is useful for organizing the task-product transition network and in determining the task and data characteristics of the human-centered domain model.

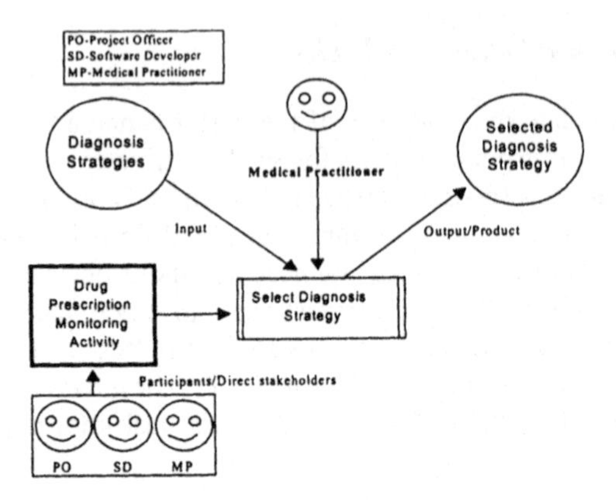

Figure 6.5 Human-Task-Tool Diagram for Select Diagnosis Strategy Task

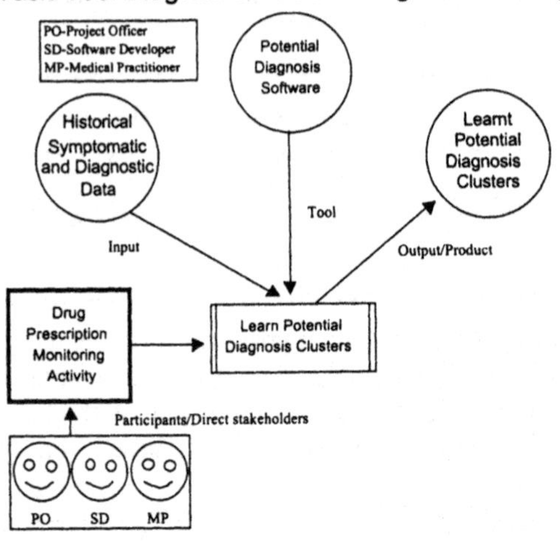

Figure 6.6 Human-Task-Tool Diagram for Learn Potential Diagnosis Clusters Task

6.2.5. *Task Product Transition Network*

The task-product transition network of the drug prescription monitoring activity is shown in Figure 6.7. In order to keep the network simple we have not shown the preconditions and postconditions associated with each task-product or product-task transition. A sample product-task transition with preconditions and postconditions is also shown in Figure 6.7. The product-task transition preconditions help us in defining the assumptions under which the task will be accomplished. For example, the preconditions shown in Figure 6.7 for the task "Learn Potential Diagnosis Clusters" implies that a large amount of historical symptom and diagnosis data should be available for the task of learning potential diagnosis clusters. The postcondition reflects not only the new product state but also the level of competence required from the method or algorithm used for accomplishing the task "Learn Potential Diagnosis Clusters" after the precondition has been satisfied. For example, the clusters learnt by say Kohonen's self-organizing maps should be cross-validated by relevant test data and also validated by the practitioners as viable potential diagnosis clusters.

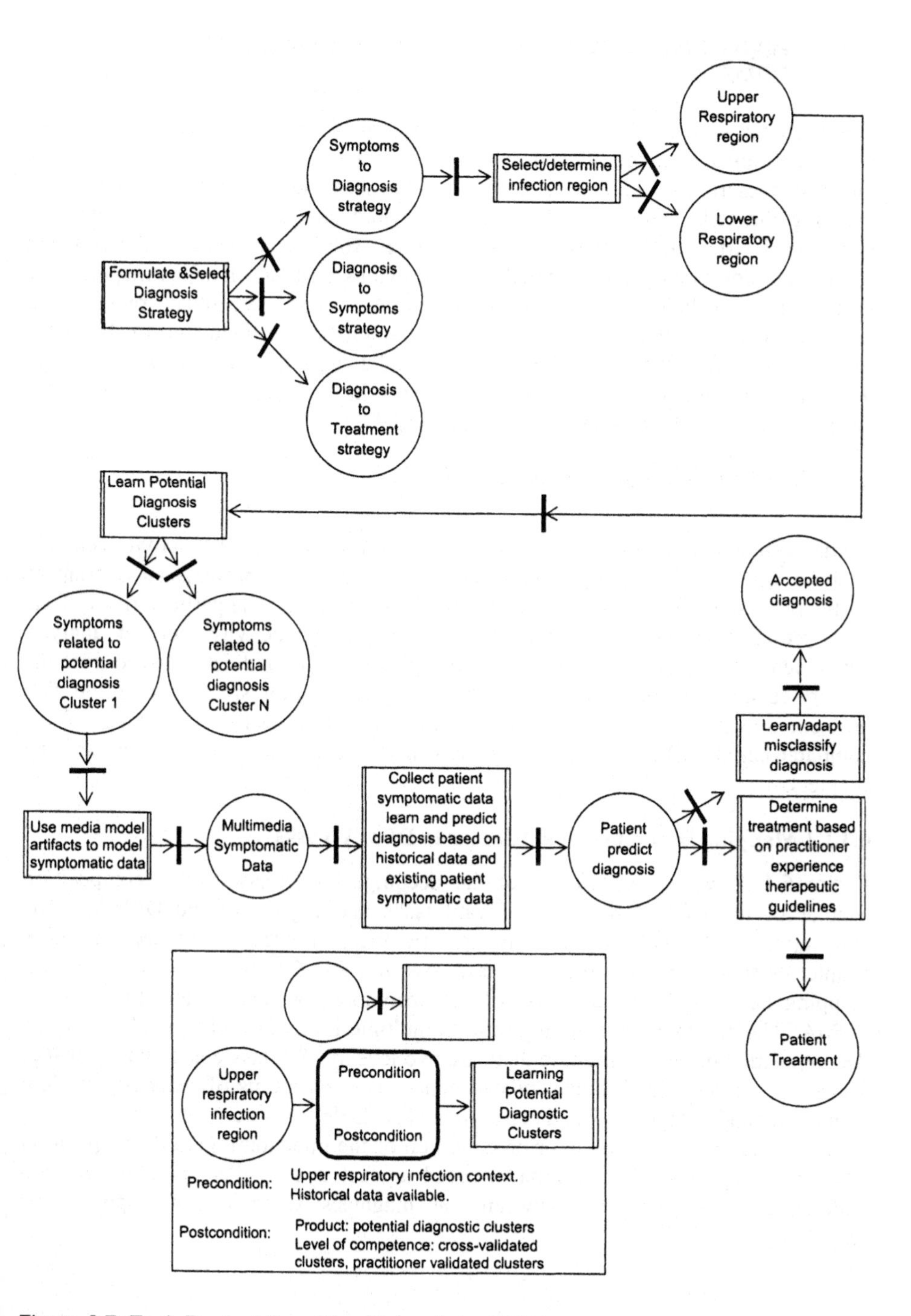

Figure 6.7: Task-Product Transition Network for DPMA

6.3. Application of HCVM Problem Solving Ontology in DPMA

The problem solving ontology component described in the chapter 5 is used to develop a human-centered domain model for the alternate computer-based system in the drug prescription monitoring activity. This is achieved by systematizing and mapping the tasks defined in the activity-centered analysis stage using persistent generalized structures, namely, the five problem solving adapters of the problem solving ontology component defined in the previous chapter. Besides mapping the tasks, the stakeholder external representations (or perceptual representing dimensions) and internal representations (e.g., linguistic/non-linguistic) of data used for each task, and the stakeholder domain model used for various tasks are also determined by the problem solving adapters. In this section, we describe the development of the human-centered domain model.

6.3.1. Human-Centered Domain Model

Human-centered domain model represents an integration of the activity-centered analysis with generalized problem solving structures. It represents an overlap of abstraction aspects of the five problem solving adapters with the particularities of the drug prescription monitoring activity modeled by the activity-centered analysis component. The integration is shown by mapping the tasks and data derived from the activity-centered analysis with the task and representation signatures of the five problem solving adapters. The rest of this section shows the mapping of the computer-based tasks defined in the task network to the five problem solving adapters.

6.3.1.1 Mapping Decomposition Adapter to DPMA Tasks

Figure 6.4 shows an association of the decomposition phase and decomposition adapter of the HCVM with the relevant tasks and objects of the DPMA. The association is established based on the generic goals and tasks of the decomposition adapter as shown in Table 6.1. The task row in Table 6.1 shows one-to-many task mapping between the generic task of the decomposition adapter and the tasks of DPMA. The one-to-many mapping reflects multiple levels of problem solving in the decomposition phase (in this case there are two levels). The task constraints shown in Table 6.2 are high level human related conceptual constraints of reducing problem complexity and computer related constraints of scalable software design.

The underlying assumption or precondition of the infection region decomposition adapter is that user of the computer-based artifact is a medical practitioner who understands the distinction between the diagnosis strategies and upper/lower respiratory infections.

Table 6.2: Mapping Decomposition Adapter Signatures to DPMA Goals, Tasks & Reps

HCVM and Representation Goal, Tasks, Signatures	Corresponding in Drug Prescription Monitoring Activity (DPMA)
Phase: Decomposition.	Infection Region Decomposition
Goal: Restrict input context Reduce complexity.	Restrict Practitioner's diagnosis strategy Reduce respiratory infection space
Task: Determine abstract orthogonal concepts.	Determine Infection Region: • Lower respiratory • Upper respiratory. Select diagnosis strategy (symptoms to diagnosis, diagnosis to symptoms, diagnosis to treatment).
Task Constraints: Orthogonality, reliability, scalability, domain dependent	Orthogonal diagnosis strategy, orthogonal infection subspace, scalable respiratory infection subspace
Precondition	Medical practitioner user
Postcondition	Upper/lower respiratory region subspace
Represented Features: Qualitative	Upper respiratory region/lower respiratory labels, diagnostic strategy labels
Represented Features: Non-linguistic	Icons for diagnostic functions.
Domain Model: Structural, functional, casual, geometric, heuristic, spatial, shape, color, etc.	Functional (diagnostic functions) Structural (infection regions)
Psychological Scale: Nominal; Formal Property; category	Nominal – Respiratory Region category and Diagnostic Function category
Representing Dimension of Features: Shape, location, position, etc. on the nominal scale	Location (region location), Shape (region shape)
Technological Artifacts	Perceptual (shape and location based multimedia representation).

The domain model used for determining the orthogonal concepts like upper and lower respiratory regions is based on the structural model of the human anatomy. On the other hand, the three diagnostic strategies are based on the functional model for diagnosis and treatment adopted by the medical practitioners.

The representing dimensions of the upper and lower respiratory regions are location and shape respectively. Given that the task constraints are non-computational we can use perceptual artifacts, based on the representing dimensions, for problem solving. Thus we employ the representative images of the upper and lower respiratory regions (see Figure 6.15) indicating the location and shape of the two regions.

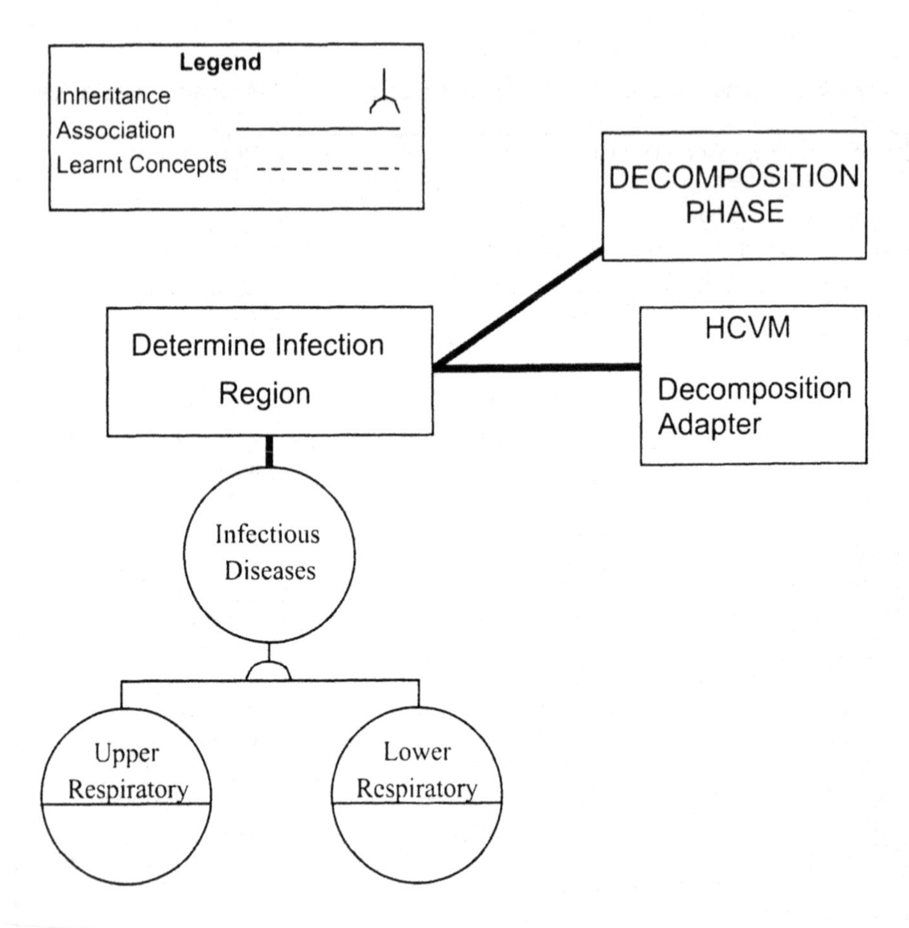

Figure 6.8:Mapping DPMA Tasks and Data Structure to HCVM Decomposition Adapter

6.3.1.2. Mapping Control Phase Adapters to DPMA Tasks

Figures 6.9 and 6.10 show the association of the control phase and preprocessing adapter of the HCVM with the relevant tasks and objects of the DPMA. The object structure shown in Figure 6.9 is extracted from Figure 6.3 based on the tasks in this phase. The association is established through the generic goals and tasks of the preprocessing adapter and control adapter. It may be noted that the preprocessing adapter can be used in decomposition, control or decision phases of the HCVM problem solving ontology.

Its application depends upon the correspondence between its goal of improving data quality and noise filtering and data conditioning tasks and the preprocessing goals and tasks involved in these phases.

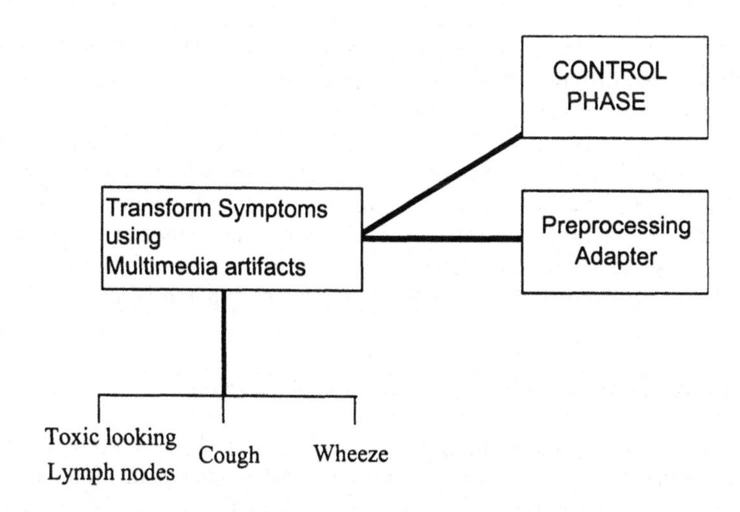

Figure 6.9: Overview of DPMA Data Structure and Tasks Associated with HCVM Preprocessing Adapter in the Control Phase

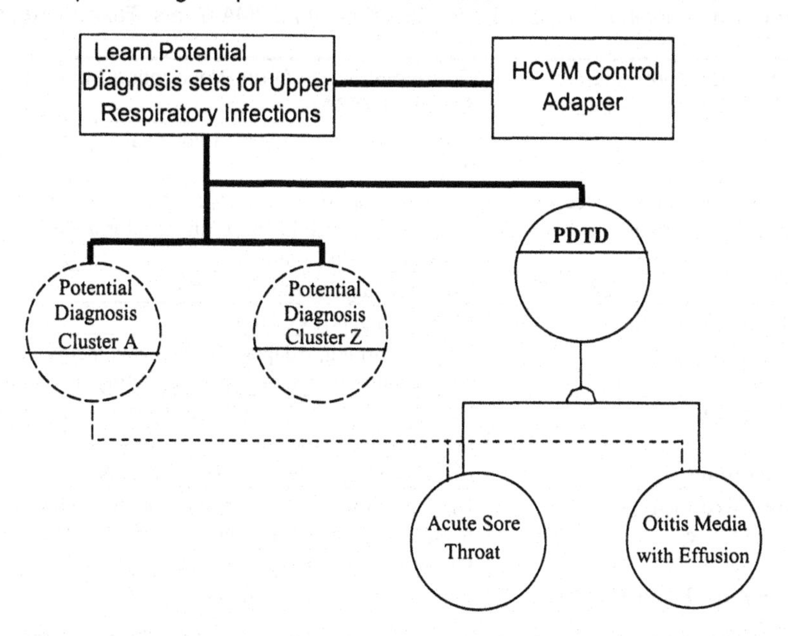

Figure 6.10: Overview of DPMA Data Structure and Tasks Associated with HCVM Control Adapter

As can be seen from Figure 6.6 the control adapter is associated with the DPMA task "Learn potential diagnosis sets/clusters". These potential diagnosis clusters represent the decision concepts used by medical practitioners for determining the actual diagnosis. However, the problems related to symptomatic data gathering highlighted in the performance and context analysis of DPMA are also applicable here. That is, the existing symptomatic data capabilities for symptoms which distinguish between different diagnostic clusters needs to be improved. This task is associated with the preprocessing adapter as shown in Figure 6.5.

The task constraints shown in Table 6.3 like learning and incomplete data influence the selection of the technological artifact like Kohonen's self-organizing maps for learning the potential diagnosis clusters. The hard computing artifacts like rules are used as control constructs for discriminating between different learnt potential diagnosis clusters. The generic task "conflict resolution" is used to resolve conflicts where multiple diagnosis related to two or more potential diagnosis clusters exist in the decision phase.

6.3.1.3. Mapping Decision Phase Adapter to DPMA Tasks

The association of the decision phase with the tasks from the task product transition network and related object structure from Figure 6.3 is shown in Figures 6.11 and 6.12 respectively. Like in the control phase, the patient symptomatic data related to a potential diagnosis cluster needs to be preprocessed in terms of transforming the

Table 6.3: Mapping Control Adapter Signatures to DPMA Goals, Tasks & Reps

HCVM and Representation Goal, Tasks, Signatures	Corresponding in Drug Prescription Monitoring Activity (DPMA)
Phase: Control.	Upper Respiratory Infection Potential Diagnosis Phase.
Goal: Establish domain decision control constructs	Decision Control Constructs for distinguishing between potential diagnosis clusters for upper respiratory infections
Task: Decision level concepts	Potential diagnosis sets (clusters)
Task: Conflict resolution	Multiple diagnosis conflicts
Task Constraints	Cluster learning, Adaptability to noisy, incomplete data, scalability to new upper respiratory infections
Precondition	Upper respiratory infection region, Historical patient symptom and diagnostic data available
Postcondition	Cross-validated clusters, viable clusters
Represented Features: Qualitative	Distinguishing patient symptoms related to upper respiratory infection clusters (e.g. wheeze, toxic looking lymph nodes)
Represented Features: Non-linguistic	Patient symptomatic patterns
Psychological Scale:	Nominal, Ordinal, Ratio
Representing Dimensions	Location, Color, etc.
Technological Artifacts	Hard computing artifact (rules), Soft computing artifact (Kohonen's self-organizing maps)

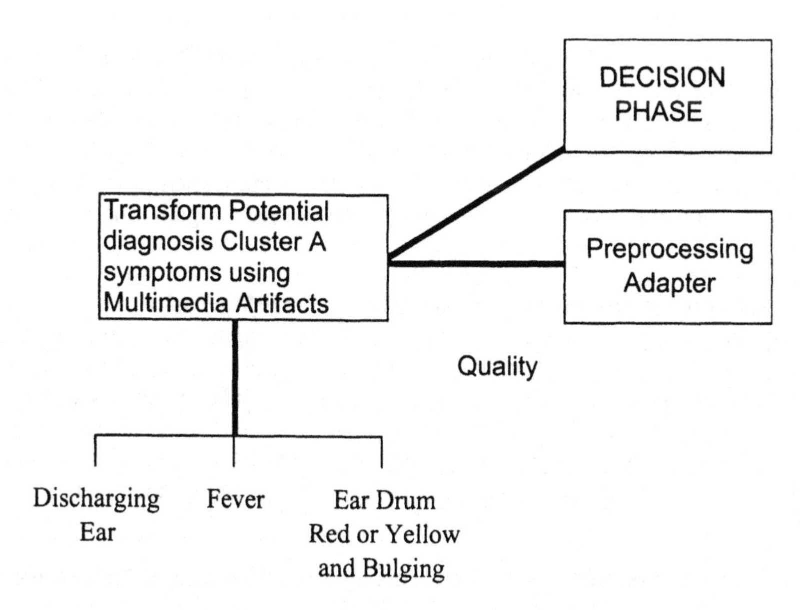

Figure 6.11: Overview of Symptoms related to Acute Ottis Media Infection, related Tasks and their Association with the Preprocessing Adapter

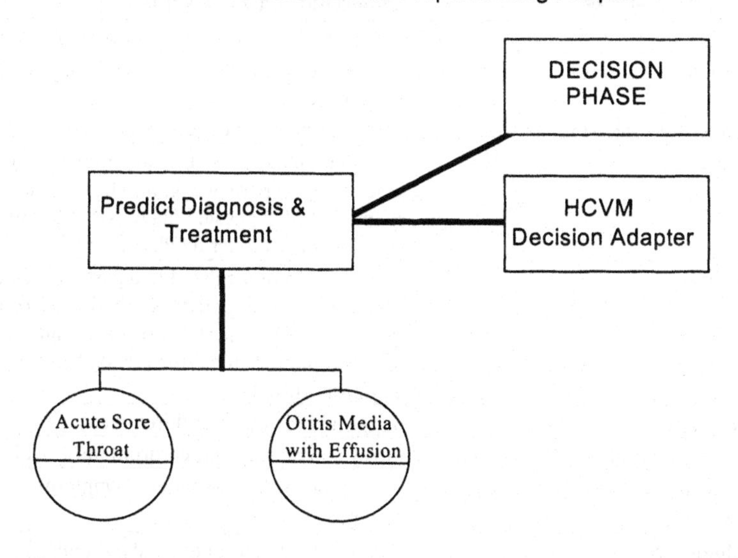

Figure 6.12: Overview of DPMA Objects and Tasks Associated with HCVM Decision Adapter

existing representation shown in Figure 6.2 to the one that employs multimedia artifacts. The association of such a task (which is also shown in the task product transition network) with the preprocessing adapter is shown in Figure 6.11. The generic task of the decision adapter shown in Table 6.3 is to determine a decision instance. In this case the decision instance is a suggested patient diagnosis and treatment in a potential diagnosis cluster based on practitioner. These are learnt using soft computing artifact like backpropagation neural network with capabilities to correct misclassified diagnosis and treatment based on practitioner's feedback.

The application of psychological scale information and representing dimensions in Table 6.4 will be described in multimedia interaction component section 6.5. The scale information and representing dimension will be used for gathering symptomatic data.

6.3.1.4. Mapping Postprocessing Phase Adapter to DPMATasks

The postprocessing phase is used to validate predicted diagnosis and treatment based on practitioner experience. The predicted diagnosis in the decision phase is validated by the medical practitioner. The predicted treatment is validated using rules based on therapeutic guidelines set by the medical experts. In case of inconsistency between predicted treatment and one recommended by the guidelines, the inconsistency is recorded. Based on the above, mapping of the postprocessing adapter is shown in Table 6.5.

Table 6.4: Mapping Decision Adapter Signatures to DPMA Goals, Tasks & Reps

HCVM and Representation Goal, Tasks, Signatures	Corresponding in Drug Prescription Monitoring Activity (DPMA)
Phase: Decision.	Suggested Diagnosis and Treatment Phase.
Goal: Provide user defined outcomes	Patient diagnosis/treatment based on practitioner experience
Task: Determine decision instance	Predict patient X's diagnosis and treatment.
Task constraints	Learning diagnosis and treatment based on practitioner experience, generalization, and adaptation of suggested diagnosis and treatment based on practitioner feedback.
Precondition	Patient symptoms related to a particular decision cluster, Historical practitioner based patient symptom, diagnosis and treatment data for upper respiratory infections related to a potential diagnosis cluster
Postcondition	Suggested diagnosis and treatment
Represented Features: Qualitative	Fuzzy (fever, sore ear, eardrum mild reddening).
Represented Features: Non-linguistic	Continuous values to patient symptomatic data.
Domain Model:	Functional
Psychological Scale:	Ordinal (fever, eardrum mild reddening).
Representing Dimension of Features:	Location, shape, color, texture
Technological Artifacts/ methods:	Soft computing artifacts (backpropagation neural network)

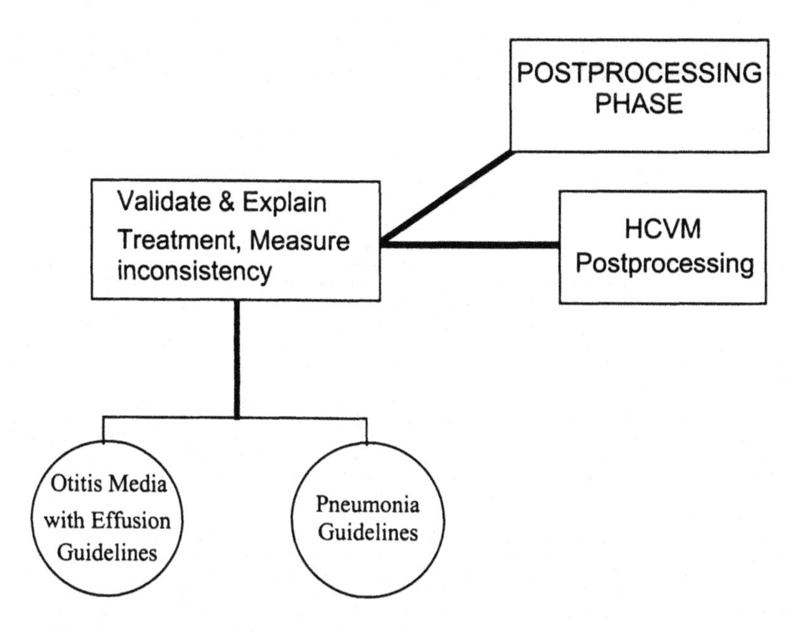

Figure 6.13: Association of DPMA Tasks and Objects with HCVM Postprocessing Adapter

Table 6.5: Mapping Postprocessing Adapter Signatures to DPMA Goals, Tasks & reps

HCVM and Representation Goal, Tasks, Signatures	Corresponding in Drug Prescription Monitoring Activity (DPMA)
Phase: Postprocessing.	Treatment Validation
Goal: Establish outcomes as desired outcomes, satisfy user	Establish correctness of treatment
Precondition	Predicted diagnosis and treatment data, therapeutic guidelines
Postcondition	Validated treatment or inconsistency recorded
Task Constraints	Proven and accepted guidelines for treatment prescription
Task: Decision instance result validation	Validation of treatment against recommended guidelines and measurement of inconsistencies
Domain Model	Heuristic (expert guidelines on treatment prescription)
Represented Features: Qualitative	Treatment model/rules based on guidelines.
Represented Features: Non-linguistic	Data measuring consistency between prescribed treatment and recommended treatment
Psychological Scale:	Nominal, ordinal, interval and ratio
Representing Dimension of Features:	Shape, color, length, etc.
Technological Artifacts/methods:	Hard computing artifacts (e.g., rules),

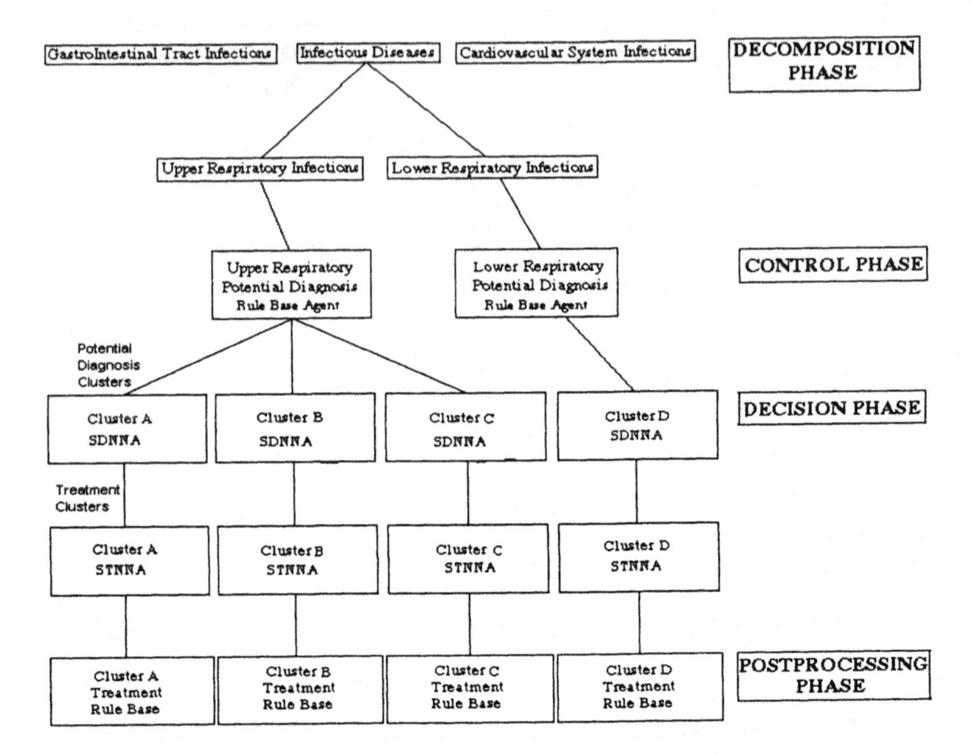

Figure 6.14: Partial Computational Overview of a Medical Practitioner's Decision Path Based on Selection of Symptom to Diagnosis and Treatment Strategy

6.4. Intelligent Diagnosis and Treatment Support Problem Solving Agents

The aim of the transformation agents is to map or transform a human-centered solution for diagnosis and treatment support into a computer-based artifact using the generic agents defined in the four agent layers of the HCVM. Thus the human-centered domain model developed in the last section is now used for defining the diagnosis and treatment support transformation agents. The transformation agents for diagnosis and treatment support include decomposition, control, decision and postprocessing problem solving agents corresponding to the problem solving agent layer, intelligent rule base (expert system) agent and self-organized or unsupervised and supervised neural network agents corresponding to the intelligent agent layer, and media agents and distributed processing and communication agents corresponding to the software agent layer. These transformation agents for diagnosis and treatment support represent a component-based software approach at the computational level.

At present, we have implemented simpler models of intelligent artifacts like neural networks and expert systems. We not implemented imprecision and fuzziness in the symptomatic data into more sophisticated models like hybrid fuzzy-neural network (fusion) agent corresponding to the intelligent hybrid agent layer.

In Figure 6.14 we show a partial computational overview of a medical practitioner's decision path based on selection of symptom to diagnosis and treatment strategy. It shows some of the transformation agents used in the four phases, namely, decomposition, control, decision and postprocessing. These phases can also be seen as representing various levels of abstraction of the decision path. The abbreviations SDNNA and STNNA stand for Supervised Diagnosis neural Network Agent and Supervised Treatment Neural Network Agent respectively.

In Figure 6.15 we show three decision paths and their corresponding phase sequences. These decision paths can also be seen to reflect different levels of expertise and different patient-practitioner situations.

In the following subsections we outline the transformation agent definitions of diagnosis and treatment support problem solving agents and their implementation results.

The media agents (software agent layer) supporting various problem solving agents are defined in section 6.5 related to multimedia interpretation component. Aspects related to the object layer have been outlined in the last section and thus are not covered here.

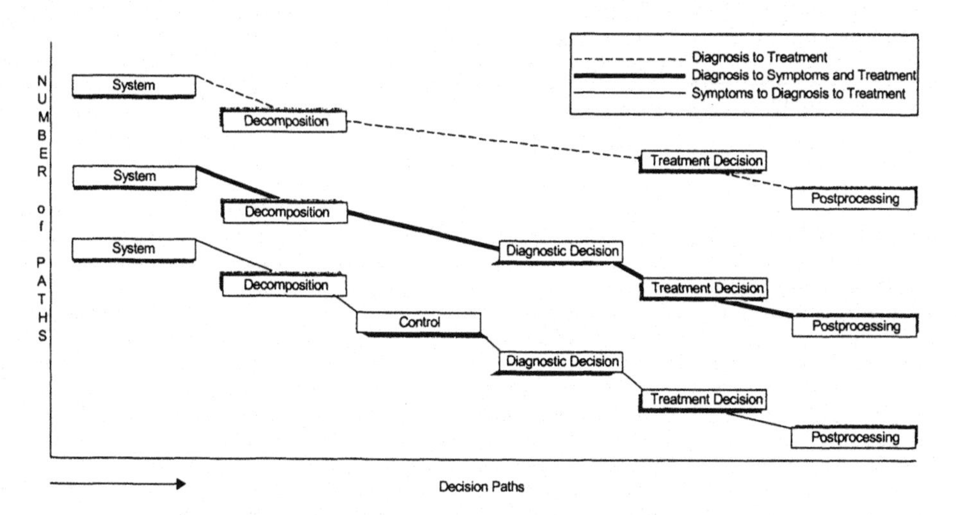

Figure 6.15: Decision Paths Taken by Practitioner and HCVM Phase Sequence

6.4.1. Etiology Selection Decomposition Agent

The Etiology Selection Decomposition Agent definition is shown in Table 6.5. It integrates the adapter definition with the generic agent definition. Table 6.5 defines the various actions for accomplishing the tasks mapped in the decomposition adapter definition in section 6.3.1.1. It also defines the communication constructs, the agents it communicates with and the external and internal tools or technological artifacts used for realizing the goals of the Etiology Selection Decomposition Agent.

In the agent definition in Table 6.5 we have also added the invoking condition for the decomposition agent. It represents the dynamic conditions of the environment under which the Etiology selection decomposition agent will become active. Figure 6.17 shows how the representing dimensions of shape and location and the psychological scale information of the upper and lower respiratory infection concepts is being used for perceptual problem solving. For the sake of simplicity, the psychological scale information and representing dimensions and internal state are not shown in the agent definition.

Table 6.5: Definition of Etiology Selection Decomposition Agent

Name :	Etiology Selection Decomposition Agent
Parent Agent	HCVM Decomposition agent
Goals:	Reduce medical diagnostic complexity Establish independent infection regions
Tasks:	Determine upper and lower level infection regions, select diagnosis strategy
Task Constraints:	Diagrammatic reasoning
Invoking Condition:	Assistance requested for diagnostic and treatment support
Precondition:	Trained Medical practitioner or medical student, solving Phase: decomposition
Postcondition:	Problem Solving Event: Control, upper and lower respiratory agents activated
Communicates with:	Upper and lower respiratory control agents, Etiology selection media agent
Communication Constructs:	Command: Upper & lower respiratory control agents
Linguistic/non-linguistic features:	Inputs from user
External Tools:	Software: Graphic object, Button Object Domain: Infectious Diseases
Internal Tools:	
Actions:	Display upper and lower respiratory etiology Activate upper/lower respiratory potential diagnosis control agent.

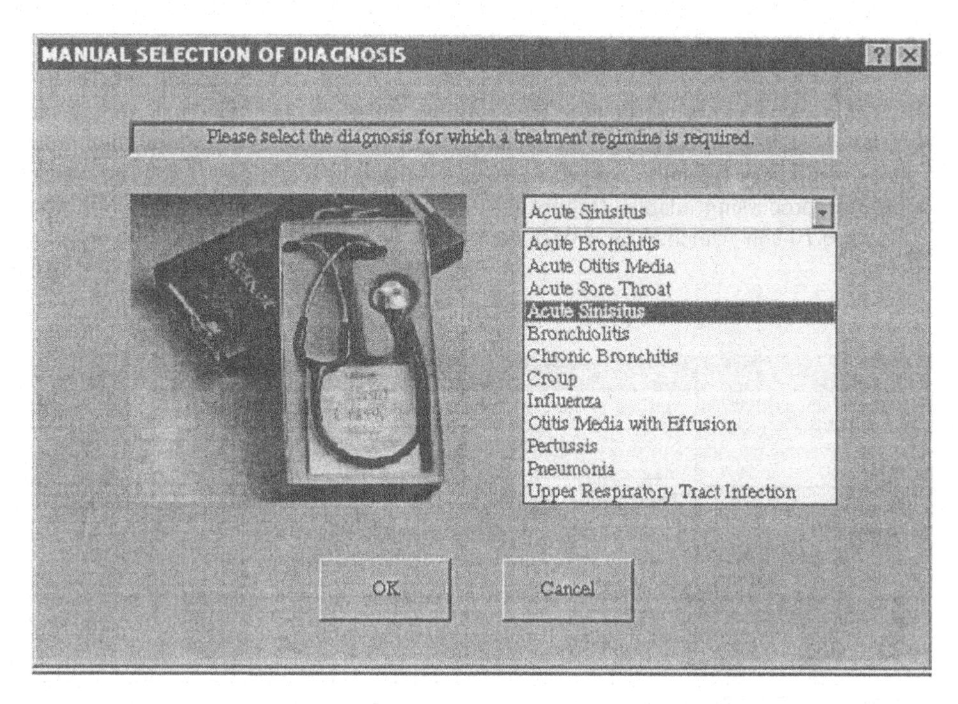

Figure 6.16: Bypassing Diagnosis – Selecting Diagnosis for which Treatment Support is Required

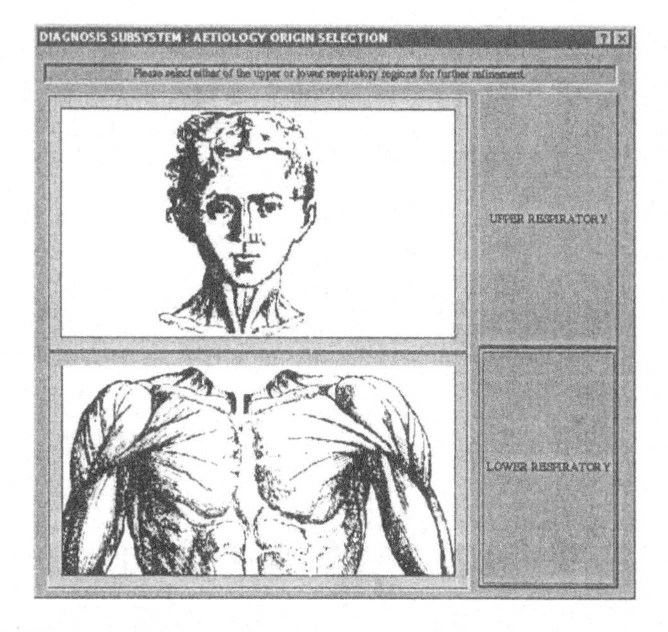

Figure 6.17 : Etiology selection

6.4.2. Upper Respiratory Control Agent

Table 6.6 shows some of the actions, internal and external tools, and communication constructs used for accomplishing the goals and tasks identified in the upper respiratory infection control adapter definition in section 6.3.1.2. The control phase preprocessing adapter (Figure 6.9) as well as and control adapter definitions (Figure 6.10 and Table6.3) have been merged in the agent definition in Table 6.6.

Table 6.6: Definition of Upper Respiratory Control Agent

Name :	Upper Respiratory Control Agent
Parent Agent:	HCVM Control agent, HCVM Preprocessing agent
Goals:	Successful determination of control constructs for each potential diagnostic cluster in the upper respiratory region
Tasks:	Transform symptoms using multimedia artifacts Learn potential diagnostic clusters from cases Conflict resolution between decision agents in case of multiple diagnosis
Task Constraints:	Shape, color, etc. of symptoms, cluster learning, adaptability to noisy, incomplete data, scalability to new upper respiratory infections, inductive and deductive reasoning
Invoking Condition:	Upper respiratory etiology selected
Precondition:	Historical symptomatic data and diagnostic data available, trained medical practitioner, Problem solving phase : upper respiratory control,
Postcondition:	Learning and Activation of potential diagnosis decision cluster/s agent/s
Communicates with:	Etiology decomposition agent, Potential diagnosis decision agents, Upper respiratory media agent, Rule base (Expert System) agent, Self-organized (unsupervised) neural network agent
Communication Constructs:	Command: Potential Diagnosis decision agents Request: Upper respiratory rule base agent Request: Self –organized (unsupervised) neural network agent
Linguistic/non-linguistic features:	Upper respiratory symptomology
External Tools:	Self –organized (unsupervised) neural network agent, upper respiratory rule base agent, media agents, upper respiratory infection objects
Internal Tools:	Planning and sequencing of actions related to multimedia symptom modeling, learning of potential diagnosis clusters, control knowledge and conflict resolution
Actions:	Learn upper respiratory potential diagnosis clusters from cases using self organizing map agent Activate potential diagnosis decision cluster problem solving agent using rule base agent Conflict resolution rules for multiple diagnosis

The upper respiratory infection control agent communicates with the media agent for generating and displaying the media artifacts used for symptomatic data gathering.

These symptoms are used by the rule base agent shown in Figure 6.14 for activating one or more potential diagnosis clusters shown in Figure 6.14. The cluster grid of the Kohonen's self-organizing maps used for learning the potential diagnosis sets/clusters is shown in Figure 6.18. Four clusters are defined from the cluster grid in Figure 6.18.

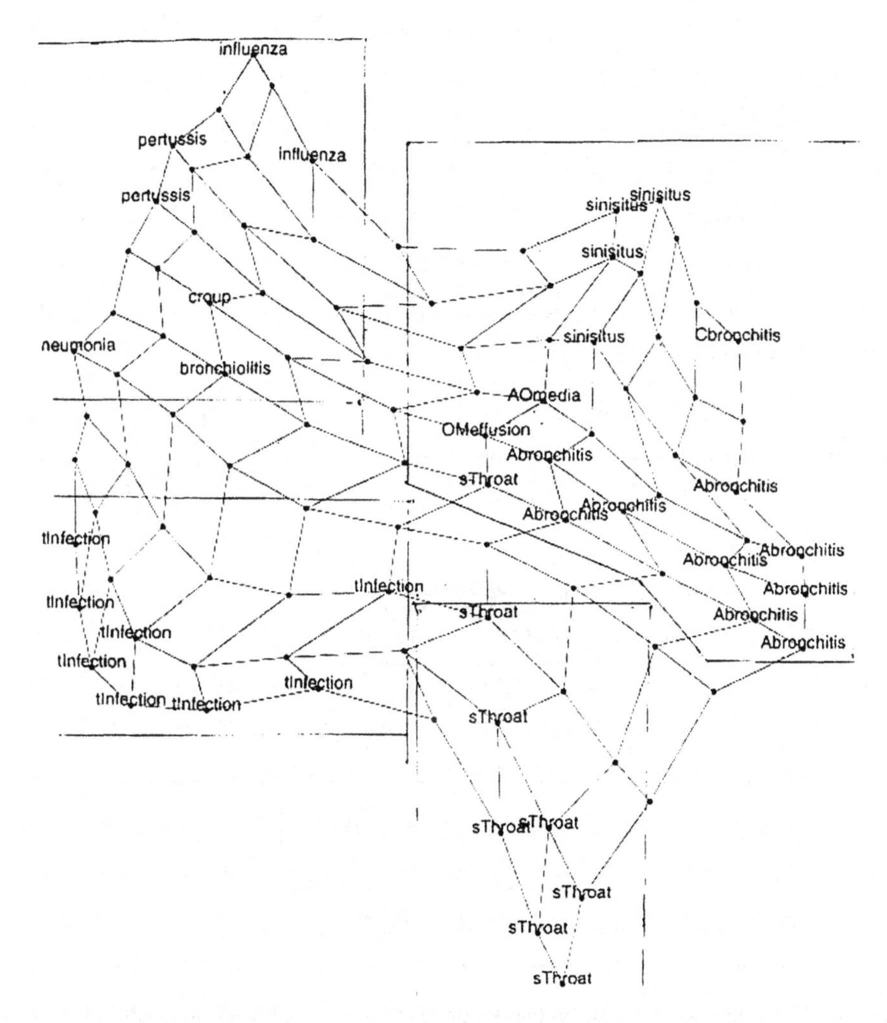

Figure 6.18: Potential diagnosis clusters using Kohonen's Self-organizing Maps

The Generalized Upper Respiratory Tract Infection (tinfection in Figure 6.18) forms Cluster A in Figure 6.14. Acute Sore throat forms Cluster B and Acute Otitis Media, Otitis Media Effusion and Acute Sinusitis upper respiratory infection are grouped into Cluster C. The other lower respiratory infection are grouped into Cluster D.

6.4.3. Suggested/Predicted Diagnosis Decision Agent

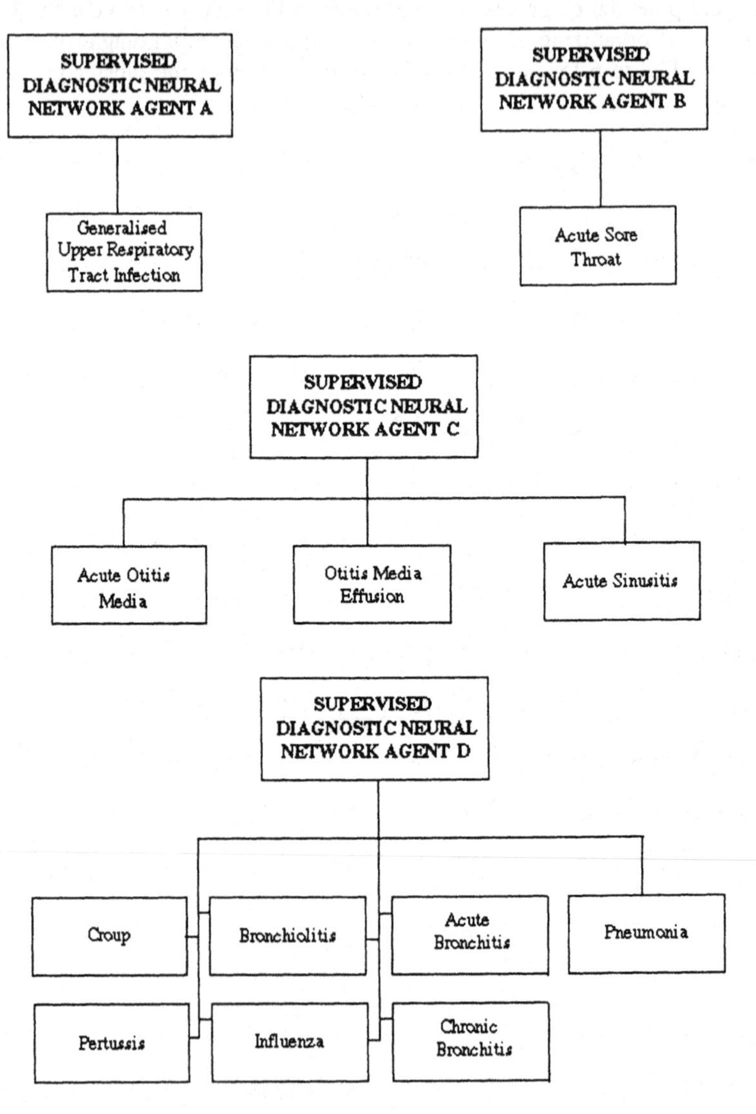

Figure 6.19: Supervised Neural Networks used in Different Clusters for Determining Suggested or PredictedDiagnosis

The potential diagnosis Clusters A, B, and C in Figure 6.14 are used as decision clusters in the decision phase for upper respiratory infections. Table 6.7 shows the transformation agent definition for predicted diagnosis decision agent for Cluster C. Based on historical data collected from medical practitioners, a supervised (backpropagation) neural network agent is used to learn the mapping between different infections and symptoms in a particular potential diagnosis cluster (Armoni 1998; Baxt 1995; Khosla and Dillon 1997; Ouyang et al. 1998, Poli et al. 1991). Four neural networks are used for learning the mapping as shown in Figure 6.14 and Figure 6.19 for upper and lower respiratory infections. In Figure 6.14 the neural network agents for predicting diagnosis in each cluster are abbreviated as SDNNA (Supervised Diagnosis Neural Network Agent). The neural network agents from the intelligent agent layer of the HCVM are used for training the neural networks.

Further, the multimedia artifacts used for gathering symptomatic data related to Acute Otitis Media diagnosis are shown in Figure 6.20. An explanation of how the various multimedia artifacts have been selected is given in the next section (i.e. section 6.5).

Based on the symptomatic data feedback, the neural network predicts/suggests a particular diagnosis. The diagnostic information related to a predicted diagnosis is shown in Figure 6.21.

Table 6.7: Definition of Upper Respiratory Decision Agent

Name :	Upper RespiratoryDiagnosis Decision Agent C
Parent Agent: :	HCVM Decision agent, HCVM Preprocessing Agent
Goals:	Predict upper respiratory infection
Tasks:	Correlate frequency of symptoms with Individual upper respiratory infections
Task Constraints:	Shape, color, etc of symptoms, learning, inductive and deductive reasoning, bottom-up knowledge engineering
Invoking Condition:	Upper respiratory potential diagnostic cluster C selection
Precondition:	Trained Medical practitioner, Problem Solving Phase: upper respiratory decision agent C, learnt potential diagnostic cluster C, historical symptomatic data for Cluster C infection diagnosis
Postcondition:	Predicted diagnosis
Communicates with:	Upper respiratory control agent, Upper respiratory treatment agent C, Upper respiratory decision media agent C
Communication Constructs:	Request: media decision agent C – Display Inform: Upper respiratory treatment agent C – diagnosis Verify: Upper respiratory control agent – multiple diagnosis
Linguistic/non-linguistic features:	Upper respiratory diagnostic cluster C symptomology data
External Tools:	Supervised Diagnosis Neural Network Agent (SDNNA) for cluster C, media decision agent for cluster C Attributes and methods of Upper respiratory objects like Acute Otitis Media, Media Effusion, Acute sinusitis etc.
Internal Tools:	Planning and sequencing of actions for symptom data gathering, learning and predicting diagnosis
Actions	Learn individual diagnosis using SDNNA Predict diagnosis.using SDNNA

Once a particular diagnosis has been suggested an option to the practitioner is to use the treatment based on practitioner's experience or based on therapeutic guidelines. The practitioner based treatment regimen is learnt using treatment neural networks shown as STNNA (Supervised Treatment Neural Network Agent) in Figure 6.14 from historical data.

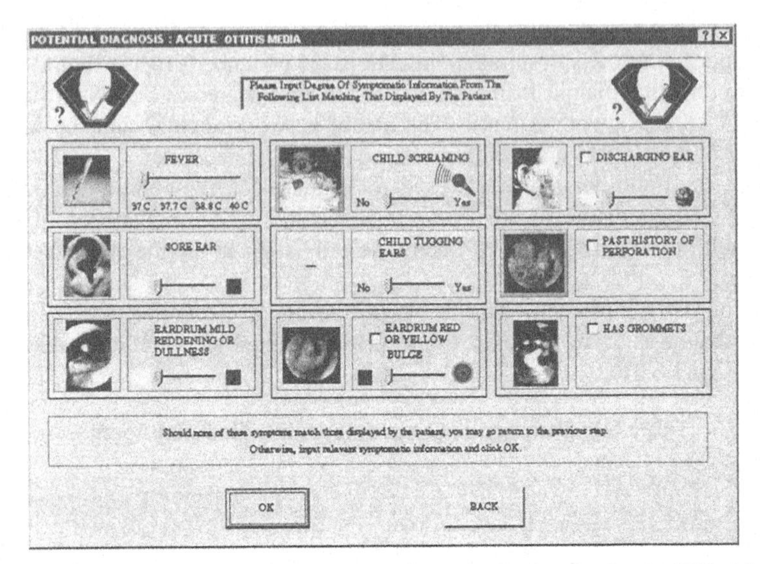

Figure 6.20: Multimedia Based Symptomatic Data Gathering for Acute Otitis Media

The predicted treatment based on practitioner experience may still be overridden by the practitioner as shown in Figure 6.22. Figure 6.22 also shows adaptation based on the accepted treatment feedback from the practitioner after the override.

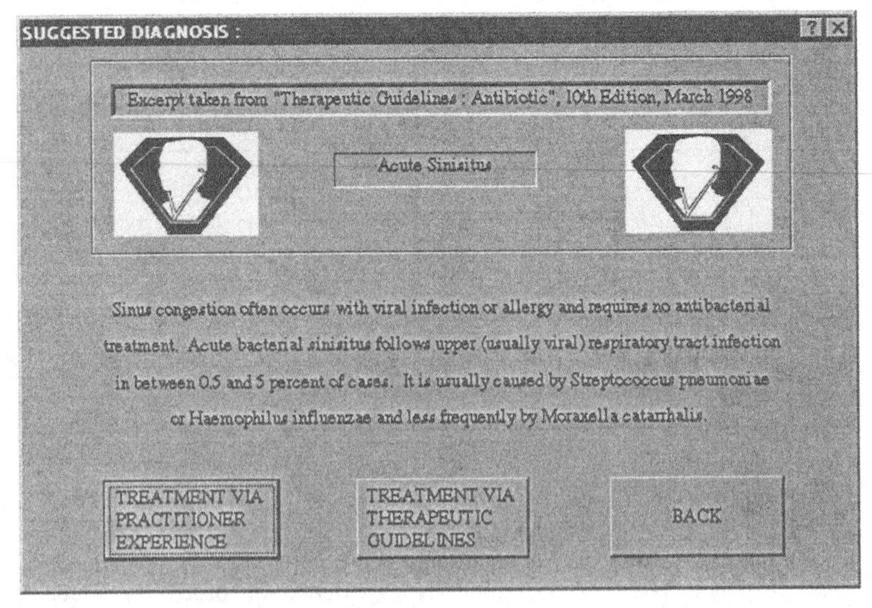

Figure 6.21 : Suggested Diagnosis Information

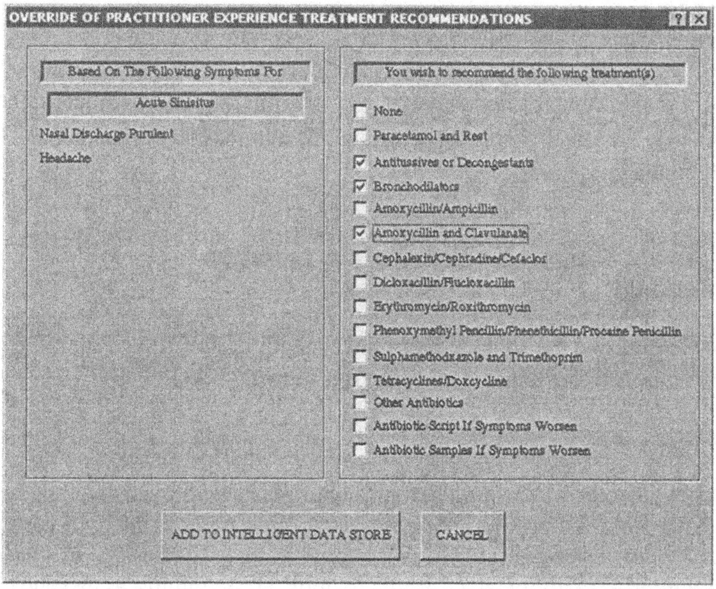

Figure 6.22: Treatment Override by Practitioner and Adaptation by Treatment Neural Network

Figure 6.23 shows the recommended treatment using therapeutic guidelines.

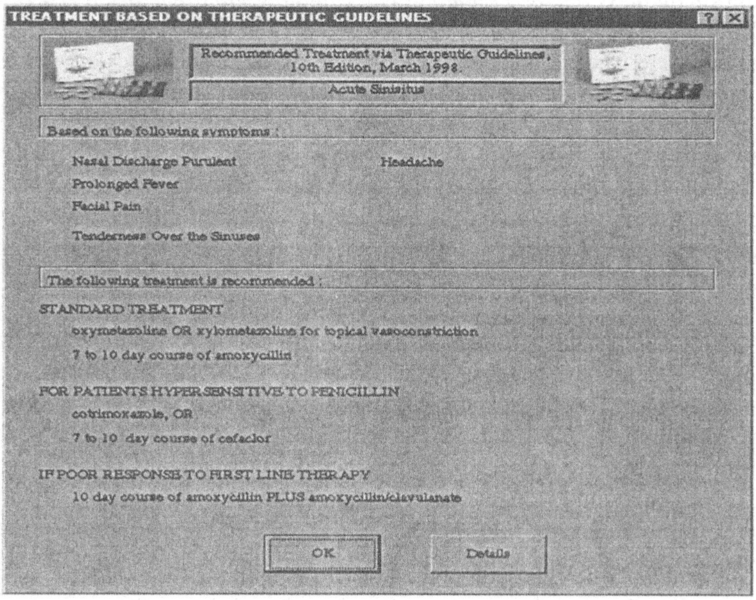

Figure 6.23: Recommended Treatment via. Therapeutic Guidelines

6.4.5. *Diagnosis and Treatment Validation Postprocessing Agent*

This agent has not been implemented as yet. Its main goal is to determine the inconsistencies in the treatments used by the practitioners compared to those recommended by the guidelines.

6.5. Diagnosis Support Multimedia Interpretation Component

In this section we describe the application of three stages of the HCVM multimedia interpretation component for the clinical diagnosis support.

6.5.1. *Symptom Content Analysis*

As mentioned in chapter 5, the psychological variables employed by humans are invariably distinct from the physical variables used for computations by a computer-based artifact. For effective patient symptomatic data gathering in the drug prescription monitoring activity, it is useful to look at symptoms as psychological variables from a medical practitioner's perspective (the person interpreting and entering the information) rather than as physical variables used for computation. In this section, we analyze the data characteristics of symptoms for Acute Otitis Media infection. We use the analysis to map the symptoms to various media artifacts.

The symptoms related to acute Otitis Media are *fever*, *sore ear*, *ear drum mild reddening or dullness*, *child screaming*, *child tugging ears*, *ear drum red or yellow and bulging*, *discharging ear*, *history of perforation*, and *has grommets*.

Data characteristics, like dimensionality, psychological scale, representing dimension, granularity, transience and urgency have been used for the analysis. The data characteristics of symptoms like *fever*, *sore ear*, *ear drum red or yellow and bulging, discharging ear,* and *has grommets* is shown in Table 6.8. The psychological scale used is based on perceptive and cognitive or interpreted representation of the symptoms by the medical practitioners in the context of the drug prescription monitoring activity. The analysis of the symptoms has been done in the context of the drug prescription monitoring activity. It is briefly discussed now.

Fever: is measured or determined on a single dimension of temperature. Medical practitioners measure the temperature in the range of 37 degrees centigrade to 40 degrees centigrade. Thus, the psychological scale information is on the interval scale, and the representing dimension is the position of mercury on this scale. The granularity of the fever symptom is considered as continuous (ranging from 37 degrees centigrade to 40 degree centigrade). There is no urgency in terms of communicating the above information in the context of the drug prescription monitoring activity and thus is classified as routine.

Sore Ear: is measured on two dimensions, namely, location (i.e. ear) and color of the ear. The representing dimension of color is density which represents different color shades like skin (normal ear) color, pink (mild sore ear) color and red (sore ear) color in a continuous range. Here the red color represents higher severity than pink or skin color in terms of magnitude. Although, perceptually the soreness is indicated by the red color (or shades of red color) of the ear, which is indicative of the ordinal scale, the psychological scale used is ratio. This is because medical practitioners interpret the degree of soreness on a continuous scale, ranging from no soreness (zero) to yes (1 - indicated by the red color of the ear), to determine the strength of the treatment. The granularity is continuous and urgency is routine.

Child Screaming: Consists of two dimensions, namely, location and density. These two dimensions are measured on the nominal and ratio scales, respectively. The location dimension relates to location of the screaming sound. The density dimension represents the screaming intensity, which is measured on a continuous scale of zero (no) to one (yes). The transient property of a scream is transient.

Ear Drum Red or Yellow and Bulging: consists of three dimensions, namely, location, color and shape. These three representing dimensions are perceived and interpreted on nominal, nominal and ratio scales, respectively. The location dimension relates to the ear drum location. The color of the eardrum (i.e. red or yellow) is perceived and interpreted on the nominal scale based on the category property. However, the bulging shape or degree of bulge of the eardrum is perceived and interpreted on ratio scale ranging from zero (flat) to 1 (bulging) with a continuous granularity.

Discharging Ear: is measured on three dimensions, namely, location (i.e. ear), color of the discharge and texture (purulent or clear discharge). These are also the representing dimensions of the symptom. The location dimension is based on the nominal scale, which includes the category property. The color dimension is based on the nominal scale and includes clear/transparent color discharge, yellow or green discharge. Finally, the texture dimension is based on the ordinal scale where the purulence of the discharge and the extent (magnitude) of discharge is determined. The granularity of this symptom is continuous and urgency is routine.

Has Grommets: represents two dimensions, namely, location and shape. These two dimensions are based on the nominal scale. The medical practitioner is looking for absence or presence of grommets only. Given the nominal scale on both the dimensions, the granularity is discrete.

In this section we have described the characteristics of a subset of symptoms used for diagnosing acute Otitis Media. The characteristics of these symptoms have been analyzed based on characteristics like dimension, psychological scale, representing dimension, granularity, and transience. These characteristics are most relevant to the Acute Otitis Media symptoms. A similar analysis has been done for treatment data and other data used by decomposition, control, decision and postprocessing agents. Additional data characteristics like volume, and urgency have also been used in the analysis of treatment and other data. For example, the volume of symptoms is mostly represented by single facts (e.g. *sore ear*) and thus is *singular*, whereas, volume of treatment text based on therapeutic guidelines is *much* and requires elaborate text description.

Table 6.8: Data Characteristics of Acute Otitis Media Symptoms

Data Characteristics	Fever	Sore Ear	Child Screaming	Ear Drum Red or Yellow and Bulging	Discharging Ear	Has Grommets
Dimension	1D	2D	2D	3D	3D	2D
Psychological Scale	Interval	Nominal,	Nominal and Ratio	Nominal, Nominal and Ratio	Nominal, Ordinal and Ordinal	Nominal and Nominal
Representing dimension	Position	Location and Color	Location and Density	Location, Color and Shape	Location Color and Texture	Location and Shape
Granularity	Continuous	Continuous	Continuous	Continuous	Continuous	Discrete
Transience	Dead	Dead	Transient	Dead	Dead	Dead
Urgency	Routine	Routine	Routine	Routine	Routine	Routine

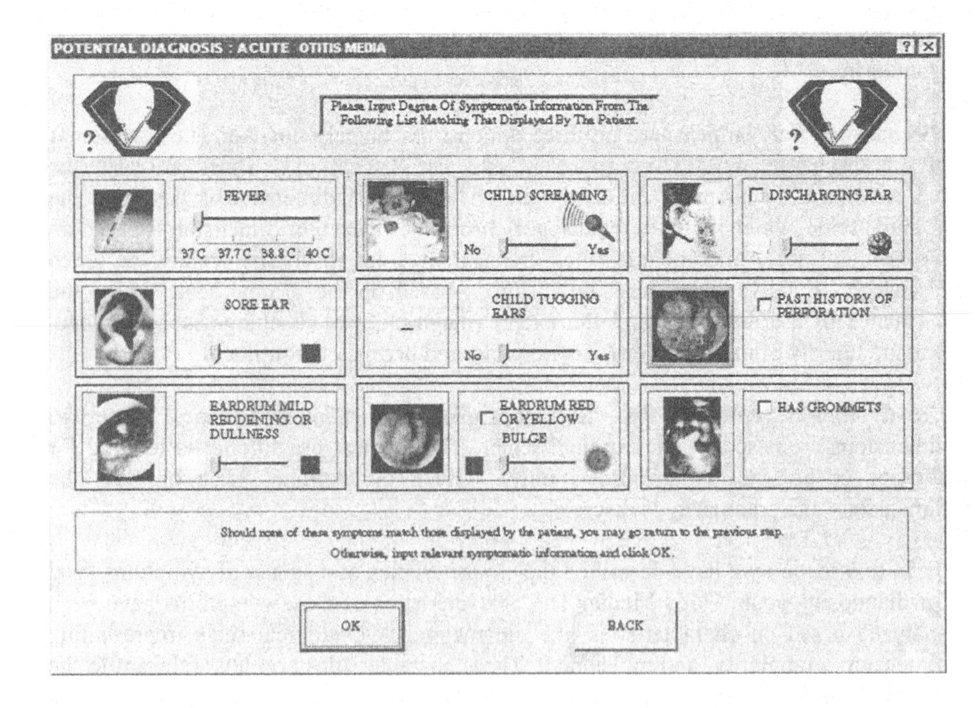

Figure 6.24: Multimedia Based Symptomatic Data and Gathering for Acute Otitis Media

6.5.2 Media, Media Expression and Ornamentation Selection

The characteristics of the symptoms outlined in the last section are used to select various media artifacts. Further, the modality or level of abstraction of various media is selected to facilitate complementation rather than duplication of media. In this section, we outline mapping of symptom characteristics to media types and the use of different levels of media expression.

Fever: A combination of text, image icon and a temperature sliding scale has been used to represent fever. A sliding temperature scale with an interval range of 37degrees to 40 degrees represents the single dimensionality and interval scale. The sliding scale in Figure 6.24 is used as a media substrate for determining patient's temperature. The scale pointer is the information carrier through which the actual physical value is recorded internally. The temperature sliding scale also represents continuous granularity of temperature.

As shown in Table the text, thermometer icon and temperature scale represent different levels of media expression, which complement each other. The thermometer icon is an abstract image icon of temperature and complements the temperature scale. The temperature sliding scale represents an elaborate level media expression of temperature and the word "fever" is a representative textual concept for temperature.

Sore Ear: Text, image and a sliding color scale are three media types employed to represent the two dimensions of the *sore ear* symptom (Table 6.9). The degree of soreness or density is represented using a sliding color scale. It is used as the media substrate for measuring the degree of soreness. The color interval ranges from normal ear to a red sore ear. The image of the red sore ear represents the location dimension as well as an elaborate level of media expression. The sliding color scale also represents a representative level of media expression for the degree of soreness. The red sore image also complements the sliding color scale representation.

Table 6.9: Definition of upper respiratory control agent

Media/Data Characteristics	Fever	Sore Ear	Child Screaming	Discharging Ear	Has Grommets
Media Type	Text, image, sliding scale	Text, image, sliding scale	Text, image and audio	Text image, sliding scale	Text image Check box
Media Expression	Text - representative Image - abstract Sliding temperature scale - elaborate	Text - representative Image - representative Sliding color scale - representative	Text - representative Image - elaborate Audio - elaborate Sliding scale - representative	Text - representative Image - elaborate Sliding texture scale - representative	Text - representative Image - elaborate Check box - representative

Ear Drum Red or Yellow and Bulging: is represented on three dimensions, namely, location, color and shape. The location dimension on a nominal scale is represented by the bulging eardrum image. The color dimension on the nominal scale is represented by the check box as well as the bulging ear drum image. The shape dimension on the ratio scale is represented using the bulge sliding scale ranging from a flat ear drum (indicating a physical value of 0) to bulging oval shaped ear drum (indicating a physical value of 1).

Child Screaming: has been shown in Table 6.9 for its variation in terms of use of text image and audio media artifacts. The image and audio artifacts shown in Figure 6.24 are an elaborate expression of a child's scream in the two media types. Although, they are at the same level they tend to complement rather than duplicate each other. Further, unlike perceptually oriented sliding scales used for other symptoms, a no/yes sliding scale has been used for this symptom. The aural nature of this symptom restricted us somewhat in providing a more perceptually meaningful sliding scale.

Discharging Ear: is represented using text, image, a texture based color sliding scale and a check box. The nominal scale of location dimension is represented using an elaborate level of the image artifact. The check box is used to confirm or negate presence or absence of the discharge. If the discharge is present the texture and color of the discharge is determined on a texture based color sliding scale. It may be noted that the sliding scale does not start with zero (0). The transparent or clear discharge on the left end of the sliding scale in Figure represents a physical value of 1 whereas, the thick greenish discharge on the right end of the sliding scale represents a physical value of 5. Further, combining color and texture dimensions into one representation in Figure is based on the assumption that color and texture vary concurrently and are interpreted together (rather than in isolation) for the purpose of determining severity of symptom and strength of the treatment.

Has Grommets: is represented using the image shown in Figure 6.24. The image is employed to represent the location and shape dimensions on the nominal scale. The check box is used to confirm or negate presence or absence of grommets.

The main purpose of this section has been to enhance the precision or quality of symptom data gathering. In this light, the multimedia representations shown in Figure 6.24 provide a richer medium for effective symptom data gathering than the PDTD forms shown in Figure 6.2. These representations, among others aspects, have been based on the psychological scales and representing dimensions employed by medical practitioners for determining the diagnosis and treatment of upper respiratory infections. The multimedia representations are expected to assist in a more clear explanation and detection of the differences and inconsistencies in the treatments prescribed by different medical practitioners.

In Figure 6.24 one can also notice two human faces with a question mark in the upper right and upper left-hand corners, respectively. These graphic objects have been for the purpose of situating the medical practitioners in terms of information processing and patient diagnosis and treatment in the system. The screen in Figure 6.24 shows that the computerized system is trying to ascertain symptoms related to a potential diagnosis of Acute Otitis Media. Finally, for ornamentation, cyan color has been used as a back group color of the screen in Figure 6.24.

6.5.3. Multimedia Agents

In the last section we analyzed the data characteristics and selected the media artifacts for representing the data. At the computational level, multimedia agents are associated with each problem solving agent (e.g., decomposition, control and decision) for generation, display, layout and coordination of various media artifacts at problem solving, information processing and task level. These multimedia agents are also used for recording practitioner's feedback and feeding it to the corresponding problem solving agent. The definition of these multimedia agents, like the problem solving agents, facilitates learning, reasoning with respect to generation, display, layout and coordination of various media artifacts. A sample definition of a multimedia agent is shown in Table 6.10.

Table 6.10: Definition of upper respiratory decision media agent

Name :	Upper Respiratory Diagnosis Decision Media Agent C
Parent Agent:	HCVM media agent
Goals:	Effective/accurate symptomology gathering Establish practitioner's location in the system Correlate symptomatic data to argumentative diagnosis knowledge
Tasks:	Determine symptoms Map symptom characteristics to media artifacts Determine media layout Map symptomatic data to antecedents of rules and input vector of supervised neural networks
Task Constraints:	Medical practitioner's context, symptom data characteristics, media characteristics information processing /task context, media layout (e.g. visual, text)
Invoking Condition:	Request from upper respiratory diagnosis decision agent C
Precondition:	Trained Medical practitioner, Problem solving phase: decision diagnosis , availability of image/graphic, audio and text objects
Postcondition:	Multimedia upper respiratory diagnosis symptom display
Communicates with:	Upper respiratory diagnosis decision agent C, graphic and audio objects
Communication Constructs:	Inform: upper respiratory symptomology to decision agent C
Linguistic/non-linguistic features:	Psychological scale and representing dimensions of symptoms, symptom for upper respiratory infection cluster C.
External and Tools:	**Software:** Graphic object, audio objects, text object **Domain:** Symptomology images, sound effects **Design:** Multimedia symptom analysis and design techniques
Internal Tools	Planning and sequencing media generation, display and coordination actions
Actions:	Generate and display upper respiratory symptomatic elaborate images, abstract sound effects and representative text, Generate & Display logarithmic sliding temperature scale and other perception based sliding scales for acute Otitis Media symptoms. Generate & Display abstract image icon to establish practitioner's location in the system (e.g., potential diagnosis state) Display feedforward neural network graphics for reasoning (not implemented)

6.6. Summary

Medical drug prescription regulation has become an important health issue in recent times. In this chapter we have described the application of various facets of the human-centered virtual machine in the drug prescription monitoring activity. That is, we have described the application of activity-centered analysis component, problem solving ontology component, transformation agent component and the multimedia interpretation component in developing a computer-based artifact to support the drug prescription monitoring activity. The intelligent multimedia multi-agent medical diagnosis and treatment support addresses number of issues related to symptomatic data gathering and improving diagnosis and treatment regimens. The modeling of various media artifacts for data gathering, problem solving including design of media agents has been described. The implementation results of various other aspects of the application have also been outlined. An interesting aspect of this application has been the ability of the human-centered virtual machine to facilitate modeling of several decision paths, which can be adopted by a medical practitioner while interacting with the system. These decision paths besides providing flexibility to the medical practitioner can also be seen as catering to different types of medical practitioners with different level of experience in the profession.

Acknowledgments
The authors wish to acknowledge Dr Susie Rogers, Project Officer, Inner South Eastern Division of General Practice, Alfred Hospital, Melbourne, Victoria and Dr. Ken Harvey, School of Public Health and Therapeutic Guidelines for collection of medical data and useful discussions on clinical drug prescription monitoring activity.

References

Armoni, A. (1998). *Use of Neural Networks in Medical Diagnosis*. M.D Computing. 15, 2, 100-103.

Baxt, G. (1995). *Application of Artificial Neural Networks to Clinical Medicine*. The Lancet, 356, 1135-113

Barrows, H.S and Pickell, G.C. (1991). *Developing Clinical Problem Solving Skills : A Guide To More Effective Diagnosis And Treatment*. W.W Norton, New York.

Gorbach, S, Bartlett, J and Blacklowe, N. (1998). *Infectious Diseases*. Saunders, Philadelphia.

Khosla, R and Dillon, T.S. (1997) *Engineering Intelligent Hybrid Multi-Agent Systems*. Kluwer Academic Publishers, Dordrecht, Netherlands.

Ouyang, M and Zazunobo, Y. (1998) *Using a Neural Network to Diagnose the Hypertrophic Portions of Hypertrophic Cardiomyopathy*. M.D Computing, 15, 2, 106-109.

Poli, R, Cagoni, S, Livi, R, Coppini, G and Vali, G. (1991). *A Neural Network Expert System for Diagnosing and Treating Hypertension*, Computer, March.

Sarbani, Sen. (1997). *Multimedia Interactive*. Kale Consultants, "http://kaleconsultants.com.au/mm.html"

TG (Therapeutic Guidelines), (1998), Therapeutic Guidelines for Respiratory Infections, Victorian Medical Postgraduate Foundation Inc and Therapeutics Committee.

7 A MULTI-AGENT SYSTEM FOR FACE DETECTION AND ANNOTATION

7.1. Introduction

The purpose of this chapter is to describe the application of the problem solving ontology component of HCVM in the area of face detection and annotation. computer-based aspects of an embedded feature-based multi-agent face detection and annotation system are described in this chapter. Unlike similar feature-based systems that start from searching for facial organs in the images and group them to find faces, the system solves the problem using the problem solving agents of the HCVM in a top-down fashion. The embedded multi-agent face detection system makes hypotheses of face locations and seeks evidences to verify them. The five problem solving agents follow a coarse grain to fine grain methodology and work on various features of color images. The coarse to fine top-down methodology is more akin to human perceptions than a bottom-up approach. The higher level problem solving agents like decomposition work as background knowledge for the control and decision agents resulting in improvement of the accuracy and speed of detection. It is also distinct from many other top-down systems in that no re-scaling of the images is needed in the searching process. To improve the detection rate, a unique iterative region-partitioning algorithm is developed in this multi-agent system. The problem solving agents model aspects related to skin-tone region segmentation, noise reduction, candidate face regions location and facial feature extraction/face detection, and annotation. The detection algorithm is invariant to scale and rotation to some degree. The performance of the multi-agent system has been tested with a large number of images and the results obtained thus far are very encouraging.

This chapter starts with an introduction to image annotation, followed by descriptions of some previous work done in the area of face detection and the approach adopted in developing the multi-agent systems. A tasks-product transition network is then drawn to show the various tasks and elementary, intermediate and final products. The task-product network is used to map various tasks to the five HCVM problem solving agents at the computational level. Finally, implementation results related to face detection are reported and future extensions are discussed.

7.2. Image Annotation

Automated image annotation has recently become an interesting research issue. This is partly motivated by the increasing need for efficient management of computerized images, which is growing explosively. Image annotation is a high-level image understanding procedure, attempting to extract semantic features from images and describe them in human languages.

Despite the huge amount of research work in this area, image annotation technique is still far from being mature because of the difficulty in extracting high-level descriptions from images. However, we could start by limiting the high-level descriptions to some specific aspects. Since the presence or absence of people in an image provides important information for the image, the ability to classify images as *people* or *non-people* will be very helpful in image annotation. Moreover, with the number and density of people, the images can be further annotated as *single person photograph*, *two persons photograph*, *group photograph* or *crowd photograph*. If this can be done successfully, more detailed image annotation could be accomplished gradually by considering other aspects.

Although many techniques may be used to judge whether an image contains people or not, the most reliable one is by face detection. We choose face detection as the heart of the image annotation not only because of its usefulness in this system, but also its potential importance to other applications. This importance stems from the fact that faces are our identities and signals communicated by our faces regulate our social interactions. Much of the early work on face detection was primarily driven by the needs of the face recognition systems and consequently assumed simple background in images. However, the recent work in this area has focussed attention on images with complex background as new applications for face detection, e.g. surveillance and tracking, have emerged. This is a difficult problem because of scale, occlusion, the possibility of multiple faces and background clutter.

We describe a multi-agent top-down face detection system for image annotation in the context of large image and video database. The multi-agent system has been applied to images with very complex backgrounds. After outlining some existing face detection systems, the details of the multi-agent system are discussed.

7.3. Some Existing Face Detection Systems

The existing approaches to face detection can be classified by gray level and by color and motion. Some existing systems with these approaches are briefly described in this section.

7.3.1. Face Detection by Gray-Level

Most digital images were in black/white format in the early years, resulting in the majority of face detection systems working with gray-level of images. Those techniques can be categorized as feature-based approaches or subspace approaches.

Feature-based approaches require priori knowledge. Certain properties of the domain must be known in advance and some heuristic measures need to be taken. This scheme is very intuitive since the features extracted are often closely related to

Feature-based approaches require priori knowledge. Certain properties of the domain must be known in advance and some heuristic measures need to be taken. This scheme is very intuitive since the features extracted are often closely related to human perceptions. Its disadvantage is that incomplete or wrong domain knowledge may have bad effect on the results. Systems in this scheme usually follow either of two models, namely bottom-up or top-down.

- Bottom-up systems start by searching for possible facial features and use their spatial relationship to form face candidates. System described in You et al (1994) is an example. A spatial filter is applied to detect possible feature points and edge maps are examined to verify them. They are grouped to form face candidates with some constraint, and the likelihood of a candidate being a face is evaluated by a probabilistic work. Leung et al Leung et al (1995) convolves the image with a set of Gaussian derivative filters to extract candidate features and used a statistical model to find possible face locations. Govindaraje et al 1989 used the knowledge obtained from the captions as an aid to locate human faces in newspaper photographs. The edge maps of images are searched for a model of face profile and other features. In a later work (1990), they developed a model-based approach. It extracts features from the edge map and groups them with cost functionals, thus gives the face location. Jeng et al (1998) proposed an approach that extracts facial features from a preprocessed image using geometrical configuration information and gave an evaluation function of the probability of the presence of a face based on the extracted features.
- Top-down strategies search the image in a coarse-to-fine style. Matches at a coarse scale are subject to verification at a finer scale. Yang et al (1994) developed a three-level method; the first two of which search the mosaic images at different scales to find face candidate positions. In the third one, edge information is used to decide whether the candidate is a face.

Subspace approach is essentially considering all images of size M*N as an M*N dimensional space, and attempting to identify the subspace where face images distribute. Rowley et al (1998) used neural networks to classify standard-sized subimages as face or non-face. An arbitrator is needed to merge the detection so that overlapping results could be eliminated. Sung et al (1987) partitioned face pattern distribution into clusters. Subimages at each position are matched against each cluster to determine the presence of faces. The advantage of the above two approaches is obvious: those systems require less priori domain knowledge and are relatively easier to be adapted to detect other targets. However, to obtain invariance to scale, images need to be searched for faces at different resolutions because the vectorized subimages are of a fixed size while faces in an image could be of any size, which will bring in extra computation expenses. And since this kind of systems work with high dimensional input data, the training set should be large although some measures could be taken to reduce it. Moghaddam et al (1998) proposed a subspace method. PCA is performed to reduce the space dimensionality. The target probability density of subimages centered at each spatial position is computed and targets are located by the maximum likelihood detection.

7.3.2. Face Detection by Color and Motion

Owing to the rapid development of hardware equipment, especially the availability of inexpensive digital color cameras and large, fast storage, color images and digital videos are becoming prevalent. This motivated the attempts to use color and motion information as cues for face detection. The system (1996) designed by Dai et al is such an example. It detects faces based on face-texture model, combined with skin region segmentation in YIQ color space to enhance the performance. Saber et al (1998) also used color to extract regions corresponding to skin-tone as the first step to find facial features, differing in that YES color coordinate is used. Zhang et al [18] converted images to the HSI color coordinate to delimitate the face of the speaker in videophone scenes. Motion information is a powerful cue for face detection in video. Since human faces may move from time to time, the searching process usually can be restricted to moving objects. Moving objects can be segmented by computing the difference between two frames. Kapfer et al (1997) investigated a multi-resolution method in videophone QCIF sequences, which used color and motion to segment the foreground and a shape model is developed for face detection. Although the approaches based on color and motion can simplify the problem significantly, two drawbacks exist. First, skin-tone extraction by color is sensitive to the chromatic characteristics of illuminance. Secondly, as the number of moving objects in the video increase, the motion-based methods lose their advantage.

7.4. Task-Product Network of Face Detection and Annotation Activity

The task-product transition network of the face detection and annotation activity is shown in Figure 7.1. The input to the system are color images and final products are annotations as *people* or *non-people images* at the simplest level, and a *single person photograph, two persons photograph, group photograph or crowd photograph* at a complex level. The tasks shown in Figure 7.1 are described in detail in the rest of this section. These tasks are based on human perceptions involving a top-down coarse to fine approach.

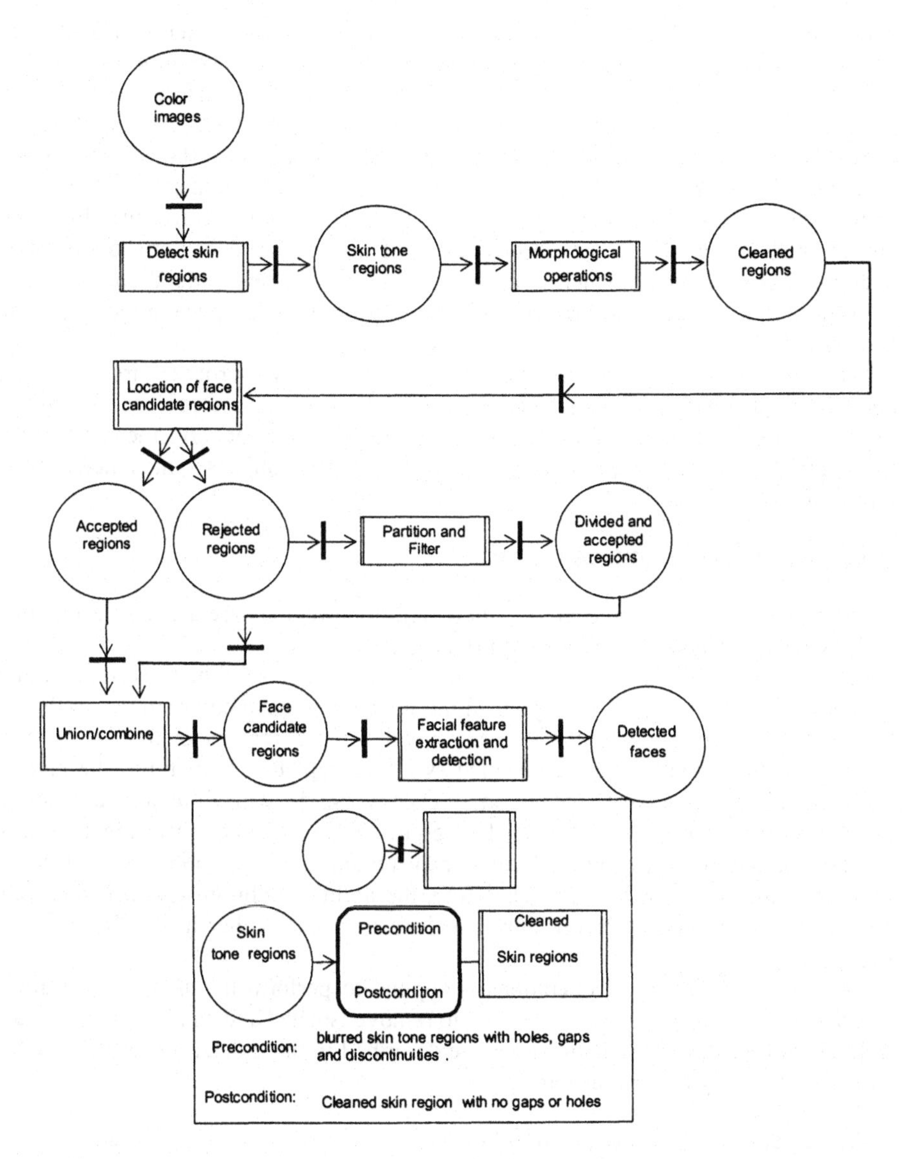

Figure 7.1: Task-Product Transition Network for Face Detection and Annotation

7.4.1. Detection of Skin-tone region

Most human face areas are of skin except those parts corresponding to eyes, eyebrows and so on. Considering the pixel representation in color images as a 3-D space, skin-tone pixels distribute within a bounded subspace of it. By identifying this subspace, a color image can be segmented as *skin-tone* or *non-skin-tone* regions. Although this subspace could be located in any coordinate model, a wise selection of the coordinate will simplify the problem. RGB coordinate is widely used in image representation, but it is not a good model for skin-tone region extraction because illuminance information is not isolated. Zhang worked in HSI coordinate to perform segmentation by color. In this process all three components are involved and thus the complexity is relatively high. The same problem exists in the use of YUV and CMY color models.

In Dai (1998), YIQ representation is used to perform skin region extraction as the I component includes color ranging from orange to cyan, which correlates closely to human skin. It has been observed that skin-tone distribution does not show obvious patterns in Q component, which can thus be ignored. However, the Y component is relevant to skin-tone. Therefore, the skin-tone distribution subspace in the Y-I plane is determined.

The skin tone regions extracted through the above process are a coarse approximations based on the hypotheses that all skin regions are face candidate regions. This is obviously very inaccurate and requires further refinement. More so, the color/YIQ images may contain clutter or non-face objects which need to be removed.

7.4.2. Cleaning of skin-tone regions

As mentioned in the preceding section, skin-tone regions are a coarse estimation and need to be refined further based on shape and size.

The purpose of this task is to remove the background noise. The background noise in terms of clutter or non-face objects in the image may have color similar to that of human skin. This often results in many small isolated regions or narrow belts of several pixels in width in the segmented skin-tone regions. Sometimes the narrow belts may be connected to a face region, or even worse, they may link a face region to another skin-tone regions. As will be described later, further refinement of face locations is based on a connected component region analysis, which may reject the regions in the above two cases. Therefore, the ability to eliminate small blobs and narrow belts and detach weakly linked regions from each other in the first place is desirable.

A set of morphological operations is applied to perform the skin-tone cleaning. These morphological operations are used to remove isolated and small regions that are unlikely to be faces. These operations are described in more detail in section 7.6.2.1. The number of regions is thus reduced.

7.4.3. Location of candidate face regions

In this task region size and shape are used to verify whether the cleaned skin tone regions are accepted as face candidate regions or rejected. Regions with reasonable size and shape similar to that of a face are accepted. To improve the detection rate, a partition algorithm is applied to the remaining rejected regions. The divided regions are checked in the same way to generated new face candidates. The new face candidates along with the accepted face candidate regions are subjected to further examination.

7.4.4. Facial feature extraction, face detection and Annotation

Each face candidate region is analyzed for the presence of certain facial features. The end product of this task is a set of successfully detected faces. After the faces are located, the number, sizes and the relative positions of them are fed to the classifier to determine the image annotation that should be applied.

7.5. HCVM Problem Solving Ontology and Face Detection and Annotation Activity (FDAA)

The purpose of this section is to show and describe the mapping of the problem solving adapters in the multi-agent face detection and annotation system. The mapping is based on the task-product transition network shown in Figure 7.1. It is also shown that the application of the preprocessing adapter is not limited to one phase but can be used in different phases.

7.5.1. Decomposition Phase Adapter Definition in FDAA

Unlike the chapter 6 where the preprocessing adapter was used in the control and decision phases of the Drug Prescription Monitoring Activity, in FDAA the preprocessing adapter is used in the Decomposition and Control phase, respectively. The association of the preprocessing adapter with the FDAA task is shown in Figure 7.2. The mapping of the preprocessing adapter signatures with FDAA is shown in Table 7.1. The goal of preprocessing is to improve data quality by isolating the illuminance property of the color image. This is done by transformation of the RGB image to a YIQ representation. The task constraints, pre/post conditions and represented features used are shown in Table 7.1. The ratio scale is based on Y and I component values. The orientation-representing dimension is indicative of the transformation from a RGB coordinate system to YIQ system.

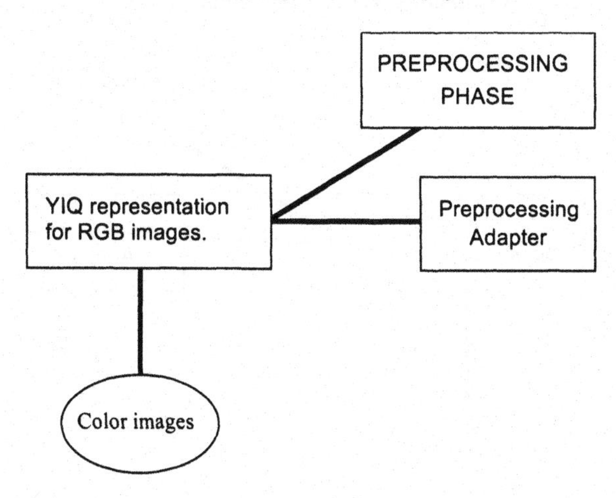

Figure 7.2: Associating Preprocessing Adapter with FDAA Task

Table 7.1 Mapping Preprocessing Adapter Signatures to FDAA

HCVM and Representation Goal, Tasks, Signatures	Corresponding in Face Detection and Annotation Activity (DPMA)
Phase: Preprocessing	Color image preprocessing
Goal: Improve data quality	Improve image quality to isolate illuminance property
Task: Data transformation	Transform RGB color image to YIQ image
Task Constraints: domain dependent	3D space
Precondition:	Color RGB people image
Postcondition:	YIQ image with Y and I component suitable for detection of skin regions
Represented Features: Qualitative or Linguistic	none
Represented Features: Non-linguistic	Color (RGB) pixel data.
Domain Model: Structural, functional, casual, geometric, heuristic, spatial, shape, color, etc.	Geometric/mathematical
Psychological Scale:	Ratio
Representing Dimension of Features:	Orientation
Technological Artifact	Mathematical techniques

The YIQ images are then used for determining the skin/non-skin region subspace as shown in Figure 7.3. The psychological scale information perceived by humans related to skin/non-skin regions is nominal and the representing dimension is color (i.e. skin/non-skin color). The task constraints shown in Table 7.2 are computational (unlike DPMA decomposition adapter definition in the previous chapter where the nature of the constraints was perceptual or non-computational in the context of DPMA). That is, in the context of the FDAA the skin/non-skin regions in the human images have to be computed by a technological artifact using an internal representation of skin/non skin color that is pixel data.

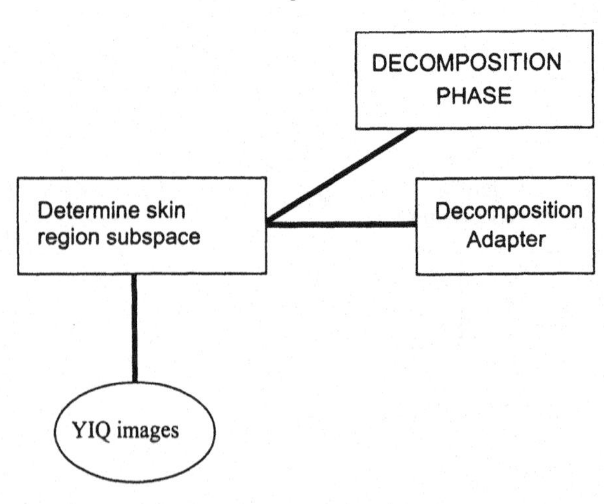

Figure 7.3: Associating Decomposition Adapter with FDAA

Table 7.2: Mapping Decomposition Adapter Signatures to FDAA

HCVM and Representation Goal, Tasks, Signatures	Corresponding in Face Detection and Annotation Activity (FDAA)
Phase: Decomposition.	Skin region decomposition
Goal: Reduce complexity.	Delineate skin regions from the YIQ image
Task: Determine abstract orthogonal concepts.	Determine skin/non-skin regions
Task Constraints:	Human images with clutter/complex background
Precondition	Transformation from RGB to YIQ representation
Postcondition	Clear skin and non-skin regions established
Represented Features: Linguistic	Skin color
Represented Features: Non-linguistic	Color (YI) pixel data
Domain Model: Structural, functional, casual, geometric, heuristic, spatial, shape, color, etc.	Color model
Psychological Scale:	Nominal (skin/non-skin color)
Representing Dimension of Features:	Color
Technological Artifact	Thresholding techniques

7.5.2. Control Phase Adapter Definitions in FDAA

The control phase definitions as shown in Figures 7.4 and 7.5 employ a preprocessing adapter and a control adapter. The mapping of the preprocessing and control adapter signatures with the tasks like cleaning skin tone regions and determining face candidate regions in FDAA is shown in Table 7.3 and 7.4 respectively. It may be noted that the task of the preprocessing adapter in Table 7.3 is noise filtering which is in contrast to the data transformation (or input conditioning) task in the preprocessing adapter shown in Table 7.1.

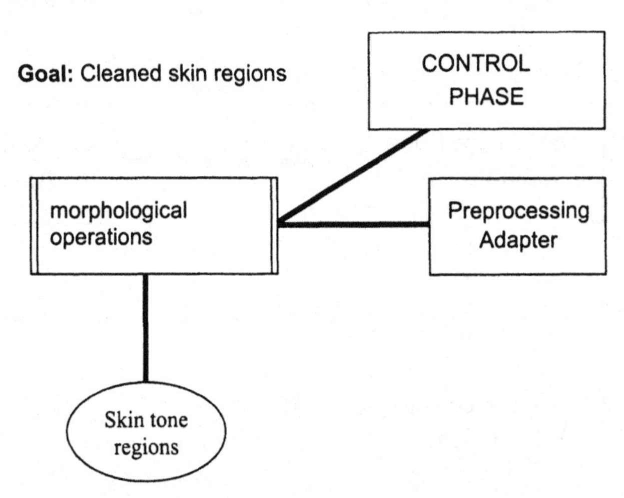

Figure 7.4: Associating Preprocessing Adapter in Control Phase with FDAA

Table 7.3: Mapping Preprocessing Adapter Signatures in Control Phase with FDAA

HCVM and Representation Goal, Tasks, Signatures	Corresponding in Face Detection and Annotation Activity (FDAA)
Phase: Preprocessing	Control phase Cleaning skin region preprocessing
Goal: Improve data quality	Improve skin tone region quality.
Task: Noise filtering	Remove background clutter/non-face objects
Task Constraints: domain dependent	
Precondition:	Blurred skin region image
Postcondition:	Cleaned skin tone region image
Represented Features: Qualitative or linguistic	
Represented Features: Non-linguistic	YI pixel data.
Domain Model: Structural, functional, casual, geometric, heuristic, spatial, shape, color, etc.	Mathematical/thresholding
Psychological Scale:	Nominal
Representing Dimension of Features:	Color
Technological Artifact	Thresholding techniques

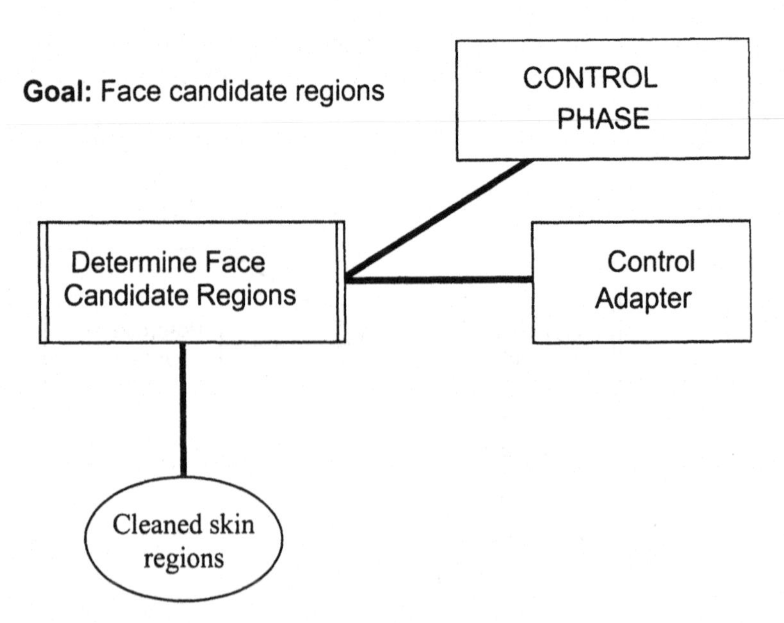

Figure 7.5: Associating Control Adapter with FDAA

Table 7.4: Mapping Control Adapter Signatures in Control Phase with FDAA

HCVM and Representation Goal, Tasks, Signatures	Corresponding in Face Detection and Annotation Activity (FDAA)
Phase: Control.	Face Candidate Region
Goal: Establish domain control constructs	Control Constructs for determining face candidate regions
Task: Decision level concepts	Face candidate regions
Task Constraints	Faces with different sizes, shape and orientation
Precondition	Cleaned skin regions
Postcondition	Face candidate regions
Represented Features: Linguistic	Small or large oval face shape
Represented Features: Non-linguistic	Pixel data of cleaned skin regions
Psychological Scale:	Ordinal, interval
Representing Dimension of Features:	Shape, size, location, orientation
Technological Artifacts:	Area and shape based filtering techniques

7.5.3. Decision Phase Adapter Definition in FDAA

The decision phase as shown in Figure 7.6 is concerned with extraction of facial features from the face candidate regions and uses these feature classify faces in the face candidates. It also annotates the faces in the original image. The decision phase adapter definition for these tasks is shown in Table 7.5.

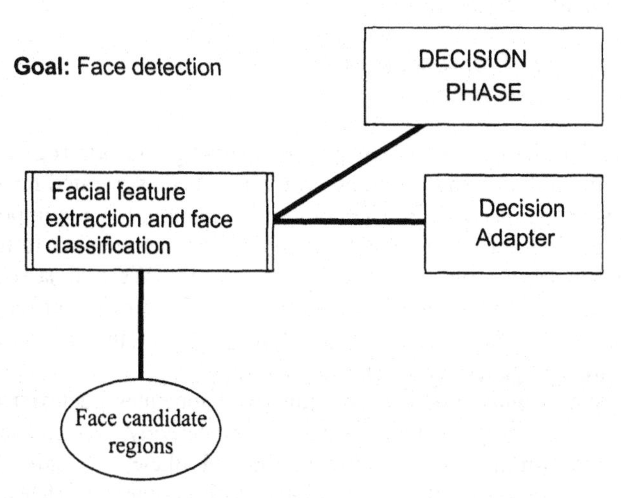

Figure 7.6: Associating Decision Adapter in Decision Phase with FDAA

Table 7.5: Mapping Decision Adapter Signatures in Decision Phase with FDAA

HCVM and Representation Goal, Tasks, Signatures	Face Detection and Annotation Activity (FDAA)
Phase: Decision.	Face Detection and Annotation
Goal: Provide user defined outcomes	Detection and annotation faces in an image.
Task: Determine decision instance	Classify face using facial features in a face candidate region
Task Constraint	Area, shape and color of facial features in a face candidate region
Precondition	Face candidate region
Postcondition	Detected and annotated faces
Represented Features: Qualitative or linguistic	Large, small and medium size of eyes, mouth, etc.
Represented Features: Non-linguistic	Pixel data related to different features/organs.
Domain Model:	Spatial, area models of facial features
Psychological Scale:	Ratio scale for size.
Representing Dimension of Features:	Size
Technological Artifacts	Soft computing artifacts (e.g. Area based thresholding techniques)

7.6. Face Detection Agents and their Implementation

This section outlines the agent definitions of the color image preprocessing, skin region decomposition, face candidate region control, and face detection and annotation agent respectively. We specifically focus on the computations associated with various technological artifacts employed. We also show the results of implementation of different agents.

7.6.1. Color Image Preprocessing and Skin Region Decomposition Agent

The agent definitions of color image preprocessing and skin region decomposition agent shown in Table 7.5 and 7.6 respectively have been separated for simplicity.

In the YIQ model used for conversion/transformation, the Y component represents the intensity while I and Q are related to colors. The conversion formula from RGB to YIQ is taken from Gonzalez and Woods (1992). We manually extracted skin-regions from a large collection of images with people of different racial background. By mapping the Y and I values of skin-tone pixels to the Y-I plane, a skin-tone distribution graph is generated as Figure 7.7.

In the graph, a white point at position (y, i) indicates occurrences of skin-tone pixels with Y and I component values y and i, respectively. We can see that skin-tone pixels distribute within a half ellipse in the Y-I plane. A half ellipse model is constructed to approximate the distribution. With another set of testing images, we adjusted the parameters of the half ellipse, including the centroid location, orientation, major axis and minor axis, to obtain optimal performance in skin-tone region segmentation. Given a pixel of a color image, if its Y and I values fall into the bounded half ellipse, it is considered to be a skin pixel and marked as white in the output binary image. Otherwise the corresponding output position will be black.

In the resulted binary image, single white pixels surrounded by black pixels are often caused by noise and so are single black pixel surrounded by white pixels. We performed an averaging operation followed by a thresholding operation to remove them. Figure 7.8 and 7.9 illustrate the process of color image preprocessing and skin-tone region extraction.

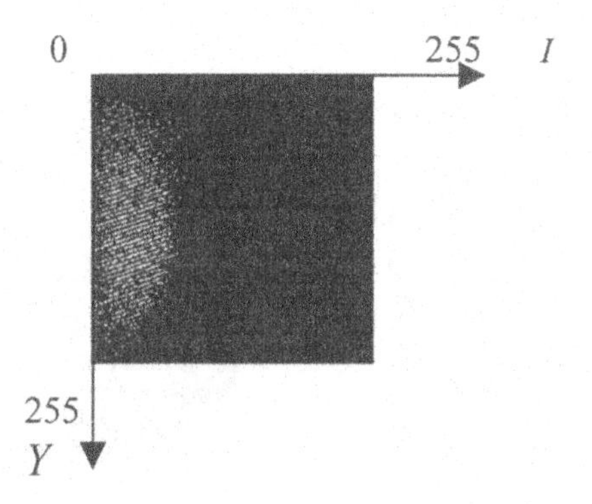

Figure 7.7: Distribution Graph of Skin-tone pixels in Y-I plane

Table 7.6 Color Image Preprocessing Agent Definition

Name	Color Image Preprocessing
Adapter Agent:	HCVM Preprocessing Agent
Goals:	YI representation for every image. This improve data quality for skin/non-skin region segmentation
Tasks:	YIQ representation for the RGB images.
Task constraints:	Color RGB images
Precondition:	People Image
Postcondition:	YI images
Communicates with:	Decomposition agent
Communication Constructs:	Inform - YI image to the Decomposition agent
Representing dimension:	Color (RGB images)
Linguistic/non-linguistic features	Color (RGB) pixel data. Each image is a matrix in which every element is a pixel having three components - one for red, one for green and one for blue.
External and Internal Tools:	Visual C++ environment, images. Algorithms for image representations
Actions:	Apply conversion formulas for Y & I components.

(a) Original Image

(b) I component of YIQ space

(c) Skin-tone pixels

(d) Post processed skin-tone regions

Figure 7.8: Color Image Preprocessing

Table 7.7: Skin Tone Region Decomposition Agent

Name	Skin Tone Region Decomposition
Adapter Agent:	HCVM Decomposition Agent
Goals:	Delineation between skin and non-skin regions.
Tasks:	Determine skin region subspace - a half ellipse model is constructed to approximate the distribution in the Y-I plane. Given a pixel of a color image, if its Y and I values fall into the bounded subspace, it is considered to be a skin pixel and marked as white in the output binary image. Otherwise the corresponding output position will be black.
Task Constraints:	Human images with complex background
Precondition:	Transformation from RGB to YIQ representation.
Postcondition:	Skin and non-skin regions (or skin region subspace) established
Communications with:	Color Image preprocessing agent, Face candidate control agent
Communication Constructs:	Inform - skin region subspace to skin control agent
Perceptual features:	color (skin/non-skin color
Linguistic/non-linguistic features:	color (YI) pixel data. Each image is a matrix in which every element is a pixel corresponding to some values for Y and I.
External and Internal Tools:	Visual C++ environment, images. Algorithms for detecting skin and non-skin regions
Actions:	Apply half ellipse model using centroid, location, orientation to extract skin-tone pixels

(a) Original Image

(b) Skin-tone regions

Figure 7.9: Skin and non-skin tone regions

7.6.2. Face Candidate Region Control Agent

As mentioned above, we make hypotheses of face locations with skin-tone regions. This is a very coarse estimation and should be refined by further analysis of each region based on shape and size. The agent definition of the face candidate region is shown in Table 7.8. The two main tasks of the control agent are cleaning the skin regions and determining the face candidate regions. The actions related to these tasks and the implementation results are described in the rest of this section

7.6.2.1. Morphological Operations for Cleaning Skin Tone Regions

Face detection and annotation is to be performed on images with clutter, which may contain non-face objects and background with color similar to that of human skin. This often results in many small isolated regions or narrow belts of several pixels in width in the segmented skin-tone regions. The existence of those regions introduces unnecessary and expensive computation in region shape and area analysis. Moreover, those regions may adversely affect the detection rate. Sometimes the narrow belts may be connected to a face region, or even worse, they may link a face region to another skin-tone regions. As will be described later in section 7.6.2.2, further refinement of face locations is based on a connected component region analysis, which may reject the regions in the above two cases. Therefore, the ability to eliminate small blobs and narrow belts and detach weakly linked regions from each other in the first place is desirable.

A set of morphological operations are applied to perform skin-tone cleaning. The basic morphological operations include *erosion* and *dilation* and other more sophisticated ones are essentially combinations of these two operations. In this stage the *opening* and *closing* operations are used. Opening is erosion of the regions followed by dilation and closing is performed in the reverse order. As stated in Gonzalez and Woods, opening can break narrow isthmuses and remove thin protrusions while closing can merge regions separated by thin gulfs and fill small holes. Closing operation is first applied to connect narrow gaps between skin-tone regions, followed by an opening operation to remove isolated and tiny and detach regions connected by thin strips.

To achieve good performance, the size of the structuring element should be delicately selected when applying the morphological operations. In this automatic process, it should be adaptive to the input image properties. During implementation we made it proportional to the image size while constrained by an upper and a lower limit. Figures 7.9 (b) and 7.10 (a) illustrate the effect of skin-tone region cleaning, in which 7.10 (a) is a cleaned version of 7.9 (b). Besides the elimination of small blobs, the boundaries of the regions are smoothed as well, making them more amenable. Note in this example, the face region is still connected to the hand region. Although for this image we could increase the size of the structuring element, the cleaning stage was intended to gain an overall optimal performance for all images in database rather than for a specific image. These regions are subject to a partition process to be discussed in the next section.
The removed regions are usually too small to be of interest. The number of regions is reduced after the cleaning process and the skin regions are refined in this sense.

Table 7.8: Face Candidate Region Control Agent

Name	Face Candidate Region Control
Adapter Agent:	HCVM Control Agent, HCVM Preprocessing Agent
Goal:	Determine the face candidate regions
Tasks:	Cleaning skin tone regions, determining face candidate regions, improving detection rate
Task constraints: :	blurred skin region images
Precondition:	skin/non-skin regions
Postcondition:	face candidate regions
Communications with:	Decision Face Detection agent
Communication constructs:	Detected faces and coordinates of eyes/mouth to Decision agent
Perceptual features:	shape, size, location and orientation.
Linguistic/non-linguistic features:	area and shape filters applied on the matrix of pixels.
Tools:	Visual C++ environment, images. Algorithms to determine the true face candidates (such as Hotteling transform) and to extract facial features
Actions:	- Apply morphological operations (erosion, dilation, closing and combination) to perform the skin-tone cleaning. - Apply area-based filter - Apply shape-based filter

7.6.2.2. Filtering Cleaned Regions by Size and Shape

The cleaned skin tone regions are based on the assumption that each skin-tone region is a face, which is very inaccurate. Observing that frontal view faces usually have an oval profile, we can check the shape of each region and if the deviation from an ellipse is below a threshold, the region is accepted as a face candidate region. Further, since in most cases we are only interested in relatively large faces, very small regions can also be rejected. Thus in this stage we refine cleaned skin tone regions by region sizes and shapes.

The face operator performs a connected component analysis, as described in Gonzalez and Woods (1992), to the binary image so that isolated regions are assigned a distinct label. Two filters are applied to each region in series to find face candidates, namely size filter and shape filter.

Filter by Region Size

Due to the simplicity of ascertaining the size of each region, the first filter is based on area. If the area of a region is below a threshold, it is eliminated. Figure 7.10 (d) is the result after the area filter is applied to Figure 7.10 (c), where one small region is removed. The threshold value is adaptive to the average area of all regions.

Filter by Region Shape

The objective of the shape filter is to find ellipse-like regions, which may correspond to face regions. There are various approaches to detecting ellipses in an image, among which the use of Hough Transform is the most famous and accurate. However this technique is not widely used due to its complexity. Saber (1998) used the Hausdorff distance to measure the similarity between a region and a set of ellipses. Sirohey (1993) attempted to find ellipses based on edge, wherein removal of intersection points and linking of edge segments is required. Linked edges are grouped to form possible ellipses. Although they reduced the work compared with Hough Transform, they are still computation intensive processes.

In our system, a very simple but still effective method is used to identify oval-shaped regions. Hotelling (KL) transform is applied to the regions so that they are aligned to their principal axis directions and the length of major and minor axes of the regions is calculated. Since the ratio between these two values reflects the elongation of the region, we can reject the region if it is above a threshold. Otherwise the area of the

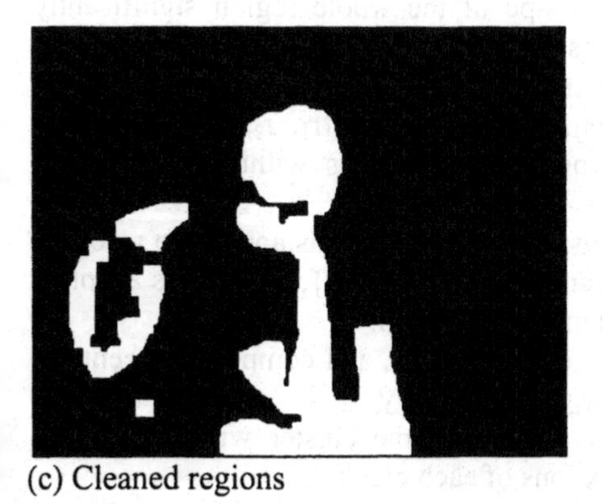

(c) Cleaned regions

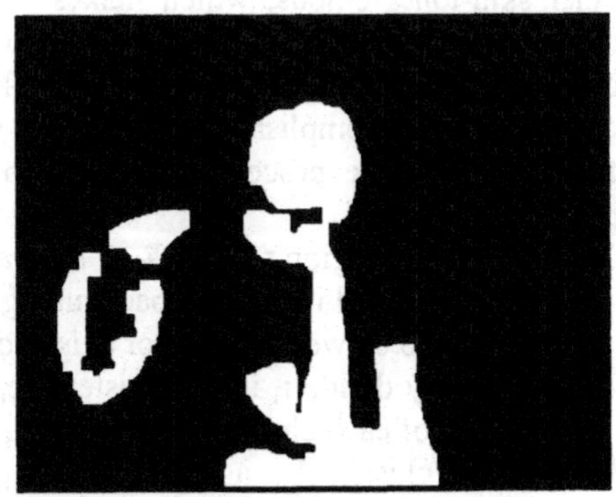

(d) Regions filtered by area

Figure 7.10: Filtering by Area of Cleaned Skin Regions

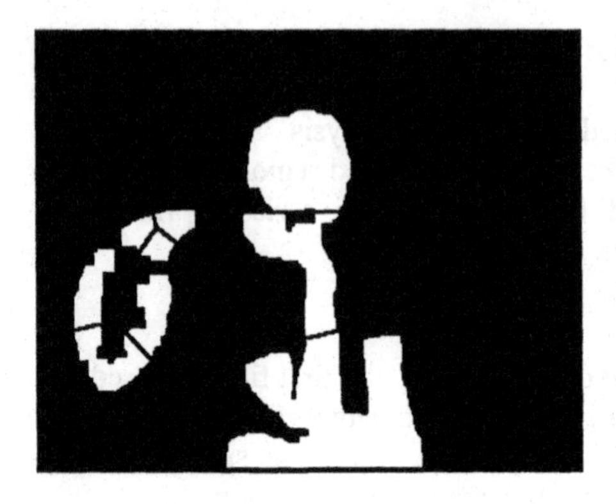

(e) Partition of rejected regions

(f) Face candidate region

Figure 7.11 Extraction of Face Candidate Region

ellipse with the same major and minor axes is computed. This area is compared with the actual area of the region and the ratio is considered the measurement of the similarity between them. Although this measurement is coarser than the above two, it works well in this situation because the regions have been smoothed by the foregoing morphological operations. If the similarity value is higher than a threshold, the region is accepted as a face candidate

7.6.2.3. Improving the Detection Rate

Thus by shape filtering, we classify all regions into two classes, namely accepted and rejected. Now what shall we do with the rejected regions? If they are simply thrown away, as is done in most other related works, problems will arise. This is illustrated in our example in Figure 7.10 (d). The shape filter rejects the face region because it is connected to the arm. This results in a low detection rate.

The problem originates from the fact that face regions are sometimes connected to other skin-tone regions, which makes the shape of the whole region significantly deviate from ellipse. If the rejected regions can be divided into smaller units, face candidate regions might be found among them. A unique region partition algorithm is developed to accomplish this. For the rejected regions $\boldsymbol{R}=\{r_1, r_2, \ldots\ldots, r_m\}$, the following iterative process is applied to pick up subregions within it that could possibly be faces.

1) Choose any region r_i from $\boldsymbol{R}$. A k-means clustering process is applied to r_i so that it is partitioned into n compact subregions $\{r_{i1}, r_{i2}, \ldots\ldots, r_{in}\}$, where n is adaptive to the ratio between the major axis and minor axis of Ri.

a) Randomly divide r_i into n clusters $\{c_{i1}, c_{i2}, \ldots\ldots, c_{in}\}$, and compute the centroid location of each cluster $\{(x_{i1}, y_{i1}), (x_{i2}, y_{i2}), \ldots\ldots, (x_{in}, y_{in})\}$.

b) Update clusters by assigning each pixel of r_i to the cluster with the closest centroid. Re-calculate the centroid locations of each cluster.

c) Compute d_i, the sum of displacement of new centroids from the old ones. Repeat from step a) until d_i converges to zero.

d) Let $r_{i1}=c_{i1}$, $r_{i2}=c_{i2}$, ……, $r_{in}=c_{in}$. Thus r_i is divided into a set of subregions $\{r_{i1}, r_{i2}, \ldots\ldots, r_{in}\}$, denoted as $\boldsymbol{R}_i$.

2) Apply the size filter described above to subregions in R_i to remove small patches. Let the survived subregion set be Ri'.
3) Apply shape filter to R_i', wherein elliptical regions are accepted as face candidate regions while the remaining are denoted as Ri''. Update the rejected region set: $R=(R \cup R_i'') - r_i$.
4) If R is not empty, repeat from step 1), otherwise stop.

The algorithm searches for face candidate regions by dividing rejected regions and this process is iterated until all partitioned subregions are too small to be of any interest and no new face candidate regions can be found. Figure 7.11 (e) shows the result of this process applied to regions rejected by shape filter in Figure 7.10 (d). It may be noted that all regions in Figure 7.10 (d) are rejected by shape filtering. Figure 7.11(f) is the result of face candidate region location, wherein only one region remains.

7.6.3. Face Detection and Annotation Decision Agent

The agent definition of Face Candidate Region Control Agent is shown in Table 7.9. It is used to extract facial features from a face candidate region and annotate the original image. The existence of face features like eyes, eyebrows, mouths and so on are evidences that the candidate is indeed a face.

Areas corresponding to the facial features are usually darker than the other parts of the face. The studies of the face patterns have shown that the portion of area that those regions occupy on the face is almost constant, varying very little (around 18 percent). These two observations lead to a histogram-based thresholding to extract possible facial features. The histogram of the Y component of each region in the original image is computed and the threshold value is chosen such that about 18 percent of pixels have intensity below it.

Table 7.9: Face Detection and Annotation Decision Agent

Name	Face Candidate and Annotation
Adapter Agent:	HCVM Decision Agent
Goal:	Detection and annotation of faces in an image
Tasks:	Classify face using facial features in a face candidate region; Annotate faces in an image
Task constraints: :	Area, shape and color of facial features
Precondition:	Face candidate region
Postcondition:	Detected and annotated faces
Communications with:	Face Candidate Region Control Agent Face Candidate Postprocessing Agent
Communication constructs:	Inform: Face Candidate Region Control Agent Validate: Face Candidate Postprocessing Agent
Linguistic/non-linguistic features:	Large, small size of eyes, pixel data of facial features
External and Internal Tools:	Soft computing artifacts
Actions:	Histogram based thresholding operation

By applying the threshold to the region in Y component, part of the region will become dark while the rest remains white. This process is shown in Figure 7.12 (g). The darkened parts may correspond to some facial organs. Since human faces share similar spatial organization, the relative position of the dark parts within the face candidate region as well as their shapes can be considered to classify them as eyes, eyebrows or other facial features. For example, a horizontal dark stripe above the centroid of the region might be an eye while a similar stripe under the centroid is considered as a mouth. Then the spatial relationships between facial features are examined to form feature groups, e.g., two eyes with similar sizes and vertical coordinate form an eye-pair. With feature groups, the face detector tells whether the region is truly a face or not. Because eyes are the most salient features on faces, the decision is chiefly based on the detected eye pairs and other features are used to confirm or negate the judgement if only one eye can be detected. The white box in Figure 7.12 (h) indicates the detection of a face.

(g) Facial features

(h) Detection result

Figure 7.12: Facial Features and Face Detection

Once faces are detected, the annotation work is relatively simple. The number of faces is counted, with which images are classified as *people* or *non-people*. People pictures are annotated by the number of faces. For further classification, the distance between faces are computed as well as the density of faces within the image so that the people group can be depicted as crowded or sparse.

(a) One person with skin-tone background

(b) Group people with complex background

(c) Group people

Figure 7.13:Some More Examples of Face Detection Results

7.7. Experimental Results and Future Work

The performance of the system has been tested using a large number of images, selected to represent varying degrees of difficulty in terms of the different factors of the problem, e.g. background objects having skin-tone, faces at different scales and variation in pose. Several results are given in Figure 7.13, in which (a) is a single-people picture with most background in skin-tone. (b) is a group-people picture in complex background. (c) was downloaded from the online demo gallery of system in Rowley et al (1998), wherein they gave very similar result (failing to detect the same face) except that there is a false positive in their system. The shading of the hat caused the missed face. The multi-agent system has been implemented on a Pentium II 266 PC and the processing time for a 256 by 256 image is less than one minute. Most images were submitted to the online demo of Rowley et al (1998) and the results obtained thus far are very encouraging in terms of precision and recall.

Like all existing systems, our face detector has false positives and negatives because of the inherent difficulties of the problem. Our future extension of this system will go in two directions. First, more powerful techniques will be incorporated to improve the performance of the face detector. We observed that most errors were introduced in the face candidate location in the control phase and facial extraction features in the decision phase. More sophisticated techniques will be investigated in the future to reduce the errors. Secondly, the multi-agent system will be extended to video. Motion information will be used to speed the searching process.

7.8. Summary

Face detection and annotation systems have applications in a number of industries. Banks can use such an embedded system for customer identification and to reduce fraud. Similarly, police can use these embedded systems for matching suspected criminals with their criminal image database. This chapter describes the application of the problem solving ontology component of HCVM in developing a multi-agent face detection and annotation system. The problem solving agents use a top-down approach to define color image preprocessing agent, skin region decomposition agent, face candidate region control agent and face detection and annotation decision agent respectively. The top-down approach is more akin to human perceptions than other approaches. The multi-agent system makes hypotheses of face locations and seeks evidences to verify them. The higher level, agent like color image preprocessing agent and skin region decomposition agent work as background knowledge for lower level agents such as face candidate region control agent and face detection and annotation decision agent in order to improve the reliability of the whole system. Further, in order to improve the detection rate, the face candidate region control agent employs a unique region-partitioning algorithm. The future applications of the multi-agent face detection and annotation system will be extended to images in motion like in video.

References

Dai, Ying et al, (1998) "Study of facial expression recognition using the Hopfield model", *DSP 90-7,* pp. 37-42 (in Japanese)

Dai, Ying et al, (1996) "Face-Texture Model Based on SGLD and its Application in Face Detection in a Color Scene", *Pattern Recognition*, Vol. 29, No. 6, p1007-1017.

Gonzalez, Rafael C., Woods, Richard E., (1992) *Digital Image Processing,* Addison-Wesley Publishing Company.

Govindaraju, V.., Sher, D. B., Srihari, R.K., and Srihari, S. N., (1989), "Locating human faces in newspaper photographs", *Proceedings of IEEE Computer Society Conference on Computer Vision and Pattern Recognition*, San Diego, CA, June.

Jeng, Shi-Hong et al, (1998) "Facial Feature Detection Using Geometrical Face Model: an Efficient Approach", *Pattern Recognition*, Vol. 31, No. 3, p273-282.

Kapfer, M. et al, (1997) "Detection of human faces in color image sequences with arbitrary motions for very low bit-rate videophone coding", *Pattern Recognition* Letters 18 , p1503-1518.

Leung, T. K. et al, (1995) "Finding Faces in Cluttered Scenes using Random Labeled Graph Matching", *Fifth Intl. Conf. On Computer Vision*, Cambridge, MA, June.

Moghaddam, Baback et al, (1998)"A Subspace Method for Maximum Likelihood Target Detection", *IEEE International Conference on Image Processing*, Washington DC, October.

Rowley, Henry A., et al, (1998),"Neural Network-Based Face Detection", *IEEE Transactions on Pattern Analysis and Machine Intelligence,* Vol. 20, No. 1, January, pp23-37

Saber, Eli and Tekalp, A. Murat, (1998) "Frontal-view Face Detection and Facial Feature Extraction using Color, Shape and Symmetry Based Cost Functions", *Pattern Recognition Letters*, V0019 N8, June, p669-680

Sirohey, Saad Ahmed,(1993) "Human Face Segmentation and Identification", *Master's Thesis*, Center for Automation Research, University of Maryland, MD, Nov.

Sung, Kah-Kay et al, (1998) "Example-Based Learning for View-Based Human Face Detection", *IEEE Transactions on Pattern Analysis and Machine Intelligence*, Vol. 20, No. 1, January

Sung, Kah-Kay, "Learning and Example Selection for Object and Pattern Detection", *Ph.D. dissertation*, Artificial Intelligence Lab., Massachusetts Institute of Technology

Yang, Guanzheng et al, (1994) "Human Face Detection in a Complex Background", *Pattern Recognition*, Vol. 27, No. 1, pp53-63.

Yow, Kin Choong, et al, (1997) "Feature-based human face detection", *Image and Vision Computing* 15 713-735

Zhang, Y. J. et al, "Automatic Face Segmentation Using Color Cues for Coding Typical Videophone Scenes", *SPIE Vol. 3024*

8 MODELING HUMAN DYNAMICS AND BREAKDOWNS – INTELLIGENT AGENTS FOR INTERNET GAMES AND RECRUITMENT

8.1. Introduction

In the last two chapters we have looked into the application of four components of the HCVM. In this chapter we wish to demonstrate two other facets of the human-centered virtual machine, which have not been specifically looked at until now. These are the role of computer-based intermediaries (for back up support in problem solving), distributed communication in human-centered multi-agent systems and using computer-based artifacts to model breakdowns in human decision making. The role of intermediaries and distributed communication is illustrated by way of an intelligent distributed multi-agent multimedia euchre application on the net. In addition, aspects of a sales recruitment and benchmarking application are described to illustrate how computer-based artifact is used to overcome breakdowns in the sales, customer service and telesales recruitment decision making. In fact the modeling of the breakdowns is the prime motivator for the sales managers and sales personnel to use the computerized system for management of recruitment and benchmarking. The sales recruitment system described in this chapter (and developed by the first author) has been used in the industry for approximately the past three years.

To facilitate illustration of the above two facets, this chapter is divided into two parts. The first part briefly outlines various components of the Net Euchre card game. It goes on to describe the role of computer-based intermediaries and distributed communication aspects of the game. In the second part we initially outline the various components of the sales recruitment activity. We undertake activity-centered analysis of the sales recruitment activity in order to demonstrate the role of the computer-based artifact as well as human breakdown situations in the recruitment decision making. Based on activity-centered analysis, we briefly describe part of design and implementation of the sales recruitment system which addresses the abovementioned problem in sales recruitment.

8.2. Net Euchre Game Application

The Net Euchre game application is described in three sections. The first section introduces the card game. The second section describes the card game structure. The third section describes the interaction between backup support game agents and human players, and distributed communication between game agents.

8.2.1. Net Euchre Card Game

Euchre is a trick-taking card game for four players. Net Euchre exists for two reasons:

a) to allow multiple players on the internet to all join in a game of euchre, and

b) to use different technologies, including Intelligent technologies, to assist the player to play the game.

The six components of Net Euchre and their scope are shown in Figure 8.1. The rules of the game of Net Euchre follow closely the rules of the game of euchre as traditionally played. Four players sit in a circle, with players seated opposite to each other partners. An initial dealer is chosen, and scorekeepers for each team are decided. For each round of play, the dealer shuffles the deck of 24 cards (Nine through Ace of each suit), deals five cards to each player, places the four remaining cards face up in front of him/her, and flips the top card of the four face-up. The suit of this up card is the proposed trump suit.

The player to the left of the dealer begins to bid. That player decides if he/she wants the up card's suit to be the trump. If the player likes the suit, he or she may call it. Otherwise, the player passes. If the opportunity to call trump gets to the dealer, and the dealer does not want the up card's trump, then the up card is flipped over, and another round of bidding begins, where each player may call any suit except for the one declined in the first round. If the bidding again makes it around to the dealer, and the dealer declines to call trump again, it is a misdeal, and the dealer passes deal to the player on his/her left. If, however, a trump is called, then the dealer picks up the up card and discards a card face down, but only if the up card's suit was called.

The ranking of the cards is slightly unusual. For non-trump suits, the highest-ranking card is the Ace, followed by the King, Queen, Jack, Ten, and Nine. For the trump suit, the highest card is the Jack of the suit (called the "right bower"). The next-highest card of the trump suit is the Jack of the other suit of the same color (the "left bower"), followed by the Ace, King, Queen, Ten and Nine (e.g. if Hearts were trump, the highest card would be the Jack of Hearts, followed by the Jack of Diamonds, Ace of Hearts, etc.). Notice that the left bower is a trump in all regards. If, for example, Hearts were trump, and Diamonds were lead, a player whose only Diamond was the Jack would not be required to play it, as it is actually of the trump suit.

The round begins. The player to the left of the dealer is the first to lead with a card. The three other players, playing in a clockwise order, must each play a card on the lead card. Each player *must* play a card that matches the suit of the lead card if possible. (If they fail to do so, they have *reneged* and the other team gets two points toward the game.) If a player does not have a card that matches the suit of the lead card, the player may play any card they desire that they have in their hand. After each player has played a card, if any trump cards have been played in the trick, then the player who played the highest-ranking trump wins the trick. Otherwise, the player who played the highest-ranking card of the same suit of the lead card wins the trick. The player who won the trick leads the next one. To win a round, a team must win three of the five tricks.

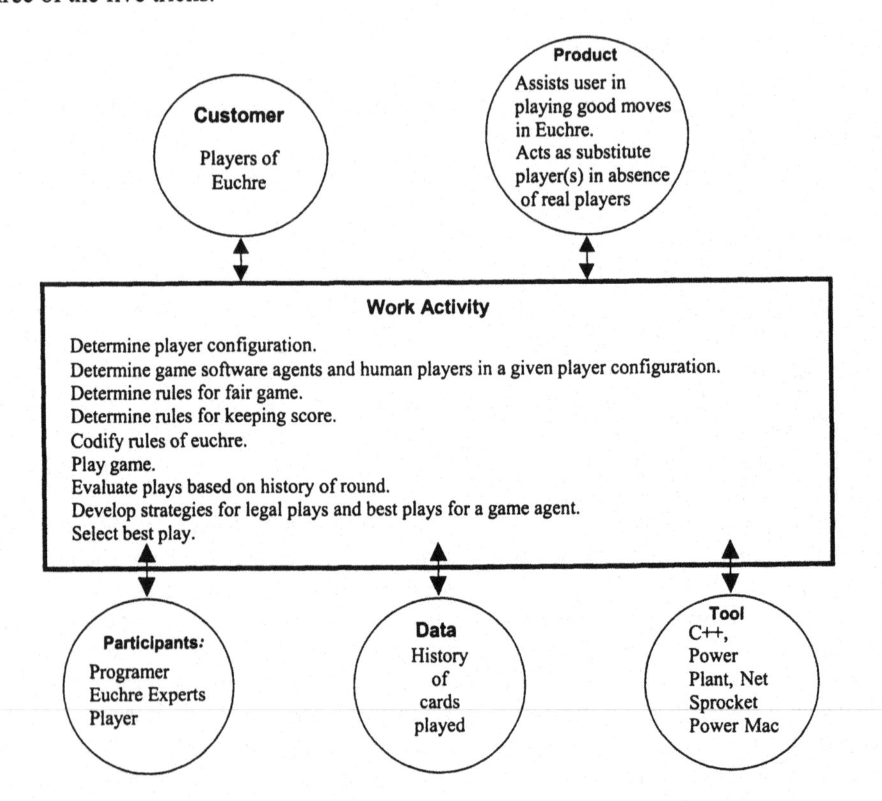

Figure 8.1: Six Components of the Net Euchre Game

Winning a round wins points for a team. If a team calls the trump, and then proceeds to take all five tricks in the round, the team earns two points. If the team called trump but won three or four tricks, they earn one point. If a team calls trump but does not win at least three tricks, they have been *euchred*, and their opponents win two points. The first team to earn a predetermined number of points wins; the goal score is usually ten points. Deal then passes to the player to the left of the dealer.

When deciding trump, a player may believe that he/she will be able to win all five tricks with the hand they were dealt. In that case, the player may call their desired trump "alone". ("Going alone" is not currently supported in Net Euchre.). If a player *goes alone*, their partner places their hand face down on the table and does not participate in the round. If the player who went alone succeeds in winning all five tricks, his/her team earns four points. If the player wins four or three, the team gets only one point, and if the player wins fewer than three, then the opposing team gets their euchre bonus of two points.

8.2.2. Distributed Net Euchre Game Agents

A Game Agent in Net Euchre is used to provide intelligent back up support to a human player if a player desires assistance. However, the Game Agent is actually capable of playing the game autonomously in the absence of inadequate human players. Figure 8.2 shows a four player Net Euchre game configuration supported by four game agents. The UI in Figure 8.2 represents the four machines used by four players on the internet. The dotted lines between different Game Agents represent distributed communication or dynamics of the game.

The Game Agent architecture for each Game Agent is based on the three problem solving adapters, namely, decomposition, control and decision. The association between some Net Euchre game tasks and decomposition, control and decision adapters is shown in Figures 8.3, 8.4 and 8.5, respectively.

The decomposition and control adapter definitions are implemented by the Game Agent which hosts the game. The decision adapter definition is implemented by all the game agents distributed across four machines. Further given the fact that different players are not present on the same machine, Net Euchre must communicate with different instances of itself to provide the players with the impression of a seamless game among four people. If there are not enough human players to play the game, additional Game Agents are created to fill the gap. All the Game Agents in a particular game communicate with each other to provide the illusion of a consistent game, despite the fact that the game is actually a cooperative effort of up to four computers.

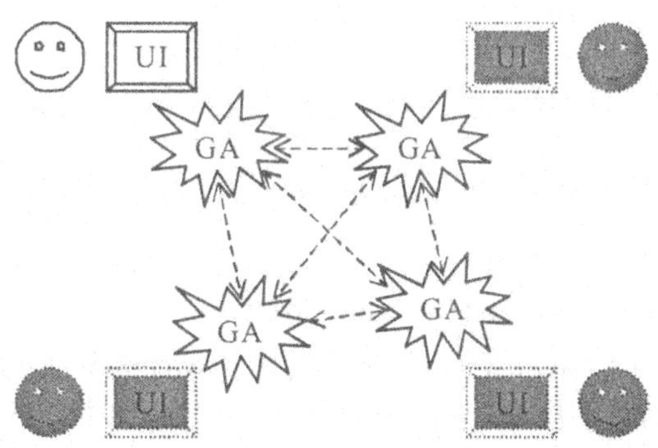

Figure 8.2:Game Agents(GA), UIs, Human Players and Distributed Communication

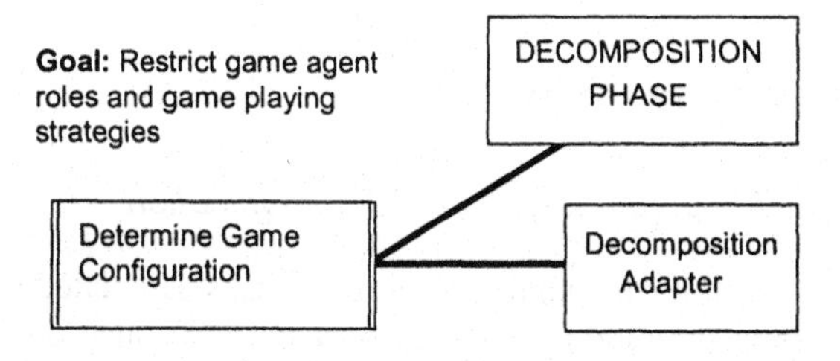

Figure 8.3: Net Euchre Tasks and Decomposition Adapter

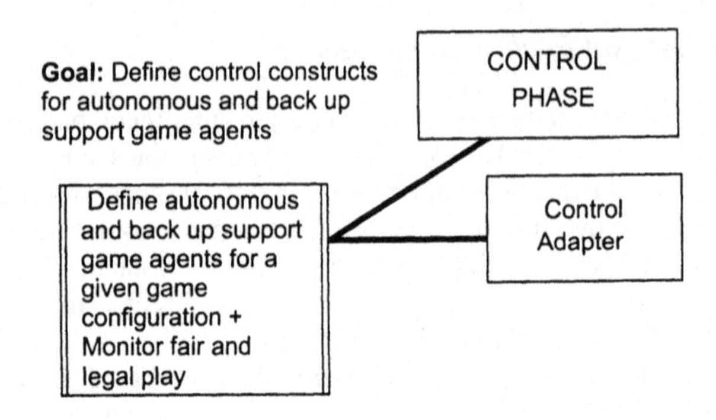

Figure 8.4: Net Euchre Tasks and Control Adapter

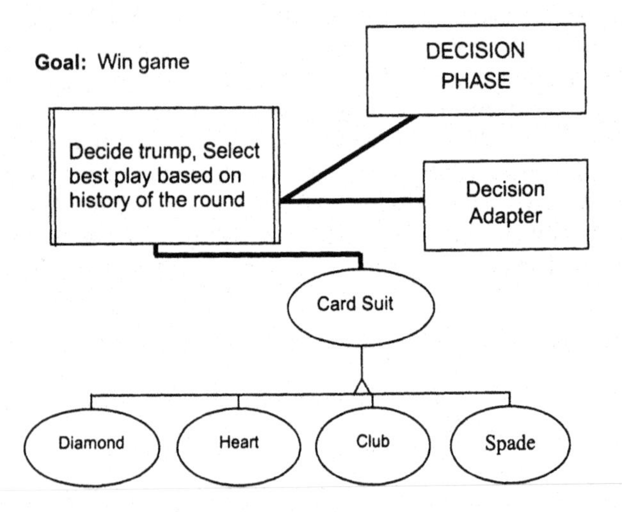

Figure 8.5: Net Euchre Tasks and Decision Adapter

The Game Agent is given a name "Pat" and a visual representation, as shown in Figure 8.6. Pat appears in the window, and just needs to be clicked by the player when he/she desires advice.

Figure 8.6: Pat – The Game Agent

8.2.3. Distributed Implementation and Communication

In this section we describe the implementation of tasks associated with decomposition and decision adapter definitions of the game agent. We also outline the interaction between different game agents and between game agents and the human players.

8.2.3.1 Game Configuration and Back up Support

The game configuration consists of two parts, namely, determining/gathering the human players and deciding the first dealer. The gathering of the players is implemented using the NetSprocket networking toolkit on Power Mac, so the user interface for that is common and well understood by players of the game. The physical architecture is host/client—one player acts as the host, and hosts the game (see Figure 8.7), while the other players get a list of hosted games and join in (as mentioned earlier, decomposition and control adapter definitions are, at the moment, implemented on the host machine).

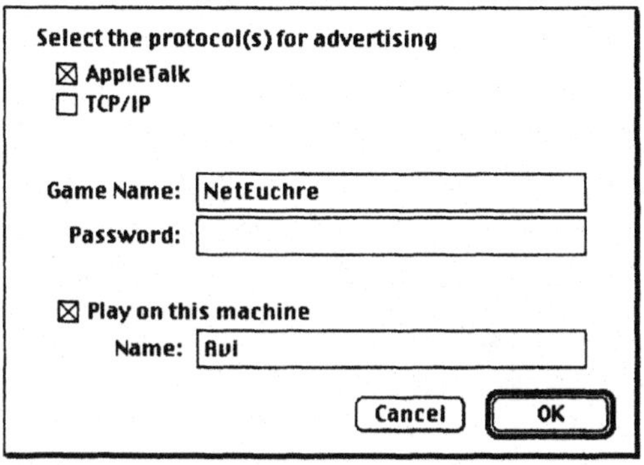

Select the protocol(s) for advertising
☒ AppleTalk
☐ TCP/IP
Game Name: NetEuchre
Password:
☒ Play on this machine
Name: Avi
Cancel OK

Figure 8.7: Hosting a Game

Then, the other players who choose to join a game get a list of games available in the zone. They select a game from the list, and join as shown in Figure 8.8.

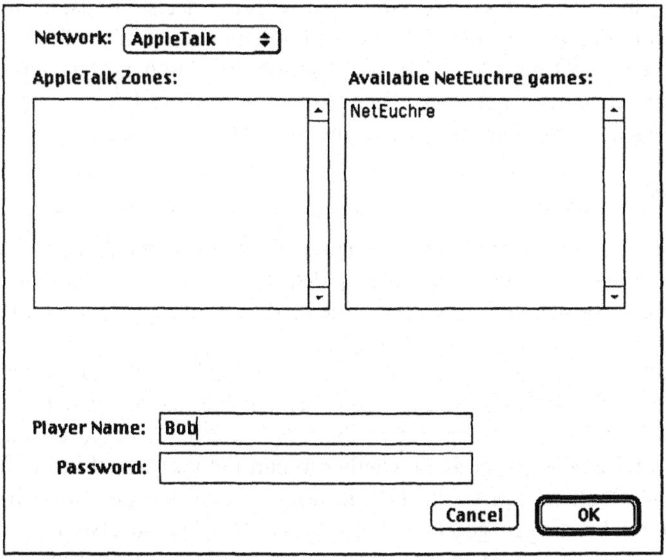

Network: AppleTalk
AppleTalk Zones:
Available NetEuchre games:
NetEuchre
Player Name: Bob
Password:
Cancel OK

Figure 8.8: Joining a Game

Once all the human players have joined in the game, one clicks the "Begin Game" button in the window as shown in Figure 8.9. The host of the game then creates enough game agents to take the place of missing human players (none in the present implementation) is needed. In this implementation all the game agents are used as back up support agents.

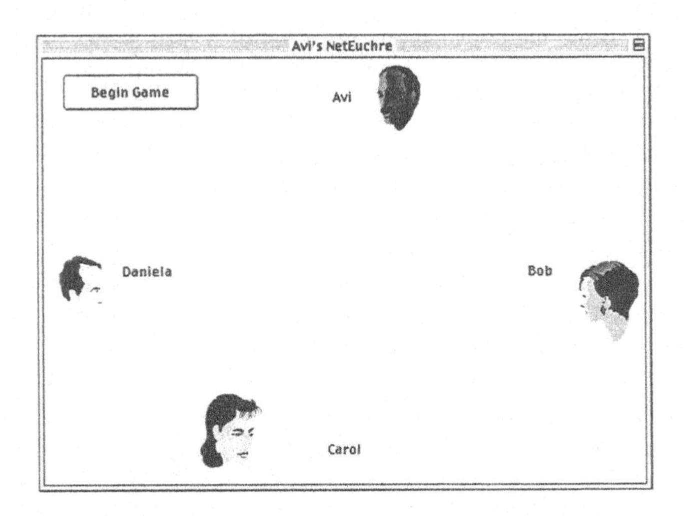

Figure 8.9: Game Configuration Before Playing

8.2.3.2 Play Game

Play game includes tasks like deciding the trump, determining best play and playing the hand. These tasks are associated with the decision adapter of the Game Agent shown in Figure 8.5. We illustrate the implementation of the decide trump task with backup support from Pat. Then we describe the play hand task and summarize the messages exchanged between Game Agents providing backup support to four human players who are distributed across the internet.

Decide Trump

Once the round is started, the dealer automatically shuffles and deals the cards. Each Game Agent, when receiving the cards, display them in the window above the player's name as shown in Figure 8.10. The cards are displayed only on the UI (User Interface) of the human player (e.g. Daniela).

Now, starting with the player to the left of the dealer and proceeding clockwise, the players have the option of calling trump. This is the first appearance of Pat (as the previous stages need no intelligence in their execution). If asked by the player for advice, Pat advises the player as to whether to call the up card (if in the first bidding round) or what suit to call (if in the second). A learning algorithm like a neural network would be a very good implementation for Pat's decision process in this phase, since there is a very clear feedback of win/lose. However, currently, the decision is implemented using a simple heuristics related to ranking of cards.

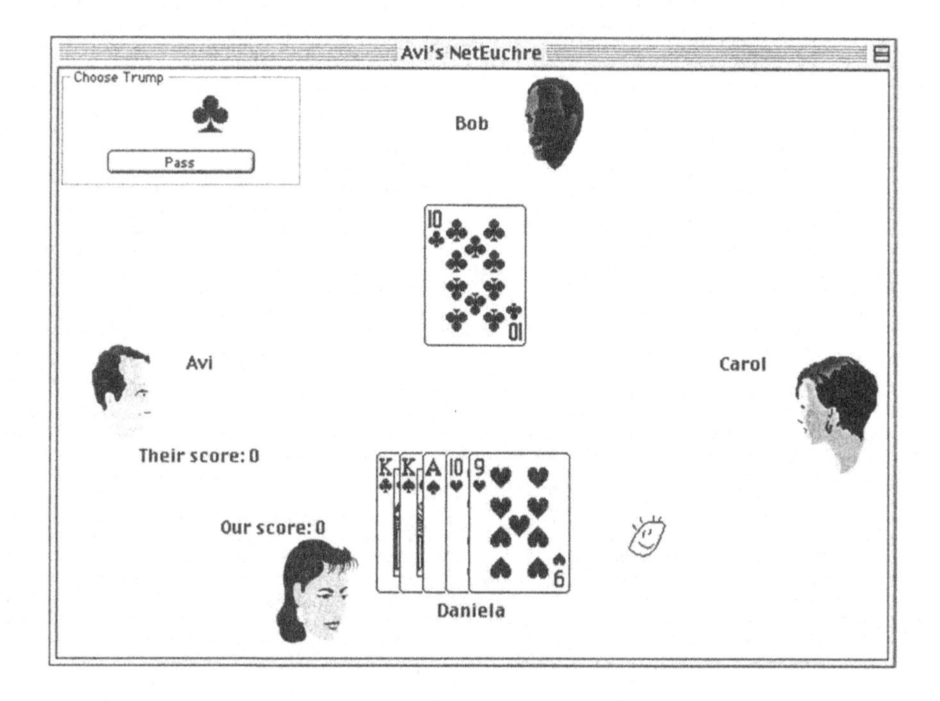

Figure 8.10: First Round of Trump Calling

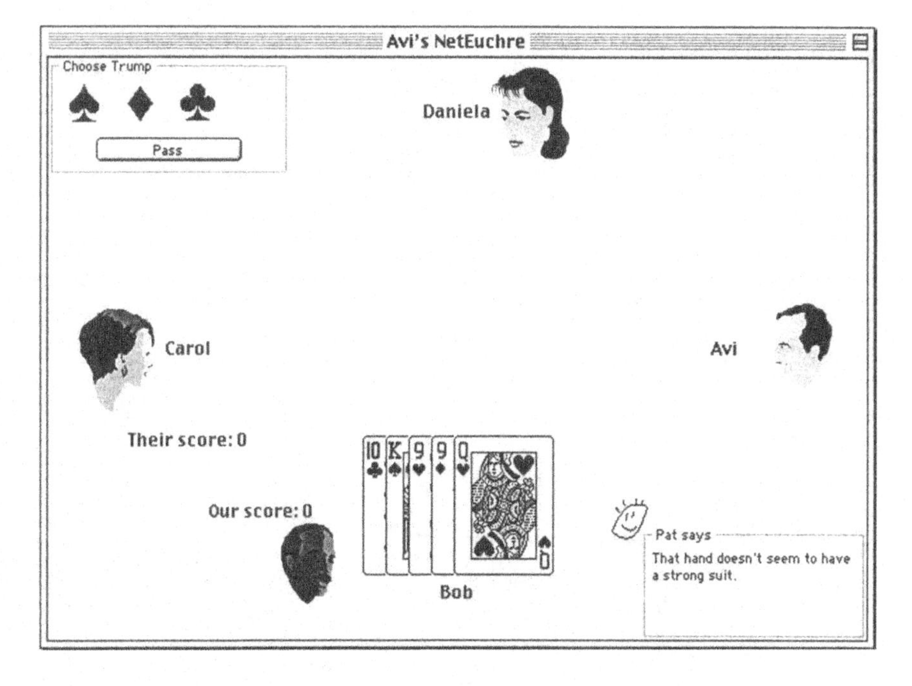

Figure 8.11: Pat's Advice to Bob for Deciding Trump

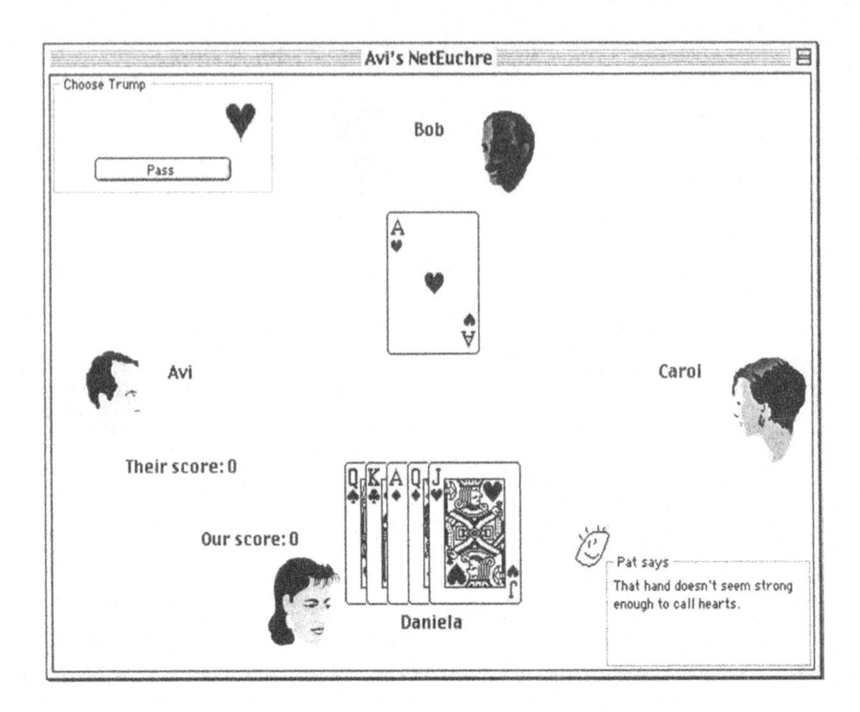

Figure 8.12: Pat's Advice to Daniela and Avi for Deciding Trumps

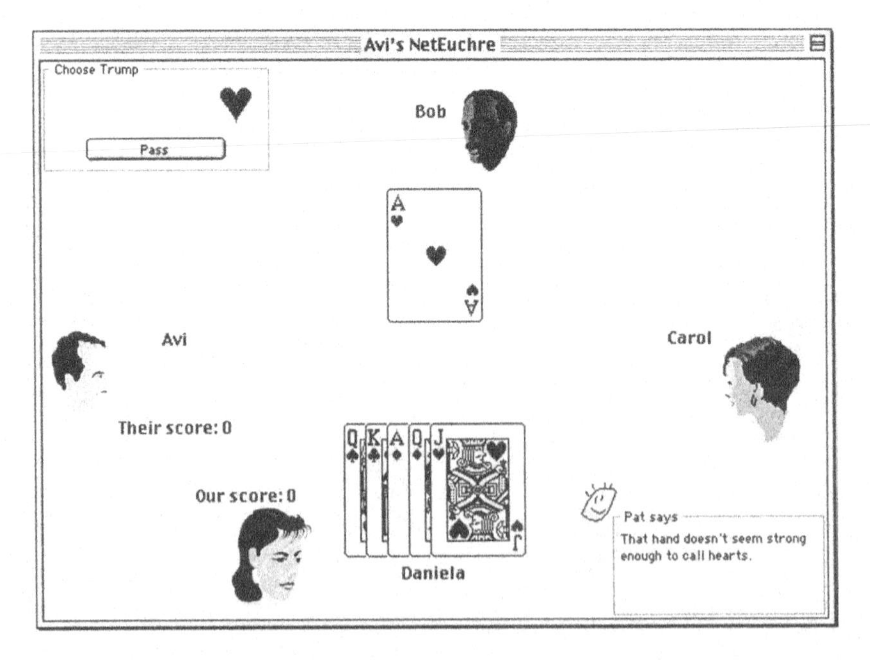

Figure 8.13: Advice from Pat to Daniela for Deciding Trumps "That hand doesn't seem strong enough to call hearts".

It may be noted in Figure 8.10 that Pat has no commentary. Unlike some "user agents" of other applications (e.g. Microsoft Office 97/98), Pat does not offer any advice unless explicitly asked. The reason that Pat behaves this way is that the "user agent" must be considered polite. If Pat offers unsolicited advice that Daniela considers obvious, then Daniela may feel Pat believes that she is an idiot. While Pat clearly has no such beliefs, the appearance that it does should be avoided. Figures 8.11, 8.12 and 8.13 show solicited advice offered by Pat to Avi, Daniela and Bob respectively.

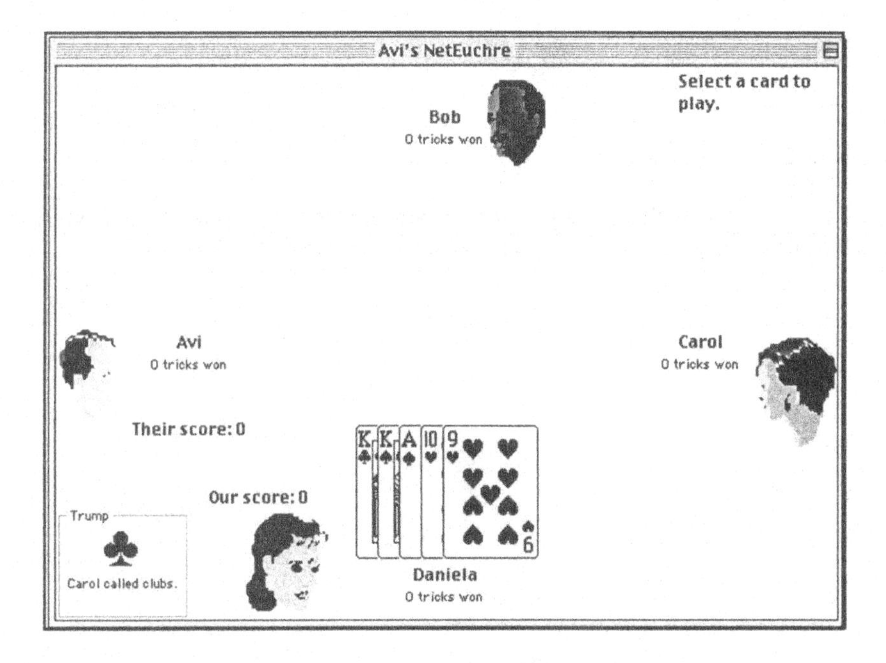

Figure 8.14 Dealer Exchange of Cards

Once a player chooses a suit, then if the up card's suit was chosen the dealer must exchange one of his/her cards with the up card (see Figure 8.14). Once that is done, or if the trump was chosen in the second round of trump calling, the Play Hand phase begins. The chosen trump is displayed in the lower-left corner of the window for the remainder of the round.

Play Hand

The player to the left of the dealer begins play by leading a card (Figures 8.15 and 8.16, respectively). Each player, in clockwise order, plays a card, following suit.

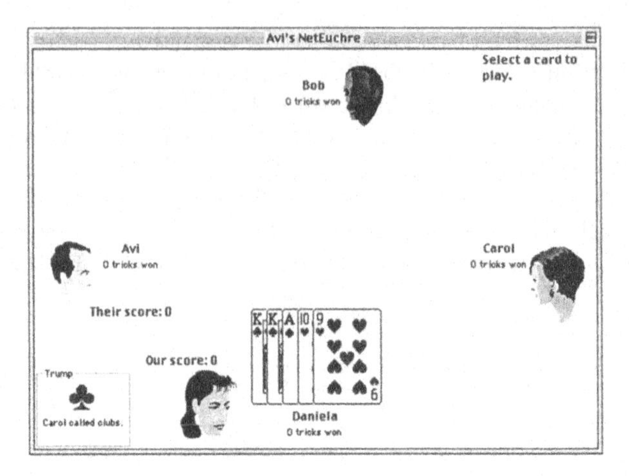

Figure 8.15: Leading a Card

After the trick completes, the player who wins it claims it, and leads the next trick. This repeats until each player has exhausted his or her hand of cards.

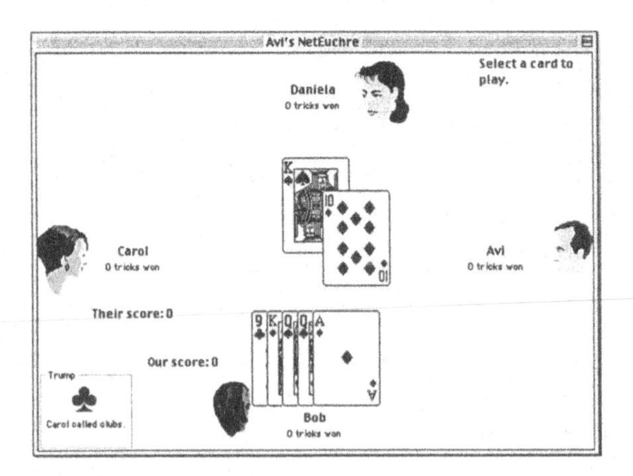

Figure 8.16: Playing a Game

Once all the cards have been played, each team counts up the tricks they took and determine whether or not they won. If they did, they get their point, the next dealer deals the cards, and the three repeating phases repeat again.

From the screenshots (Figures 8.15 and 8.16), it can be seen that Pat is not present during the playing of the game. Pat, however, will in the future be added to the game-playing phase for selecting best play strategy. A game agent could give advice or play for the player (if asked to). The most likely implementation of Pat for game playing would be heuristics. There are plenty of rules of thumb by which Pat could play, and the heuristics come with different settings of aggressiveness. Further, Game Agents do not currently take the place of human players who are absent and the card that the player plays is not checked for legality.

Distributed Communication

The different instances of the Game Agent communicate with each other through messages in a distributed manner. These messages are based on various communication constructs like broadcast, inform, suggest, request and tell. The different messages hold different pieces of information, and allow the Game Agents to keep the common game state coordinated. The communication constructs and content, source and destination agents and the parameters exchanged are shown in Table 8.1.

We now look into aspects related to the second part of this chapter, namely, using computers to model human breakdowns in the sales recruitment activity.

Table 8.1: Distributed Communication between Game Agents

Communication Construct: Content	Source agent	Destination agent	Parameters
Broadcast: I Roll A	every	all	random number
Sent by every game agent at the beginning of a game. The highest roller will assume the role of dealer.			
Broadcast: Begin Deal	dealer	all	none
Sent by the DealerGame Agent to indicate the start of the dealing.			
Inform: You Are Dealt	dealer	player	card
Sent by the dealer to each agent dealt a card.			
Inform: Up Card Is	dealer	all	card
Sent by the dealer, revealing the up card.			
Suggest:You May Call Up Card	player	player	none
Sent by a player to another to pass calling rights.			
Broadcast: I Call Up Card	player	all	none
Sent by a player if the up card is called.			
Suggest: You May Call Trump	player	player	none
Sent by a player to another to pass the right to call trump.			
Broadcast: I Call Trump	player	all	suit
Sent by a player to call a suit as trump.			
Broadcast: Misdeal	dealer	all	none
Sent by the dealer in case of a misdeal. (The only misdeal that should be possible is no trump call.)			
Tell: You Deal	old dealer	new dealer	none
Sent from the dealer of a finished round to the dealer of the next round.			
Request: You May Lead	dealer	player	none
Sent by a dealer to the player on its left to tell them they may start playing.			
Broadcast: I Lead	player	all	card
Sent by a player to lead a card.			
Request: You Go	player	player	none
Sent by a player to the next player to tell them to play a card.			
Broadcast: I Play	player	all	card
Sent by a player to play a card.			
Broadcast: I Take This Trick	player	all	none
Sent by the player who takes the trick.			
Broadcast: We Win	team scorekeeper	all	number of points won
Sent by the scorekeeper of a team when a round of play is won.			
Broadcast: We Won Game Thank You For Coming	team scorekeeper	all	none
Sent by the team scorekeeper once the team reaches ten points.			

The order in which these messages are sent is fairly self-explanatory.

8.3. Sales Recruitment

Human resource management function is one organizational area where information technology and computerization have not had much impact. Conventional and traditional methods and techniques are still preferred to automated or computerized techniques in a number of work activities (e.g. recruitment, etc.) in this important organizational function. One of reasons for the resistance to computerized techniques is the fact that work activities such as recruitment involve analysis of human behavior which is considered too complex for any computerized system to model. This resistance also highlights a fact that if computer-based artifacts are to find a place in an area like human resource management they must provide enough motivation to the stakeholders in a human resource work activity for their use. We look at human breakdowns in work activities like recruitment as a means of motivating the stakeholders to use computer-based artifacts. That is, we use computer-based artifacts for modeling the human breakdowns in these work activities.

In the rest of this chapter, we describe how human breakdowns in sales recruitment have been modeled using computer-based artifact (Sales Recruitment System or SRS). SRS has been used commercially for the last 3 years, primarily because of its ability to model these breakdowns.

In the next section, we introduce the sales management function and highlight some of the problems with the existing recruitment procedures. We then outline the performance and context analysis of the recruitment activity to highlight the goals and tasks of SRS. Finally, we show how the human breakdowns have been modeled and implemented in SRS.

8.3.1. Managing Salespersons

Sales management among other responsibilities includes forecasting demand (sales), managing salespersons, and establishing sales quotas. Managing salespersons involves such activities as recruiting of salespersons, supporting the salespersons in their work, meeting with customers, establishing territories and evaluating performance. Recruiting the right type of salesperson to match the organizational needs and ingraining them with proper selling attitudes has a critical impact on the performance of the sales force, sales manager, and the organization as a whole.

Most organizations rely on interviews as the main strategy for recruiting salespersons. Product knowledge, verbal skills, hard work, self-discipline, and personality are generally assumed to be well taken care of in the interview process. However, it is difficult to objectively determine a candidate's selling behavior during an interview.

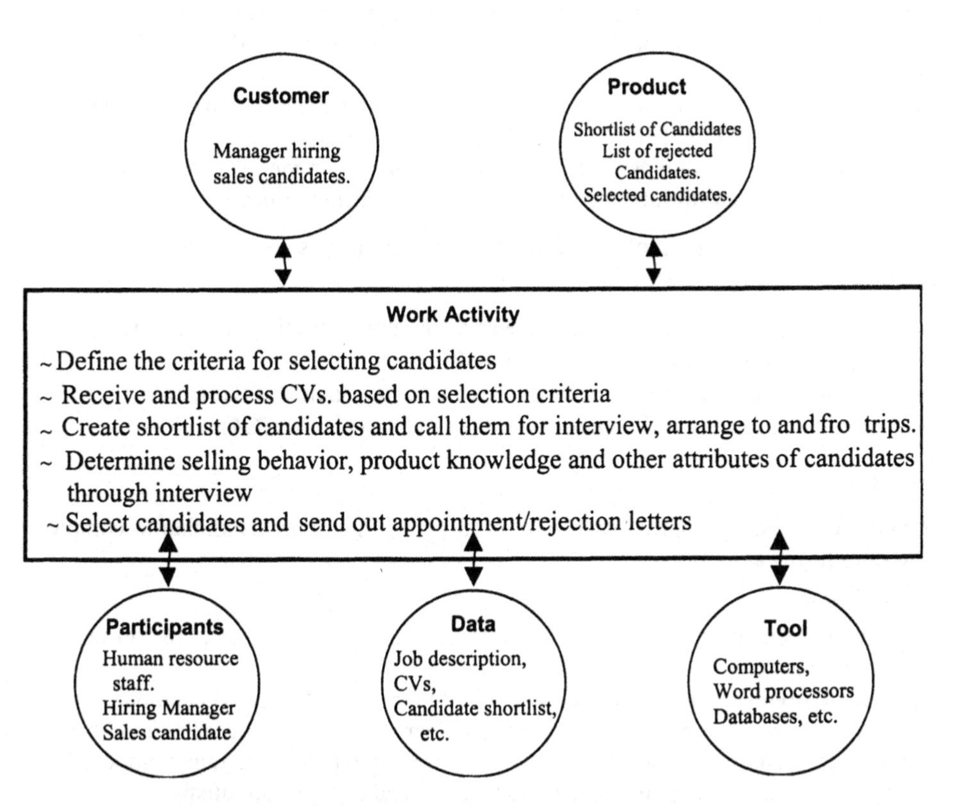

Figure 8.17: Scope of the Six Components of the Existing Sales Recruitment Activity

As a result, the existing recruiting procedures though useful have met with limited success. The high salesperson turnover and stress levels on sales managers while on the job are good indicators of the limited success of these procedures. An alternative system is sought to enhance the interview process and to address the existing problems, whilst improving the overall effectiveness of the recruitment activity. In order to determine the role, goals and tasks of the alternative computer-based system, performance and context analysis of the existing sales recruitment activity is undertaken.

8.3.2. Performance Analysis of Sales Recruitment Activity

The purpose of the performance analysis is to identify the role and goals of the computer-based artifact in the sales recruitment activity. In this section, we briefly outline the performance analysis of the relevant parameters related to the six components of the existing sales recruitment activity shown in Figure 8.17.

Product

Cost: Direct and indirect costs in terms of senior management time, to and fro expenses of candidates, etc. run into hundred of thousands of dollars.

Quality: approximately 30% of selected sales candidates leave the organization within 2 years of employment. There are no objective benchmarks for selecting the best candidate.
Reliability: Decision to shortlist candidates for interview is entirely based on information provided in their CVs, which can be unreliable
Conformance to Standards: Because of heavy reliance on the interview process it is difficult to ensure uniformity and objectivity in the selection process

The performance analysis of the product component in this section identifies a need for computer-based artifact with an aim to

- reduce direct and indirect cost of the sales recruitment activity (Goal 1, G1),
- improve recruitment decision quality (G2),
- improve reliability of information related to sales candidates (G3), and
- provide uniformity and objectivity in the selection process (G4).

Work Activity

Cycle time: It takes approximately 10 to 15 person hours per candidate to complete the sales hiring activity.

Participants

Skills: Generally hiring managers have 15 to 20 years of experience in the profession, which is considered to be satisfactory
Degree of Involvement: Given the high (30%) turnover of selected candidates in the sales profession, the time hiring managers can commit to the selection process diminishes. More so, the stress resulting from the time constraints imposed on managers in make hiring decisions leads to a reduced level of job satisfaction.

Data

Quality: The methods used to determine the accuracy of the candidate data based on their CV are inadequate. The hiring decision in the sales activity involves, among other aspects, data related to the selling behavior of a candidate. This data is not readily available in the existing sales recruitment activity.
Accessibility & Presentation: The only accessible data is the job criteria and appropriateness of a candidate to suit that criteria. Other important data is not accessible in the exiting system, such as selling behavior data.

Tool

Functional Capabilities: The functional capabilities of the computer tools are limited to writing selection and rejection letters and storing candidate information in a database.

Some of the additional goals for the computer-based artifact based on the performance analysis of the work activity, participant, data and the tool are:

- reduce cycle time (G5),
- improve data quality and accessibility (G6). This will also assist in reducing the stress levels of the hiring manager as well as reducing salesperson turnover.

8.3.3. Context Analysis of the Sales Recruitment Activity

In order to understand the context in which the goals can be effectively realized, a context analysis of the six components of the sales recruitment activity is undertaken. This context is analyzed in terms of social (participants and tools) organizational (culture), data structure and security context, product substitution, and new emerging tools context. The outcome of the context analysis is a set of context based tasks for an alternate computerized system.

Participant Goals and Incentives:

Hiring Manager: The hiring managers at present do not have an objective means of benchmarking the sales, customer service and telesales candidate's suitability against the organization's existing successful salespersons, customer service, sales support and telesales personnel. This is a human breakdown in the decision making process for sales recruitment. The hiring managers are prepared to support the use of a computer-based artifact for improving recruitment decision-making if it could support benchmarking. The computer-based system should provide a means for benchmarking candidate behavior profiles with existing profiles of successful salespersons, customer service, sales support and telesales personnel.

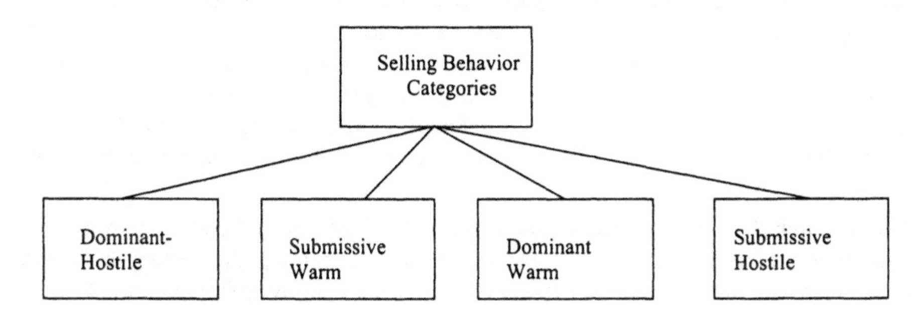

Figure 8.18: Selling Behavior Categories

Further, for improving data quality and accessibility, and reducing cycle time, computer-based artifact should:

- relate behavioral category of sales candidates on areas directly related to the selling profession rather than use indirect methods of behavioral analysis. The behavioral analysis is based on four behavioral categories; dominant hostile, submissive warm, dominant warm, and submissive hostile. These four categories are based on two dimensions, namely, "Submissive------Dominant, and Warm-----Hostile". These two dimensions are the two most significant dimensions in which selling behavior is expressed. More details on the behavioral dimensions and some earlier work done in the development of SRS is reported in (Khosla and Dillon 1992; Khosla and Dillon 1993; Khosla, Dillon and Parhar 1994).
- to support ease of use and understanding, the language used for evaluating a sales candidate on various areas should reflect the language used by them in their day-to-day activities

- be able to identify their suitability to other sales related jobs such as customer service, and sales support. At present this information has to be evaluated by the hiring managers during the interview process.
- be able to provide not only the behavioral category but also the breakup of the behavioral profile based on various areas related to the profession, such as competition, customer, product, decisiveness, success and failure, rules and regulations, job satisfaction. Overall, seventeen areas have been identified.

Sales Candidate: The sales candidates are ready to support development of a computer-based recruitment decision support system if it asks them direct sales or customer service related information rather than indirect psychological questions (e.g., Myer's Briggs behavior profiling systems) which they do not understand.

Work Activity

Practitioner Cultural issue: Traditionally, hiring/sales managers do not have confidence with using computerized sales recruitment systems, especially those based on indirect methods (e.g., Myer's Briggs behavior profiling system). Further, although sales managers do class salespersons in different behavioral categories, the selection of a sales candidate is determined by the function of the behavioral category, culture of the organization and organizational policies for sales and marketing (e.g., high promotion aggressive strategy or sit back strategy). The culture of the organization can be interpreted in terms of what type of sales and customer service personnel are currently successful in that organization. Thus, here again the capability of an alternative system to benchmark a sales candidate against the existing successful salespersons is a motivating factor for using a computerized sales recruitment system.

Practitioner Concerns: Practitioners are concerned about how a computerized system processes a candidate's information for determining the selling behavioral category of a sales candidate. In other words, the selling behavior model should correspond to the one used by them in their training programs.

Data

The data structure of various areas that relate to the selling profession is shown in Figure 8.19. The data security issues involve access of the behavioral profiles of the candidates. The access has to be restricted to the hiring managers and senior management.

Tool

The new emerging tools are internet technologies and e-commerce. The internet can be used to provide access to a computer-based sales recruitment system for use by hiring managers at different geographic locations. It will create uniformity and consistency in the recruitment activity. It will also help to reduce costs by allowing the behavior profiling of candidates from remote locations (thus reducing the to and fro expenses as well as preparing a better quality shortlist of candidates for an interview).

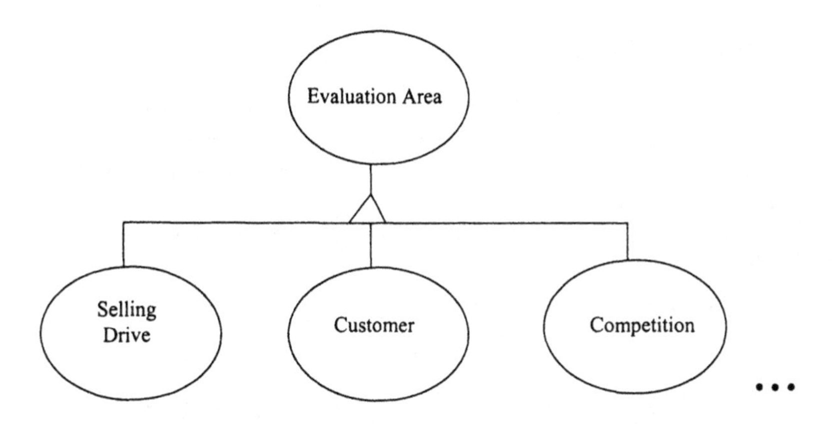

Figure 8.19: Evaluation Areas for Selling Behavior Categorization

8.3.4. Goals and Tasks

Based on the context analysis of the participant, work activity, data and tool components, a number of tasks can be identified. We list some of those tasks and cross-refer them with the goals identified in the performance analysis of the sales recruitment activity.

In order to improve the decision quality (G2), the following tasks need to be modeled. We list some of those tasks and cross-refer them with the goals identified in the performance analysis of the sales recruitment activity.

- **T1** The selling behavior profile of a sales candidate has to benchmarked against the behavior profiles of existing successful salespersons in the organization
- **T2** The degree of fit of a candidate's profile to a frontline sales, customer service and sales support position has to determined
- **T3**: The training needs for frontline sales, customer service and sales support have to be determined

In order to improve the reliability of the information about sales candidates, data quality and accessibility, the following tasks need to be modeled:

- **T4** The behavior categorization has to be based on selling related areas, as shown in Figure 8.19. The language used for evaluating the areas shown in Figure 8.19 has to reflect the language used by salespersons and managers in the profession.

In order to reduce costs (G1), cycle time (G5) and improve conformance to standards (G4) the following task needs to be modeled:

- **T5** Emerging internet technologies should be used to facilitate behavior profiling of candidates from remote locations. Through consistent use of the computerized sales recruitment system, it will also assist hiring managers in different geographical locations to conform to recruitment standards laid down by the organization.

8.3.5. Human-Task-Tool Diagram

The human-task-tool diagrams in Figures 8.20, 8.21 and 8.22, respectively, show the division of labor between the computer-based Sales Recruitment System (SRS), hiring manager, and the sales candidate. Figures 8.21 and 8.22 also show the human interaction points with SRS.

In the next section we show the implementation results of some of the tasks outlined in the last section.

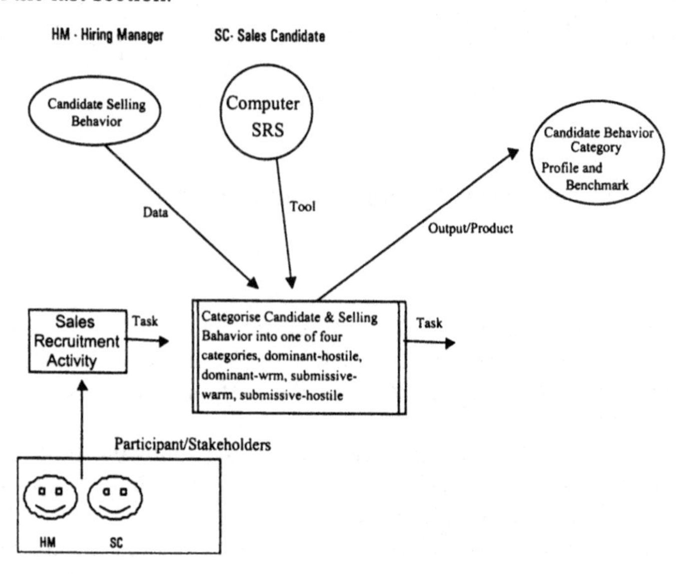

Figure 8.20: SRS Task

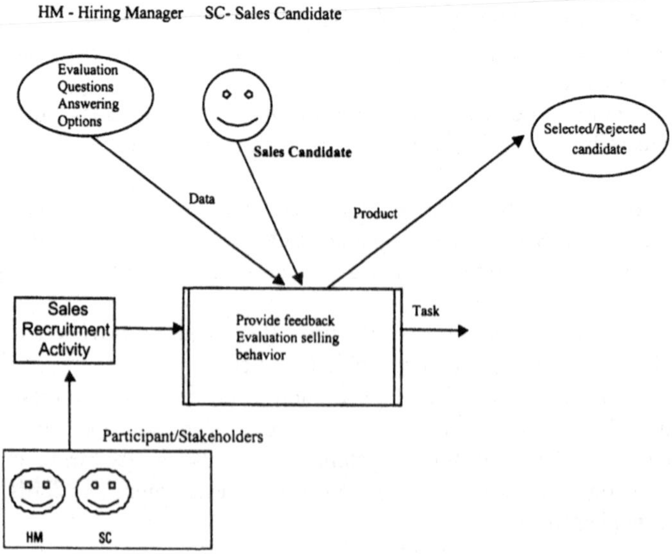

HM - Hiring Manager SC- Sales Candidate

Evaluation Questions Answering Options

Sales Candidate

Selected/Rejected candidate

Data

Product

Sales Recruitment Activity

Provide feedback Evaluation selling behavior

Task

Participant/Stakeholders

HM SC

Figure 8.21: Sales Candidate Task

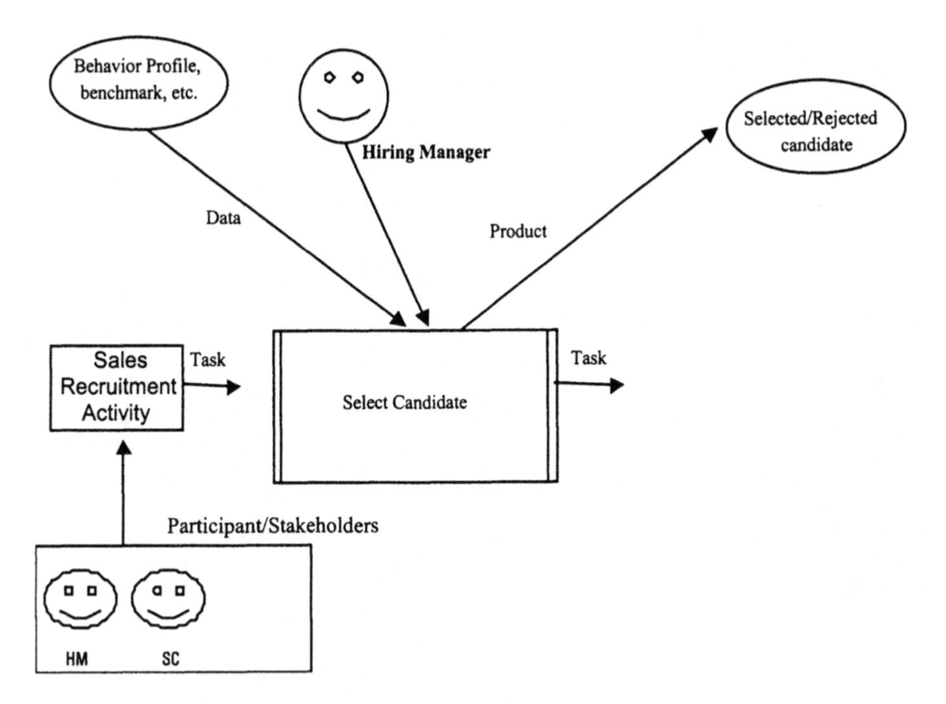

Figure 8.22: Hiring Manager Task

8.4. Behavior Profiling and Benchmarking

The intelligent selling behavior evaluation control agent shown in Table 8.2 models behavior profile and benchmarking tasks of SRS. Figures 8.24 and 8.25 show the results of implementation of the sales candidate behavior profiling and benchmarking tasks of the SRS respectively. The screen shot in Figure 8.23 shows the implementation of tasks T2 and T3. These two tasks involve two psychological variables of interest to the hiring manager, namely, degree of fit. Further, the selling behavior profile is shown at two levels of abstraction. The pie chart represents the overall distribution of four category scores, whereas, area wise behavior profile is shown in the upper right hand corner of Figure 8.23. That is, the upper right hand corner of Figure 8.23 shows the area wise breakup of a candidate's selling behavior as related to the Dominant Hostile (DH) category.

In Figure 8.24 we show a comparison of the candidate's profile (black color) with the benchmark profile (grey color) of a particular organization. The hiring manager is particularly interested in the orientation of the two profiles. That is, are the two profiles parallel or do they cross each other (as in Figure 8.24)? They are less interested in the magnitude of difference between the two profiles (which if required can be deciphered from the Y coordinate dimension of the comparison of profile graph).

Table 8.2: Selling Behavior Evaluation Control Agent

Name :	Selling Behavior Control Agent
Parent Agent:	HCVM Control Agent, Expert System Agent
Goals:	Define candidate behavioral category selection knowledge for selling behavior categories
Some Tasks:	Define conflict resolution rules for determining predominant behavioral category from four decision categories, namely, DH, DW, SH, and SW. between decision agents in case of multiple diagnosis
Task Constraints:	Learning new heuristics for behavioral category
Precondition:	Sales Candidate Behavior Category raw scores
Postcondition:	Candidate Predominant selling behavioral category, Candidate behavioral profile, training needs, etc.
Communicates with:	DH behavioral category decision agent, DW behavioral category decision agent, SH behavioral category decision agent , SW behavioral category decision agent, Behavioral profile and Benchmarking media agent
Communication Constructs:	Command: Behavioral category decision agents Tell: Behavioral profile and Benchmarking media agent
Linguistic/non-linguistic features:	Raw scores
External Tools:	Software: Graphic object, Button Object Domain: Competition, customer, product objects
Internal Tools:	Expert System and neural networks
Actions:	Rules for determining predominant behavioral category Supervised neural network algorithm for learning new heuristics

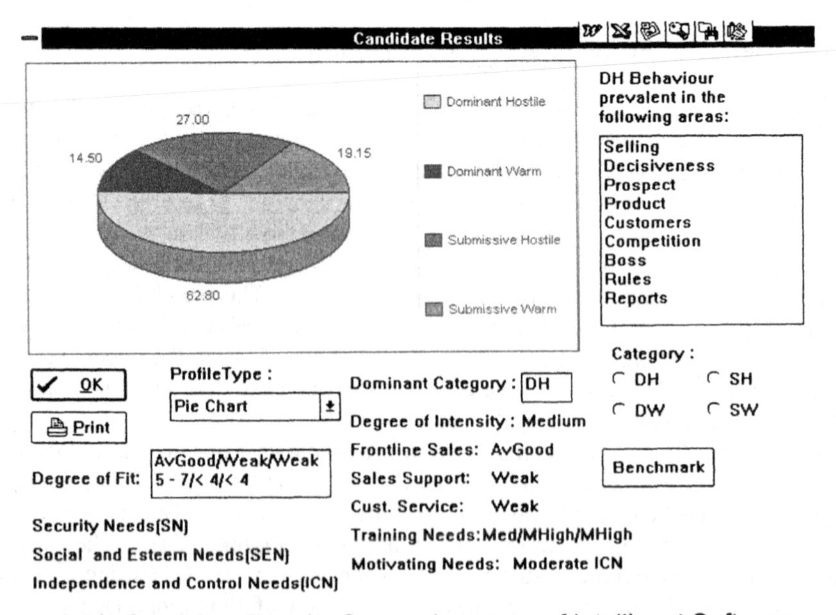

Figure 8.23: Candidate Results Screen (courtesy of Intelligent Software Systems, Melbourne, Australia)

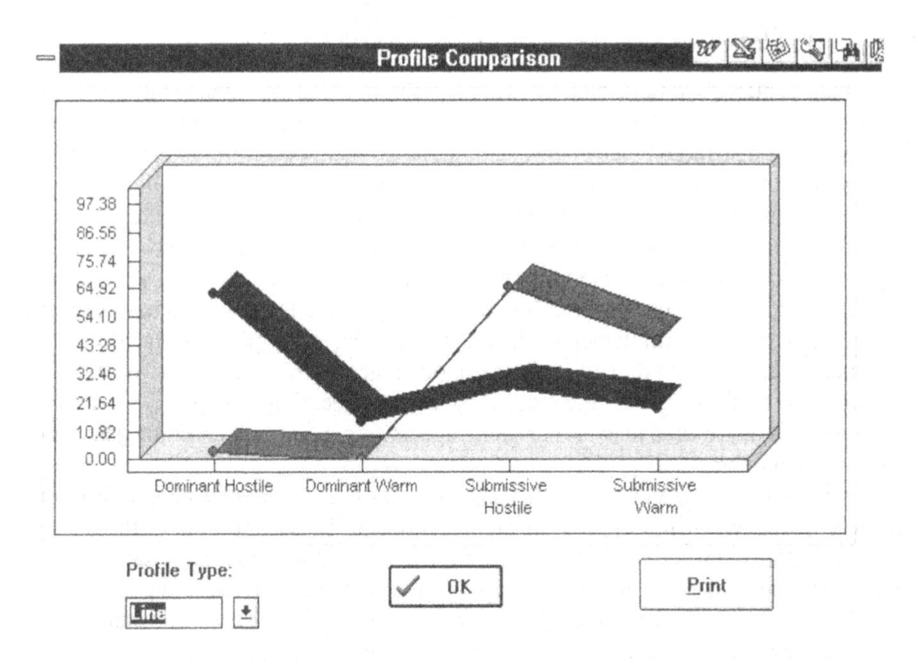

Figure 8.24: Benchmarking of a Candidate's Behavior Profile (courtesy of Intelligent Software Systems, Melbourne, Australia)

Figure 8.25 shows as an example the language used for designing the four questions related to the *competition* area. It can be seen that the tone and words used mirror the language used in the selling profession.

At present SRS is being ported on the internet to reduce recruitment costs.

SRS is being present used in the industry for recruitment of salesperson, telesales personnel, customer service personnel and sales support personnel.

1. In sales, the law of the jungle prevails. It's either you or the competitor. You relish defeating your competitors, and fight them hard, using every available weapon.	Behavioral Category:	DH
2. The best hope to outwork and outsell competitors is by keeping abreast of competitive activity and having sound product knowledge of your product.	Behavioral Category:	DW
3. You may not be aggressive otherwise, but when it comes to competition you are just the opposite. You spend good deal of your time to explain to the customer why he should not buy from the competitor.	Behavioral Category:	SH
4. You do not believe in being aggressive towards your competitors. Competitors are people like you and there is room for everybody.	Behavioral Category:	SW

Figure 8.25: Questions Related to the Competition Area

8.5. Summary

Human-centered computer-based systems should lead to enhancement of user competence. In this chapter we demonstrate two effective ways of achieving this goal. We describe a distributed Net Euchre card game application on the internet in which intelligent Game Agents interact with the human players to provide solicited advice for deciding the trumps and selecting best play strategy (the later feature has not been implemented as yet). Further, the Game Agents can also be substituted for human players to allow for different game configurations.

Distributed communication and intelligence is another feature of human-centered systems that has been implemented in the Net Euchre card game application. The four game agents communicate with each other across four different machines in various stages of the game (e.g., deciding dealer, deciding trumps, etc.). Further, intelligence is distributed across four game agents to generate a system or game level intelligent behavior.

Traditionally, computer-based artifacts have not been a popular choice in human resource management function of an organization. In the second half of this chapter we demonstrate how computers can be used as tools for modeling breakdowns in human decision making situation in a sales recruitment activity. The breakdown modeled by a computerized Sales Recruitment System (SRS) relates to benchmarking the incoming sales candidates with the existing successful salespersons in an organization. Benchmarking has proved the motivation to the hiring managers in the industry to make SRS an integral part of their sales recruitment activity.

References

R. Khosla and T. Dillon, *'An Intelligent Assistant for Improving Sales/Customer Service Performance'* - in IEEE Workshop on Customer Service and Support, San Jose, California, U.S.A, July 1992

Khosla, R. and T. Dillon , *'A Knowledge Based Approach for Recruiting Salespersons',* Sixth Artificial Intelligence Technology Transfer Conference in Industry and Business}, Monterrey, Mexico, Sept. 1993, pp.83-9

R. Khosla, T. Dillon and A. Parhar, *'Synthesis of Knowledge Based Methodology and Psychology for Recruitment and Training of Salespersons'*, in Lecture Notes in Computer Science (LNCS), Springer-Verlag , 18th German Annual Conference on Artificial Intelligence , Saarbr"ucken, Germany, September 1994

9 INTELLIGENT MULTIMEDIA INFORMATION MANAGEMENT

9.1. Introduction

In the last three chapters we have described applications of HCVM in medical informatics, image processing, internet games and sales recruitment. In these applications we have looked at multimedia artifacts in terms of how they can be used for improving the representational efficiency, effectiveness and interpretation of computer-based artifacts and also to some extent how they can be used for perceptual problem solving. However, there are interesting research issues and problems associated with management and retrieval of text, image, video, audio artifacts from single and multimedia databases. The area of multimedia information management and retrieval has in recent times become important due to the advent of the internet and World Wide Web (WWW). The solution to a number of issues in this area today lies with traditional solutions used for conventional databases. Recently, intelligent agents and human-centered information retrieval models are being proposed by researchers to address the problems in this area. In order to gain a proper understanding of the various problems and issues involved, we start this chapter by outlining the three evolution stages of this time honored area. We then describe these evolution stages in some detail. We include in this description applications and solutions proposed for multimedia information management and retrieval. We conclude the chapter by outlining open research problems in this area.

9.2. Evolution Stages of Multimedia Information Management and Retrieval

Intelligent multimedia information management and retrieval is a time-honored research area that addresses a broad range of issues spanning multiple disciplines such as multimedia data processing, computer graphics and visualization, human computer interaction, intelligent agents, knowledge representation, software design, and information retrieval. In the past few years, the increasing importance of multimedia content on the WWW has brought multimedia-related issues to the center of the interests of both industry and academia. Several textbooks have been devoted to the subject (the interested reader may want to refer to (Khosafian and Brad Baker 1996), (Maybury 1995) and their rich bibliography). Following an updated version of the classification proposed in Maybury (1995), the recent evolution of this discipline could be broadly described as involving three main stages:

Single media processing. Traditionally, largely independent research communities have focused on the automated processing of single media, including text processing (Grosz et al. 1986), spoken language processing (Waibel and Lee 1990), and image and video processing (Chen et al. 1993; Fuhrt et al. 1996).

In more recent years, several new requirements for access to global and corporate information repositories were introduced, which go beyond text retrieval and increasingly include graphics, imagery, audio (speech, music, sound), and video artefacts. The challenge posed by managing huge multimedia digital libraries turned the research community's attention toward the problem of integrated access to structured data and textual sources, as well as media, with their spatial and temporal properties (Gibbs et al. 1994). These lines of research were aimed at creating new integrated capabilities, including multimedia information browsing, search and visualization. For instance, ad-hoc indexing techniques turned out to be essential for efficient search and retrieval of information from multimedia database systems.

Traditional data indexing techniques, though they perform satisfactorily in low dimensions, were found to be unsuitable for indexing higher dimensional data encountered in multimedia database systems. Results of these research approaches produced innovative commercial applications such as video-on-demand, interactive radio, as well as content-based multimedia authoring and presentation tools (see for instance the *SlideFinder* system presented in Niblack (1999). The MUDS project at the University of Maryland seeks to develop a unified treatment of multimedia data. For a given class of media, the key functionalities of that media-type are identified. Based on this identification, indexing structures are developed that are as independent of a low-level representation as possible. Unified query languages that can be used to query wide varieties of multimedia data exploit such indexes. However, a basic difference emerged between multimedia and conventional query languages, i.e., the need to manage *probabilistic* and/or *fuzzy* retrieval. In conventional database system, the result of a query is usually a set. In multimedia database systems, *content* queries are involved that return a *ranked* result set, indicating the multimedia objects (images, audio or video data) that are closest to the user query.

WWW-based multimedia today, increasing use and expansion of the internet as the global information highway has made the World Wide Web (WWW) the ultimate, large-scale multimedia database. Finding useful information from the WWW without encountering numerous false positives (the current case) poses a challenge in the area of information indexing and retrieval. Specifically, the word *Web indexing* has acquired a new meaning as it refers to the extraction and representation of semantic content from the WWW. In this framework, *metadata* for content representation of multimedia and semi-structured sources are being actively investigated, and several approaches to medium specific metadata for audio and video information have been proposed (see for instance (Chang et al. 1997)).

In this chapter, we shall describe in some detail the evolution path sketched above, underlining the user-centred (as opposed to technology-centred) emerging requirements, such as representation of content and content-based querying. The satisfaction of these requirements, especially in the framework of the World Wide Web, is increasingly recognized as one of the main challenges facing current multimedia research. Moreover, the role of emerging metadata format based on the *eXtensible Mark-up Language* (XML) will be discussed. A detailed description of XML features has been given in chapter 2; here, it is sufficient to say that the function of the mark-up in an XML document is to describe its logical structure without having to deal with presentation and rendering. In other words, XML provides a *lightweight schema*, called the *document-type declaration*, to define the logical structure of a set of valid documents. An XML document is valid if it has an associated document-type declaration and if the document complies with the constraints expressed in it. Document-type declarations are made in a *Document-Type Definition* (DTD) file. The DTD file contains a formal definition of a particular type of document outlining the element names and the structure of the document.

The chapter is structured as follows: after the present introduction, an overview of techniques for dealing with multimedia data in the context of traditional databases is provided in section 9.3. Section 9.4 deals with approaches to content-based and multi-model query support for images, video and audio in the framework of specialized multimedia database system. section 9.5 presents some real world applications of the notions presented in the previous sections and section 9.6 briefly discusses the application of fuzzy techniques to multimedia database querying. Then, in section 9.7, representation and querying techniques for WWW data are outlined, while section 9.8 presents a flexible query support system for Web based semi-structured and multimedia information. Finally, section 9.9 focuses on some of the key remaining problems in multimedia retrieval research.

9.3. Multimedia Information Management Using Conventional Databases

Many conventional database management systems currently allow storage and retrieval of non-traditional data types, such as imagery, graphics, text, audio, speech and, more recently, video. Multimedia types are often characterised by large data volume; moreover, some of them, such as video and audio, are continuous and time dependent, posing complex problems of temporal relationship and synchronization among multiple data sources. Several issues must therefore be taken into account in order to include multimedia support in a conventional database management system:

- Storage and transfer of large objects is expensive. Transporting large multimedia objects over networks may require substantial bandwidth over an extended time period. Even with compression, these objects are larger than conventional data types and may not fit into main memory of a client.
- The size of a multimedia object may vary dynamically creating allocation challenges.
- Temporal data, such as audio and video, must be expressed as a function of time. Moreover, temporal data is delay-sensitive.

- Synchronization is required to ensure a temporal order of events.
- Speed of retrieval from disks may not be sufficient to handle the demands of a large multimedia system.

Other interesting issues that have been at least partially addressed in the framework of conventional DBMS are user interaction, the management of spatial composition and, more recently, content based retrieval. In this section, we shall present some approaches to multimedia data management by means of conventional databases.

9.3.1. BLOBs

Before 1980, most research and commercial development in document processing focused on text retrieval. In the early days of multimedia, DBMS did not include any support for multimedia data types, and text-retrieval techniques were often applied to documents containing textual descriptions of multimedia data, separately stored using a conventional file system.

The first approach chosen by DBMS designers in order to manage multimedia data was the introduction of the notion of *Binary Large OBject* (*BLOB*). Basically, BLOBs are long field data used for storing unformatted data such as text and images as indexed *chunks*. In older systems, only restricted operations (such as `blobread` and `blobwrite`) were usually permitted on BLOBs as BLOB storage was obtained by means of a conventional file-system, without integration with the rest of the database architecture. Moreover, in early proposals the DBMS provided indexing facilities only, while presentation of BLOBs was initially entirely left to application programs.

In Christodoulakis (1986), functions were presented providing multimedia presentation and browsing capabilities to MINOS, a general-purpose information system exploiting a file system for BLOB storage. The presentation and browsing capabilities of MINOS were among the first to make use of the capabilities of a modern workstation to increase the so-called "man-machine communication bandwidth".

Table 9.1. BLOB Management in Conventional RDBMSs

Functionality	*Description*
Request interpreter	Multimedia data capture, retrieval, and presentation
Request processor	Binding of components, communication control, and transaction management
Intra-relationship management	Media synchronization, spatial composition, media conversion
Mixed object management	Long field data server, video data server

Symmetric capabilities for text and voice browsing were also provided. Later, BLOB management got more and more integrated into relational and object-oriented databases; as a result of this integration, some query facilities were added. Summarizing the above discussion, it can be said that BLOB management at the RDBMS level includes the functionalities presented in Table 9.1, while reference architecture is depicted in Figure 9.1.

As an example of a rather sophisticated interface for BLOB management in the framework of a RDBMS, we shall refer to the POSTGRES95 version of the well-known POSTGRES database management system (Stonebraker and Kemnitz 1991). Originally, POSTGRES supported three standard implementations of BLOBs: as files external to POSTGRES, as UNIX files managed by POSTGRES and as data stored within the POSTGRES database. Today, POSTGRES only supports BLOBs as data stored within the POSTGRES database. Even though internally managed BLOBs are slower to access, this technique provides stricter data integrity.

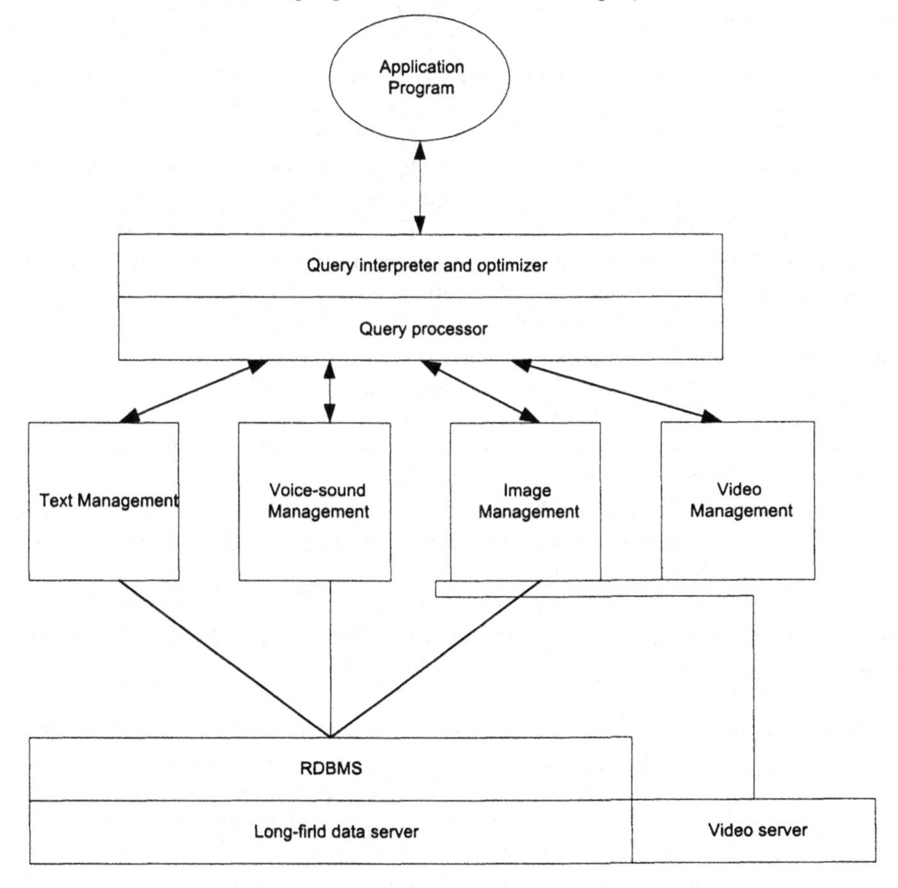

Figure 9.1. Managing BLOBs in the Framework of Conventional DBMS

For historical reasons, in the framework of the POSTGRES project, large objects are called *Inversions*; hence, we will use terms *Inversion* and *BLOB* as synonyms in this section. The Inversion large object implementation breaks BLOBs up into chunks and stores the chunks in relational t-uples in the database. A *B-tree index* guarantees fast searches for the correct chunk number when doing random access reads and writes to the database.

The POSTGRES BLOB interface was modeled after the well-known UNIX file system interface (Bach 1986), with analogues of basic file access functions such as *open(2), read(2), write(2), lseek(2)*, etc. User-defined functions call upon these routines to retrieve only the data of interest from a BLOB. For example, if a BLOB type called *cast* existed that stored photographs of human faces, then a function called *haircolor* could be declared on *cast* data.

Function *haircolor* could look at the upper third of a photograph, and determine the color of the hair that appeared there, if any. The entire BLOB value need not be buffered, or even examined, by the hair function. POSTGRES provides a fully-fledged C Application Program Interface: BLOBs may be accessed from dynamically loaded C functions or database client programs that link the C library.

In the early Nineties, as the object-oriented data model increasingly gained acceptance in the database community (Kim 1990), (Cattell 1991), (Cattell 1994). BLOB management was included as a standard extension of object-oriented data model in several O-O databases, and acquired importance in later releases of the ODMG standard definition (Cattell 1996). The rationale of these systems is that relational databases do not naturally support modeling for multimedia data types.

These data types can instead be easily defined as *User Defined Types* (UDT) at the object level, without interfering with traditional data types. They can make use of traditional Object-Oriented operations, such as *create* and *delete.*

The object-oriented model allowed the specification of BLOBs as simple objects linked by a part-of relationship to complex objects, also including text information. Text fields were thus available for multimedia data tagging and description-based retrieval. For instance, GMD-IPSI's well-known AMOS (*Active Media Object Stores*) research project deals with issues of object-oriented modeling of multimedia objects (particularly audio), taking into account synchronization, parallel processing, and issues of distributed architectures.

As a result of the powerful data modeling features of the object-oriented approach, effective query support is often provided in this framework. A query in a sample object-oriented query language is reported in Figure 9.2.

"Find the last version of all documents, dated after January 1st. 1999, containing a company logo and having the word "Athena" either as sender name or in the product presentation (which is a textual component), with the words "Personal Computer"(with a possible trailing 's) in the product description section (which is another text component), and with the word "Italy" either constituting the country in the address or contained in any part of the entire document."

```
FIND DOCUMENTS VERSION LAST
WHERE Document.Date > /1/1/1999
      AND (*Sender.Name = "Athena" OR *Product_Presentation
CONTAINS "Athena")
      AND *Product_Description CONTAINS "Personal Computer%"
      AND (*Address.Country = "Italy" OR TEXT CONTAINS "Italy")
      AND WITH *Company_Logo;
```

Figure 9.2. A Sample O-O Query

Moreover, object-oriented databases proved very effective in supporting the notion of temporal operations on multimedia objects (Aberer and Klas 1994), so that much of the research on multimedia support in conventional database systems is presently focused on the object and object-relational data model.

9.3.2. SQL Multimedia (SQL/MM)

While academic research focused on the representation of multimedia BLOBs as objects stored in object-oriented databases, it is widely recognized in the industry that today's conventional SQL-based database engines must be able to effectively address multimedia information processing. As the need for a standard technique for multimedia data representation and querying in the framework of SQL became more evident, the SQL standardization committee announced a ISO/IEC international standardization project for development of an SQL class library for multimedia applications, which was approved in early 1993. This standardization activity, named *SQL Multimedia* (SQL/MM), specifies SQL *abstract data types* (ADT) definitions using the facilities for ADT specification and invocation provided in the emerging *SQL-3* specification. SQL/MM aims to standardize class libraries for science and engineering, full-text and document processing, and methods for the management of multimedia objects such as image, sound, animation, music, and video. It is also intended to provide an SQL language binding for multimedia objects defined by other JTC1 standardization bodies (e.g. SC18 for documents, SC24 for images, and SC29 for photographs and motion pictures).

SQL/MM is a multi-part standard consisting of an evolving number of parts. Part 1 is a framework that specifies how the other parts are to be constructed. Each of the other parts is devoted to a specific SQL application package. The following Table 9.2 outlines the SQL/MM Part structure:

Table 9.2. The SQL/MM Standard

Part #	*Content*
Part 1	Framework A non-technical description of the standard.
Part 2	Full Text Methods and ADTs for text data processing.
Part 3	Spatial Methods and ADTs for spatial data management.
Part 4	General Purpose Methods and ADTs for complex numbers
Part 5	General Purpose Methods and ADTs for still images (G4,JPEG,TIFF)
Part 6	General Purpose Methods and ADTs for full motion video (MPEG)
Part 7	General Purpose Methods and ADTs for audio (MIDI)
Part 8	General Purpose Methods and ADTs for spatial 2D & 3D (IGES,PHIGS,GKS)
Part 9	General Purpose Methods and ADTs for seismic data
Part 10	General Purpose Methods and ADTs for music (MIDI, SMDL)

Even after the SQL/MM standard emerged, however, the support for multimedia data in the framework of the relational model is still being actively investigated.

For instance, in Soffer and Samet (1999) a novel method is presented for integrating images into the framework of a conventional relational database management system, storing and indexing images as t-uples in a relation. Moreover, indices are constructed for both the contextual and the spatial data, thereby enabling efficient retrieval of images based on contextual as well as spatial specifications. This technique is admittedly not entirely general, as it is applicable to a class of images termed *symbolic images* in which the set of objects that may appear are known a priori. Finally, the geometric shapes of the objects conveying symbolic information are relatively primitive. This approach is however promising, as emphasis is given on extracting both contextual and spatial information from the raw images; to this aim, a *logical* image representation that preserves this information is defined.

As we shall see, such logical representations (as opposed to raw multimedia data) are an essential prerequisite for content-based querying.

9.3.3. Storage and Retrieval Issues in Multimedia Databases

In a database system holding multimedia data, most processing has to do with storage and retrieval of BLOBs (Gemmell et al. 1995). A complete, though not recent, survey of research on this and related issues is presented in Kunii et al. (1995). Modern disks have a throughput of 5 Mbytes/sec and more; increased disk speed ensures that retrieving a few BLOBs of CD-quality audio (1.4 Mbits/sec) poses no particular problem. The same can be said for MPEG-encoded video, though in the past, broadcast-quality full-motion pictures often required ad hoc architectures. A classic approach is presented in Oomoto and Tanaka (1993), describing the OVID video database system.

Today, several commercial systems are available providing fast BLOB management. The *Oracle Media Server* (Oracle Corp. 1999), for instance, is a recent cross-platform system for multimedia data that can easily handle tens of thousands of simultaneous audio and video streams. Currently, research in this area focuses on *disk-striping* as a storage mechanism, disk-scheduling algorithms and admission-control algorithms.

Usually, an audio object is physically stored in discrete blocks, so that retrieval actually involves a series of retrievals. Retrieval time consists of the time it takes to locate, seek, read and transfer each block. All of these actions must take place at a rate that is less than or equal to the time it takes to fill a buffer.

One method of storage is *disk striping* (Berson et al. 1994), which is at the basis of standard technologies such as *RAID (redundant array of inexpensive disks*). This method involves the use of multiple storage disks as one logical disk. This technique enables a multimedia object to be stored in discrete blocks across disks in a round robin fashion, increasing transfer rate. Alternatively, it can be used to maintain a steady transfer rate by allowing simultaneous access to different portions of the object.

A potential problem with disk striping is that a single disk failure may affect the availability of many objects; additional redundancy is therefore required. Once retrieved from disks, BLOBs should be efficiently cached at the server side. Traditionally, database servers have used *data sharing* (caching data when it is known that other users have requested the same data).

Besides server side caching, buffering at the client could also be used to smooth out the presentation of the transferred blocks. At present, a dual buffering system is often employed, while the use of multiple buffers and queuing is foreseen in the future.

Disk-scheduling algorithms control the scheduling of retrieval of needed blocks of material. Two main techniques are used:

- *Earliest Deadline First (EDF)* is a widely used technique. The needed block that is closest to its retrieval deadline is accessed first. The downside of this algorithm is that it results in excessive seek time and poor server utilization because the server is looking for one block at a time.
- *Scan scheduling* is often more effective because the disk head scans the medium and retrieves all blocks within a certain deadline. It works especially well if retrieval deadlines are clustered together.

Admission-control algorithms are used to guarantee presentation in an acceptable form. When the user makes a request for presentation of multimedia material, these algorithms check to be sure that adequate resources are available for presentation. If resources are inadequate, the request for presentation is denied.

In fact, they generate strongly varying loads on the media server during the presentation of media data. Admission control approaches for limiting the number of concurrent users and thus guaranteeing acceptable service quality, currently existing, are only suited for applications with uniform load characteristics like video-on-demand.

In Aberer and Hollfelder (1999) a session-oriented approach to admission control is proposed based on the stochastic model of *Continuous Time Markov Chains*, which allows describing the different presentation states occurring in the interactive access to the multimedia database. The model is derived from semantic information on the forthcoming browsing session, taking into account the relevance of the videos to the user.

From the above considerations it should be clear how, until recently Staehli et al. (1995), Vogel et al. (1995), *Quality of Service* (QoS) for multimedia has referred to system-level parameters such as speed of transmission, error rates, and connection failure probabilities. These parameters meant to be unseen by the application, and are usually hidden from the user as well.

This view is however changing as a result of the relationship between the system and time-dependent data in distributed systems. As a consequence, although QoS was introduced as a technology-related notion, it is now seen as an *interface* providing users with choices and is increasingly defined as a function of user satisfaction.

The QoS goal is to deliver results of acceptable quality, taking into account a price/performance ratio. For instance, users should be able to choose the quality of a presentation, and be prepared to pay a fee related to their choice. Moreover, if an digital audio selection is in high quality format but the user's devices support only low quality format, or network traffic is high, it would be cheaper and faster to be able to specify that retrieval of the selection is also acceptable in a low quality format.

9.4. Content-based Retrieval in Multimedia Databases

While many conventional databases still use BLOBs to store and retrieve multimedia data, *content-based retrieval* capabilities are considered essential for effective utilization of information from multimedia repositories. Content-based retrieval is characterized by the ability of the system to retrieve relevant information based on the contents of the media rather than by using simple attributes or keywords assigned to the media itself.

Intelligent multimedia information retrieval goes beyond conventional database environments to provide content based indexing of multiple media and management of the interaction between these materials by representing and reasoning about models of the media, user, discourse and task.

As Figure 9.3 illustrates, intelligent multimedia information processing include:

- *Analysis* of the multimedia information to index or extract information from them
- *Content-based querying and retrieval* of indexed information using single and cross-media query languages
- *Presentation* of the query results.

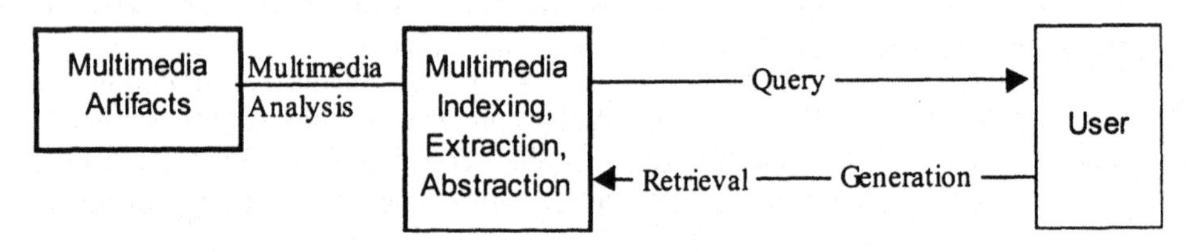

Figure 9.3. Phases of Multimedia Information Processing

Content-based querying is intended to enable, for example, users to pose queries not only by keywords but also by color, shape, and texture for imagery; by characterizing the pitch, tone, and timbre for retrieval of sound. It should be noted that content-based retrieval poses several difficult problem related to multimedia information semantics; for instance, the relevance of retrieved information is often judged differently by system users for an identically formulated query. No wonder, since the notion of relevance is subjective in nature, and is a function of both the user's retrieval need and context.

Indeed, the research community has yet to agree upon methods of describing the basic elements of imagery and audio. For example, (Blum et al. 1995) addresses automated indexing of sound based on features such as pitch, loudness, and duration. However, while such parameters are readily describable and computable, timbre (characterizing the tonality of an instrument), is primarily a perceptual concept, and thus must be derived from "primitive" acoustic properties. Similarly, elementary image properties such as color and shape are directly computable from raw image contents, while more sophisticated notions such as texture must be derived starting from other properties, such as the directionality, granularity or coarseness, and contrast of an image.

9.4.1. Content-based Access to Imagery

Many methods for automated indexing of imagery support content-related search and browsing based on complex features. In fact, in multimedia information retrieval applications, content-based image retrieval is essential for retrieving relevant multimedia documents. Following the current research approaches, the content of an image may be represented by a set of *visual objects*. Many research proposals represent image content by means of *feature vectors*, whose elements are content-related features such as texture, color, and shape.

The semantic meaning of an image is determined based on the spatial relationship between the visual objects and the heterogeneous features of each object. The similarity between two images is then determined based on the distance between features of the images in the feature space. Spatial features and features in texture, color, and shape are normally generated using different computational methods. Other metrics (Liu and Dellaert 1998) are based on spatial arrangements in 3D images. Thus, different features may have different similarity measurement. Because of this, the content-based retrieval process is normally performed on individual features; the retrieval results are often unrelated to the query semantically.

In Flickner et al. (1995) the seminal *Query By Image Content* (QBIC) system (now included in IBM's *DB2 Image Extenders*) is described, which indexes imagery and video on the basis of visual properties such as color, shape, and texture as well as motion analysis.

Another line of research in imagery indexing and retrieval requires pixel-level original images to be automatically or manually transformed into an *iconic image* containing meaningful graphic descriptions, called *icon objects*. For this format, a conventional *spatial match representation scheme* called DLT (Direction Lower Triangular) is defined, which allows it to accurately describe three-dimensional spatial relationships between icon objects. (Kanehara et al. 1995) describes the *Image Retrieval Prototype System* as an image database that can be queried using visual examples, using a simple and effective technique that can be used as an example.

Without the use of any a priori knowledge, the proposed system is capable of automatically recognizing several features of each of regions composing the overall target image; thereby allowing a user to initiate queries representing detailed demands, intentions, and viewpoints. The process employed to perform image retrieval can be outlined as follows. Each image contained in the image database (*target image*) is processed and indexed off-line using the following three steps:

1. The image is segmented into parts or regions, termed as primitives using our novel method of shape decomposition and image segmentation;
2. Several features of the image are then extracted from these primitives;
3. The extracted features are respectively translated into two descriptive bit patterns used for image retrieval, named data signatures, i.e., diverse and local properties of the target image data are encapsulated within these signatures.

More specifically, one data signature represents several attributes of each primitive, such as scale, position, slope and color; while the other represents several relations between each primitive and another one near to it, such as their relative scale and spatial position.

In this method of shape decomposition, the interior of a given shape is virtually eroded at every concavity at certain speeds, corresponding to configurations of each concavity, and the object shape is then divided only when an eroded region reaches the other side of the shape. This procedure is recursively performed.

The following description explains how queries are generated for performing image retrieval. To input either the whole or part of a particular "query image" to be retrieved from the database, the user draws its outline using polygon-like shapes sketched on the screen of our built retrieval interface.

The user can individually manipulate these primitives such that various query conditions can be generated based on geometric constraints selected among primitives of interest. Primitives that have been manipulated to form query conditions, are then used to create query signatures, which can be matched with corresponding data signatures. Thus, the matching process is simply and easily carried out. The conjunction, shift and bipartite graph matching operations are required to provide a comparison with corresponding data signatures, two query signatures for each primitive.

In Wang et al. (1999), a different though related indexing technique enables the system to avail the user of the ability to visually generate query conditions that can manipulate detailed and diverse features of the basic feature of an image to be matched with a corresponding data images.

A different, though related approach that exploits *metadata* as well as segmentation for image indexing and querying is described in Chang et al. (1997). In the same vein, (Sheikholeslami et al. 1998) describes a system for semantics-aware clustering and querying on heterogeneous features for visual data. The subject of metadata support for content-based retrieval of imagery will be explored in some detail in the following sections.

9.4.2. Representing and Accessing Video Data

While still images can be indexed *a posteriori* by segmentation techniques, it is widely recognized that in order to enable content-based representation and indexing of video data it is mandatory to define innovative metadata standards to be associated with raw data. There is currently a growing interest in organizing and querying large bodies of video data; in the VIQS project (Adali et al. 1999), for instance, a simple SQL-like video query language is proposed which extracts the relevant segments of video that satisfy the specified query condition. A standard relational database system is extended in order to handle such queries.

As we shall see in the next section, a number of approaches have considered the application of the *eXtensible Markup Language XML* and the *Resource Description Framework (RDF)* (Berners-Lee 1998) to multimedia data indexing and content-based querying. These proposals are interacting with the development effort of the MPEG-7 standard. In this subsection, three of the main requirements related to representing video information will be given, following (Hunter and Armstrong 1999):

- *Hierarchical structure definition.* The schema must be able to constrain the structure to a precise hierarchy in which complete video documents is at the top level. *Part-of* relationships should be used to represent decomposition into sequences, which contain scenes and so on. At the bottom level, video frames contain objects or actors.

- *Cardinality* within attributes should be represented, as well as the minimum and maximum number of attributes.
- *Spatio–temporal* specifications and relations. Video metadata must be able to support the specification of temporal characteristics, e.g. begin and end time of segments and their duration. Spatial relations, such as neighboring objects, and temporal relations, such as sequential or parallel segments, should be supported. Given such a relationship between two classes, it should also be possible to constrain specific attribute values of these classes. For example, the start and end times of scenes contained within a sequence must lie within the start and end time of that sequence.

9.4.3. Content-based Audio and Video Retrieval

A multimedia database management system, which would include audio capabilities, requires the integration of several database and multimedia technologies, namely an audio data type (Rakow and Lohr 1995), must include the following:

- *Temporal relationships.* Digital audio is recorded by storing samples. A sample has a *size* (16 bits for CD audio) and a *rate* (44.1 kHz for CD audio). A digital audio object (sometimes called a stream) will also have duration. If one is selecting an audio clip, there will be one or more *positions* in the audio object. There will also be methods for *record*, *start play*; *pause* and *stop play* which allow for interactive control.
- *Compression.* Sound-processing systems exploit compression techniques so that storage requirements may be minimized and faster transport enabled. The best-known standard is MPEG-Audio (Motion Pictures Experts Group). This means of compression is supposed to primarily remove material that is not perceptible to the human ear. There are problems, however. Audio data is usually manipulated in uncompressed form. Thus, editing, which is a major process in multimedia applications, requires methods for repeated compression and decompression. This results in audible loss of quality.
- *Format.* The *format* of audio data has to do with sampling rates. Higher quality results from a higher sampling rate. Choice of format depends upon amount of available storage, computing power, and type of presentation devices. High definition format requires about 20 times as much storage space as a low definition format. In building a large distributed system, a parametric system that supports multiple sampling rates, might be advised. This, in turn, requires methods for making *conversions* from one format to another.

In Blum et al. (1995) a sound classification and retrieval system is described that computes both acoustic and perceptual properties in order to provide content-based access to an audio clip base.

Multimedia Information Environments based on video are now growing rapidly. Three types of environments have been identified, namely *Stream type MM environment* by digital TV broadcasting, *Real-World type MM Environment* monitoring roads and towns, and *Network type MM environments* on the WWW, that are considered especially promising targets.

A framework for developing applications and services from these three MM environments, called *Multimedia Mediation Mechanism*, is proposed and discussed in Sakauchi (1999). Another proposal, the FRAMES project, aims to developing Video Information Systems (VIS) such as digital video libraries, video-on-demand systems and video synthesis applications. The framework includes a functional component to represent video and audio analysis functions, a hypermedia component for video delivery and presentation and a data management component to manage multi-modal queries for continuous media. A meta-model is described for representing video semantic at several levels.

As one would expect, a common feature of such frameworks is that they require a high-level representation of video sequences. General approaches to this problem envision systems for encoding the contents of video sequences based on their syntactical and semantics-aware description. Such techniques require two phases: *analysis* and *synthesis*. The analysis phase involves the automatic generation of a syntactical structure using image and image-sequence analysis, image understanding as well as text recognition.

The full representation of video sequences is then generated during the synthesis phase by combining results from the analysis phase. While several general approaches based on these notions have been proposed, research on video indexing and retrieval has sometimes started from domain-specific applications to develop more general approaches.

For instance, in Zhang et al. (1998) a video parsing algorithms for broadcast news is described which takes advantage of models of anchor and reporter shots to go beyond transition detection (e.g., cuts, fades) to classify shot segments. This research was later developed to obtain *NetView* Zhao et al. (1999), an integrated system for selecting and accessing distributed video data. Other approaches address content-based access to video via indices of associated streams of spoken and written language (e.g., closed caption text).

9.4.4. MPEG 7

In October 1996, MPEG started a new activity in order to provide a standard solution to some of the indexing and retrieval problems described above. The new MPEG standard, called *Multimedia Content Description Interface* (in short 'MPEG-7'), addresses content-based indexing by including several new data types.

While much work has been done in defining the scope of the standard and its requirements, the reader should be aware that it is likely to evolve considerably. A general description of MPEG-7 requirements can be found in (MPEG Requirements Group 1998).

The emphasis of MPEG-7 is the description of audio-visual content. Specifically, MPEG-7 defines a standard set of descriptors that can be used to describe various types of multimedia information. MPEG-7 will also standardize ways to define custom descriptors and structures (*Description Schemata*) for the descriptors and their relationships. The combination of descriptors and description schemata shall be associated with data content, to allow fast and efficient searching for material of a user's interest. MPEG-7 includes a fully-fledged language to specify description schemes, i.e. a data *Description Definition Language* (DDL).

Figure 9.4 shows the extensibility of the above concepts. The arrow from DDL to DS Repository means that the Description Schemata are generated using DDL. Furthermore, the figure shows the extensibility of MPEG-7 schemata, i.e. the fact that it is possible to build a new schema by reusing an existing one.

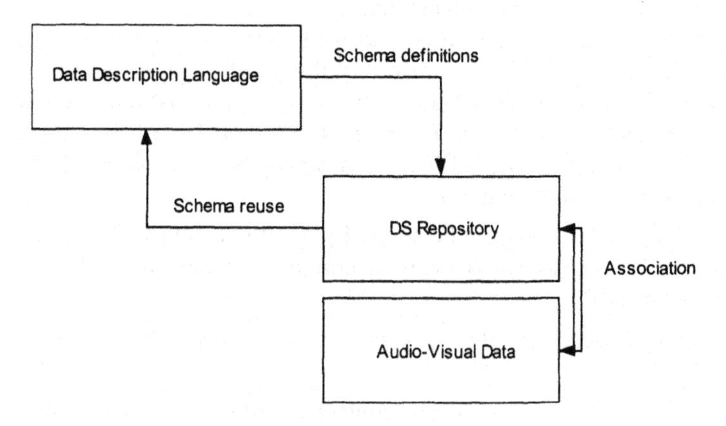

Figure 9.4. Relations between MPEG DDL and DS.

Multimedia information that can be equipped with MPEG-7 descriptions includes still pictures, graphics, 3D models, audio, speech, video, and how these elements are combined in multimedia presentations. Special cases of these general data types include facial expressions and personal characteristics.

The MPEG-7 standard builds on previous (standard) representations such as MPEG-1, -2 and -4. A main feature of the standard is to provide references to suitable portions of them. For example, perhaps a shape descriptor used in MPEG-4 is useful in an MPEG-7 context as well, and the same may apply to motion vector fields used in MPEG-1 and MPEG-2.

MPEG-7 descriptors do not depend on the coding or storage characteristics of the described content. It is thus possible to attach an MPEG-7 description to an analogue movie or to a picture that is printed on paper. Even though the MPEG-7 description does not depend on the (coded) representation of the material, the standard builds on MPEG-4, which provided the means to encode audio-visual material as objects having certain relations in time (synchronization) and space. Using MPEG-4 encoding, it was possible to attach descriptions to elements (objects) *within* a scene, such as audio and visual objects. MPEG-7 provides an even finer granularity control in its descriptions, offering the possibility of having different levels of discrimination.

Moreover, MPEG-7 descriptive features depend on the application context, i.e. they are different for different user domains and applications.

This implies that the same multimedia information can be described using different types of features, tuned to the area of application. In the case of imagery, a lower abstraction level is the description of shape, size, texture, color, movement (trajectory) and position. Analogously, low abstraction description for audio data includes key, mood, tempo, tempo changes, and position in sound space. The highest level would give semantic information, specifying *objects* in the semantic space of the application. Intermediate levels of abstraction may also exist.

The level of abstraction is related to the way the features can be extracted: low-level features can be usually extracted in fully automatic ways, whereas high level features often need more human interaction.

MPEG-7 data may be physically located with the associated multimedia information, in the same data stream or on the same storage system. When the content and its descriptions are not co-located, mechanisms that link AV material and their MPEG-7 descriptions are useful; these links should work in both directions. Besides content descriptions, MPEG-7 metadata also include other types of information about the multimedia data:

- *Form* - Such as the coding scheme used (e.g. JPEG, MPEG-2), or the overall data size. This information helps determining whether the material can be 'read' by the user;
- *Conditions for accessing the material* - Including copyright information, and price;
- *Classification* - Parental rating, and content classification into a number of pre-defined categories;
- *Links to other relevant material* – Used in order to increase search speed.
- *Context* - The occasion of the recording (e.g. Winter Olympic Games 2006, Turin, Italy)

An important distinction between MPEG-7 and other description techniques presented in the previous sections is the concept of *on-line indexing*, which means that information is associated with multimedia content while it is being captured. MPEG-7 addresses both multimedia data that can be stored (on-line or off-line) and streamed (e.g. broadcast, push models on the internet), and can operate in both real-time and non real-time environments.

Figure 9.5 shows a highly abstract block diagram of the MPEG-7 processing chain, including feature extraction (analysis), the description itself, and the search engine (application). To fully exploit the capabilities of MPEG-7 descriptions, automatic extraction of features (or 'descriptors') will be extremely useful; however, it is widely recognized that automatic extraction is not always possible, especially at the higher level of abstraction. For this reason, neither automatic nor semi-automatic feature extraction algorithms are inside the scope of the standard. Also the search engines are not specified within the scope of MPEG-7. The subject of semantics-aware searching will be addressed in the next sections.

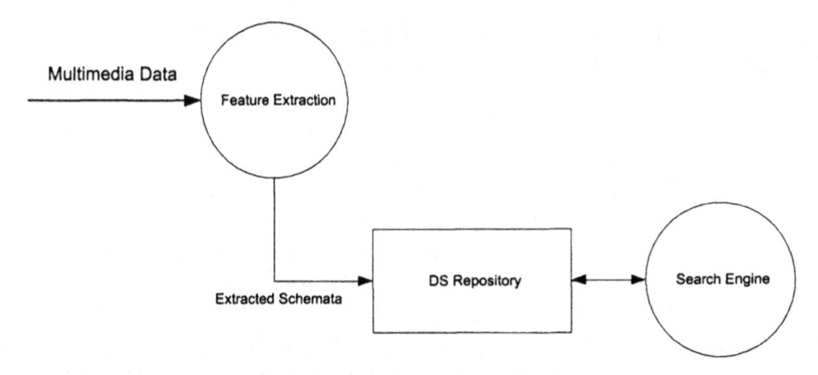

Figure 9.5. Processing Chain of MPEG-7

Figure 9.5 presents a general view of MPEG-7 operation. It should be noted that besides the descriptors themselves, the database structure plays a crucial role in the final retrieval's performance. To allow the desired fast judgment about whether the material is of interest, the indexing information will have to be structured, e.g. in a hierarchical or associative way (MPEG Requirements Group 1998). It should be noted that besides the descriptors themselves, the database structure plays a crucial role in the final retrieval's performance. To allow the desired fast judgment about whether the material is of interest, the indexing information will have to be structured, e.g. in a hierarchical or associative way (MPEG Requirements Group 1998).

Additional details regarding the MPEG-7 call for proposals can be found in the MPEG-7 Evaluation Document (MPEG Requirements Group 1998) and in the MPEG-7 Proposal Package Description (PPD) (MPEG Requirements Group 1998).

9.4.5. SMIL

The *Synchronized Multimedia Integration Language* (SMIL) (SMIL Specification 1998) is a new standard, recommended by the World Wide Web Consortium (W3C), aimed at defining synchronized multimedia presentations on the World Wide Web where audio, video, text and graphics are combined in real-time. Experience from both the CD-ROM and the Web research communities suggested a declarative format for expressing media synchronization on the Web as an alternative and complementary approach to scripting languages. The Synchronized Multimedia Working Group was established in 1997 by W3C, and focused on the design of a declarative language for multimedia presentation. The outcome of this work gave rise in June 1998 to the SMIL specification. Philipp Hoschka chaired the working group and was the editor of the specification (SMIL Specification 1998). The SMIL language is written as an XML application; it enables authors to specify temporal constraints for the presentation of multimedia information. For example, SMIL statements can control the precise time that a sentence is spoken and make it coincide with the display of a given image appearing on the screen. As a Web-oriented standard (unlike, for instance, Microsoft PowerPoint), the primary requirement underlying SMIL design was not to require full download to begin displaying multimedia elements. The SMIL file itself contains only information that describes the presentation, but not the presentation elements itself.

Another requirement taken into account by SMIL is the possibility of selecting alternative multimedia elements, depending on the available bandwidth and the language of the user. For example, a SMIL presentation may choose the most suitable among several different versions of a video clip file, according to the quality of the network link between the user and the server site.

The basic idea of SMIL is to name media components for text, images, audio and video within a *Uniform Resource Identifier* (URI) namespace and to schedule their presentation either in parallel or in sequence. The components have different media types, such as audio, video, image or text. Begin and end times of different components are specified in SMIL relative to events in other media components. For example, in a slide show, a particular slide is displayed when the narrator in the audio starts talking about it. Familiar looking control buttons such as “stop”, “fast-forward” and “rewind” allow the user to interrupt the presentation and to move forwards or backwards to another point in the presentation.

Additional functions are *random access*, meaning that the presentation can be started anywhere, and *slow motion*, i.e., playing the presentation slower than its original speed. The SMIL standard also supports hypertext links embedded in the presentation. Linking in SMIL is unidirectional: all links have exactly one source and one destination. As SMIL supports the URI addressing scheme, internet URLs are a part of this convention. SMIL allows for some flexibility in the specification of link targets, as it supports the same "#"-connector used in HTML documents. Using this symbol, it is possible to link to a part of a SMIL document.

```
<smil>
  <head>
    <layout type="text/smil-basic">
      <region id="slide-area"     left="20"  top="50"/>
      <region id="slide-comments" left="20"  top="120"/>
    </layout>
  </head>
  <body>
    <seq>
      <par id="firstslide">
        <img  href="firstslide.gif"  region="slide-area"/>
        <text href="firstslide.html" region="slide-comments"/>
        <audio href="firstslide.mp3"/>
      </par>
      <par id="secondslide">
        <img  href=" secondslide.gif"  region="slide-area"/>
        <text href=" secondslide.html" region="slide-comments"/>
        <audio href=" secondslide.mp3"/>
      </par>
    </seq>
  </body>
</smil>
```

Figure 9.6. A Sample SMIL Presentation

Figure 9.6 contains a simple example of a SMIL presentation, which displays two "slides". The slides are shown in two regions: "slide-area" and "slide-comments", as defined in the `<region>`-element. The presentation starts with the first `<par>`-element. A `<par>` element contains objects that are displayed in parallel, at the same moment. In the region "slide-area", a GIF-image shows up (as defined by the attribute loc.) and in the region "slide-comments", a html-document is displayed. At the same time, a MPEG-3 audio file starts. When the audio stops, the next slide is displayed, with the ID "secondslide". It replaces all objects by their successors.

9.5. Sample Applications to Medical Multimedia Information Processing

Medical multimedia repositories are an ideal test bed for many of the notions developed in the previous sections of this chapter. Indeed, in the past few decades, physicians have been increasingly basing their diagnoses on high-tech laboratory results.

While the tools used for diagnoses, such as *Magnetic Resonance Imaging* (MRI) scans, have improved greatly over the years, the method of storing these results into a medical record has not changed much. In fact, the medical system is still using paper-based record, and this results in inefficiency. Medical professionals are confronted with a basic issue: How should they manage the huge amount of different types of information so that they can improve medical treatment and upgrade patient care?

A medical *multimedia repository* can be defined as a facility for storage and retrieval of specialized medical experience available as multimedia data.

Medical multimedia repositories require massive storage and links to a vast selection of supporting resources enabling users to retrieve, cooperate, and learn. This section outlines the possible use of a multimedia database to create a medical information system. Though our discussion is loosely based on some advanced real world applications, such as the *National Medical Multimedia Knowledge Bank (NMKB)* (Sterling 1996) designed by Warren Sterling's group at Teradata, the Data Warehouse Solutions division of NCR, we shall not describe a single system in detail. Rather, we try to focus on some architectural and design issues widely understood to be of general applicability.

As we saw in previous sections, the core of any multimedia architecture is a fast database engine with multimedia extensions. The NMKB project relies on a high-performance parallel object/relational database, while less recent systems (Khosafian and Brad Baker 1996) exploited pure relational engines with multimedia extensions.

The database engine serves as a central data repository, capable of storing, retrieving and analyzing multimedia data such as medical images (e.g. MRI scans, digital x-rays images), 3D images for operating-room reference (Jolesz 1997), video clips showing surgical and other health related procedures and other imagery such as photographs, graphics to accompany video presentations. Of course, the central repository also stores a wealth of information in the form of text documents and audio clips. Recent systems provide a simple Web based interface for exploring the available information.

Besides the multimedia database engine, the architecture comprises other components, such as simple user interfaces for healthcare practitioners, some medical diagnosis CBR, the formatting of patient medical records to support CBR, automated analysis of magnetic-resonance brain scans, and virtual conferences.

9.5.1. Architectural Design

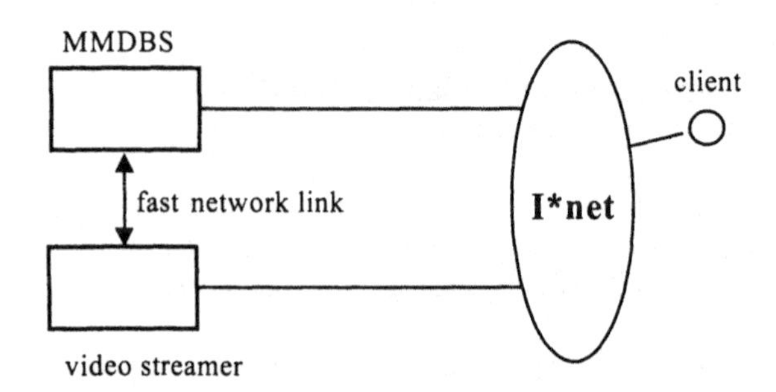

Figure 9.7. The Architecture of a Medical Multimedia Database

The design of a medical multimedia database system is often based on a three-tiered design pattern. The first tier is the web client, which uses standard browser technology and appropriate multimedia repository plug-ins for displaying various types of multimedia data. The second tier is the WWW server acting as a web interface to the multimedia repository and the application server. The application server is a software layer responsible for implementing the appropriate multimedia repository application, such as information retrieval, case based medical diagnosis or medical education. One important task of the application server is to update and maintain references to multimedia data stored in the database within web pages. The third tier is the multimedia database itself. This layer acts as a media server that streams images, video and audio to clients over the Intranet/Internet. In the remainder of this section we shall briefly outline some aspects of such a three-tiered system design especially relevant for multimedia applications.

The second-tier video support

To deliver video to clients across the Intranet/Internet with minimal latency and small client buffer space, the system's second tier may include a video streamer, a software component capable of generating concurrent streams of video to clients. Selected videos are retrieved from the database and cached on the media server, whence they are streamed to the clients. It should be noted that this technique eliminates the need to support video Quality of Service directly from the database but adds the latency of the full transfer of the video to the media server. A high-speed sub-network between the database engine and the video streamer may help to reduce this latency.

The third-tier database engine

Extremely high-performance database engines are required to service the load imposed by multimedia data. The NMKB project database engine, for instance, is based on massively parallel technology originally developed for the Teradata Database, capable to handle multi-terabyte databases in mission-critical applications (Sterling 1996). The query language for the database is a multimedia extension of SQL based on Structured Query Language Three (SQL-3) (section 9.3.2).

In such an object/relational database, t-uples contain both the built-in, or alphanumeric, attributes normally associated with relational databases, and attributes such as the video, image, audio, and document attributes as UDTs (section 9.3.1). Associated with UDTs are user-defined functions (UDF) that are used to process UDT values. For example, a UDT for Graphic Interchange Format (GIF) images may have a set of UDFs that perform standard image processing operations (rotate, crop, edge enhancement, etc.) on the images.

9.5.2. Usage Scenarios: Computer-Assisted Medical Diagnosis

Besides straightforward applications to medical training and cooperative work, usage scenarios include the promising field of computer-assisted diagnosis and surgical therapy (Jolesz 1997).

The primary technology applied for this application is case based reasoning; as we shall see in the next chapter, hybrid techniques have also been proposed.

Computer-assisted diagnosis involves retrieving medical cases from the system, which are similar to a case currently being analyzed by a physician.

In the domain of neurology, for instance, research is focusing on automating analysis of MRI brain scans to detect and characterize brain tumors, aneurysms and bleeds (Liu and Dellaert 1998) (Black 1997) and then determining how to use the derived attributes to measure case similarity.

9.6. Fuzzy Query Processing in Multimedia Databases

As we have seen in the previous sections, queries to multimedia databases involve several specific issues with respect to conventional database querying. For example, in a multimedia database it is reasonable to ask for images that are "similar to" a given reference picture, and to require the answer to be a ranked list instead of a set. In recent years several researchers, notably, R. Fagin (Fagin 1998) proposed to use fuzzy predicates (e.g. "*like*") to model multimedia data similarity, and described techniques to execute "mixed queries", i.e., logic formulae including both fuzzy predicates (on multimedia data) and Boolean predicates (on associated conventional data types).

In this setting, a multimedia database query $Q(Y_1, \ldots, Y_n)$ can be informally seen as a set of fuzzy and crisp predicates, together with constants, variables and, obviously, conjunction, disjunction and negation operators. Result to such a query is, in turn, defined as a list (and not, as usual, an unordered set) of n-tuples of the form $< X_1, \ldots, X_n >$.

In Candan and Li (2000) and Dubois et al. (1999) the problem of computing such a list is stated as follows: "Given a multimedia query that involves multiple crisp and fuzzy predicates, find an efficient way to process it".

The straightforward approach to this problem is to use fuzzy logical operators that are natural extensions of the crisp ones, dealing with ordinary Boolean predicates as if they were fuzzy predicates with 0 or 1 truth-value.

There are however some disadvantages to this approach: for instance, straightforward extensions of crisp logical operators (e.g. *min* for AND) inherit for their crisp counterparts the *absorption* property (one of the argument being zero means the result will be zero regardless of other values) and the *non-strict-monotonicity* property (an increase in one of the operands does not ensure the result to increase if the other operand does not increase as well). Other classical definitions of binary operators used in fuzzy logic (such as those based on the product of the truth-values) introduce other problems, as they lack the distributivity property and are therefore unfit for query optimization.

Some researchers proposed the solution of considering queries in disjunctive normal form and to resort to the direct evaluation of n-ary operators. The recent paper (Candan and Li 2000), for instance, contains a novel product-based definition for n-ary fuzzy logical operators that is strictly monotonic and exhibits no absorption. Moreover, the proposed operators retain good statistical properties; for instance, the proposed conjunction has a fairly steep slope (though, of course, not as steep as *min* conjunction). This confines high result values to a small region, and keeps the definition suitable for effective execution of similarity queries with threshold.

In this setting, a query evaluation algorithm for unordered predicates can be given (Candan and Li 2000) whose complexity is close to the lower bound given by Fagin (Fagin 1998) in the ordered case. The statistical properties of the operators favor query optimization.

However, this approach has its own drawbacks: either one keeps Boolean predicates together with fuzzy ones, loosing some optimization opportunities, or the proposed improvements come at some expense of generality, as fuzzy predicates about multimedia data have to be dealt with separately from crisp ones about other data.

This is perhaps not a very realistic assumption in real systems where QBE interfaces allow the user to specify both.

For instance, consider a naïve query like

(I.Date='12/12/99' AND I.Filesize=1MB AND like(I.image a.gif)) OR
(I.Date='12/12/99' AND I.Filesize=2 MB AND like(I.image a.gif))

Since like is a fuzzy predicate, we could resort to classical *min* conjunction and factor the mixed query as usual, e.g. as follows:

(I.Date='12/12/99' AND like(I.image a.gif) AND(I.Filesize=1MB OR I.Filesize=2 MB))

Using product-based semantics for conjunction, factorization is no longer possible and one needs to evaluate directly 3-ary conjunction. If the third predicate is Boolean as well (suppose for instance that instead of like (I.image a.gif), the query includes predicate I.Creationtime='12.00') either the system looses the opportunity for optimization or it passes on the query to another, conventional execution engine.

9.7. Representation and Querying of Multimedia Data on the WWW

Due to the ubiquitous nature of the WWW, more and more audio-visual information in digital form is becoming available through the internet. Before users can access such information, however, it has to firstly be located. Currently, many techniques exist allowing Web-based searching for textual information. Hundreds of search engines are available on the World Wide Web, and their huge number of visitors indicates they foresee a real demand. Identifying information is, however, more difficult for audio-visual content, as no generally recognized description of this material exists.

Multimedia database techniques described in the previous sections allow searching for pictures using characteristics like color, texture and information about the shape of objects in the picture. The huge amount of multimedia and semi-structured information published via the World Wide Web has led to a number of research efforts on techniques to index, query and restructure WWW sites' contents.

In many cases, these approaches differ significantly from those used by multimedia databases, as multimedia content representation must be human (and agent) readable as well as machine readable, making some of the indexes presented in the previous section wholly unfit to the task.

In this section we provide a brief overview of techniques for multimedia representation and retrieval on the WWW (see also (Florescu et al. 1998)). Our discussion is based on how the various approaches deal with semantics-aware representation of Web contents.

Free text indexing - No content representation

Early approaches to Web indexing were text-oriented inasmuch as they tried to collect and index title-like information about every reachable page of data on the WWW and then build Boolean keyword searches into the resulting document. Many current Web search engines are still partially based on this approach, where search results are flat lists of HTML pages, completely unrelated to the hypertext structure of the sites they come from. In the past few years, as the amount of WWW multimedia data continued to increase, users grew dissatisfied with pure keyword matching.

Nowadays, one could hardly find a Web search engine relying on terms indexing alone. Some keyword-based indexes try to complement keyword indexing by taking into account HTML tagging in order to make educated guesses about multimedia data semantics. For example, the Lycos search engine summarizes the actual content of documents by taking advantage of human-tagged information (HTML headings), of often-appearing keywords and of the introductory text that generally is positioned at the beginning of a file.

Semantics representation via taxonomies

Several search engines do not use keyword indexing but exploit a taxonomy representing sites' content. Another popular search engine, Yahoo, relies on a broad hierarchical classification system of subjects, much similar to those used by the Library of Congress or by the ACM Classification of Computer Science Topics.

Structural representation of sites

A considerable amount of research has been made on how to complement keyword-based searching with database-style support for querying the Web. Several projects addressed this problem, and several WWW query languages have been proposed so far, among them Web3QL (Konopnicki and Shmueli 1995) and WebSQL (Mendelzon et al. 1996). None of those approaches is specifically aimed at multimedia, as these languages are modeled after standard SQL used for RDBMS. WebSQL offers a standard relational representation of Web pages, such as `Document(url, title, text, type, length)`, which can be easily constructed from HTML tagging. The user can present SQL-like queries to Web sites based on that relational representation. Content-related queries (for instance: `Document.text` = "Italy"}) are mapped in free-text searches using a conventional search engine.

In addition to similar query capabilities, Web3QL offers an elementary graph pattern search facility, allowing users to search for simple paths in the graph representing the navigational structure of a Web site.

Besides query languages, other proposals aimed at Web retrieval were made over the years: the Dublin Core (Dublin Core 1998) is a fifteen-element metadata set that was originally developed to improve resource discovery on the Web.

Dublin Core elements are intended to describe document-like objects, specifying metadata such as Title, Subject, Description, Format and others. However, site semantics is only conveyed as natural language descriptions and no formal representation of such semantics is given. A richer Web object model is also described in Manola (1998), but again it lacks any representation of multimedia data semantics.

Explicit representation of data semantics

Nowadays it is widely recognized that to effectively build multimedia Web-based services, developers must be able to superimpose some sort of semantic structure upon Web sites in order to support efficient information capture. In fact, the subject of semantics representation for multimedia is currently actively investigated. A well-known technique for instance-based representation of multimedia content is semantic tagging, i.e., the use of HTML or XML tags to represent semantic information. Several systems are available to exploit semantic tagging; (Sheth and Kshah 1999) presented VisualHarness, a system aimed at searching heterogeneous digital media accessible on the internet and Intranets. Through the extensive use of metadata, VisualHarness can support keyword-based, attribute-based and content-based searches, as well as their combinations. The content-based search is particularly facilitated by the black-box approach that can use third party engines to deal with specific types of digital media, coercing them to extract metadata. Once the metadata has been extracted, the Web-based VisualHarness server can provide access to Web-accessible data using any combination of the three search alternatives and allowing the user to combine access to data of heterogeneous media.

A related line of research pursued by the same group aims at supporting the concept of information correlation involving heterogeneous digital media (called MREF) and corresponding information request processing techniques. As we will see in the following, both the Resource Description Format (RDF) and the eXtensible Markup Language (XML) standards have been exploited for MREF representation and implementation.

9.7.1. Markup-Languages and Multimedia

The use of mark-up languages in the context of multimedia content representation and retrieval long preceded the internet and WWW diffusion. For instance, the Structured Information Manager (SIM) developed at University of Melbourne and at the Royal Melbourne Institute of Technology (RMIT) was designed to manage multi-gigabyte collections of documents containing text, images and other kinds of data.

SIM provides for the representation of document structure based on logical units such as titles, chapters, paragraphs, figures, and so on, relying on special-purpose markup languages defined via the *Standard Generalized Markup Language* (SGML), the ISO standard for the representation of the logical structure of documents as well as the text contents.

In order to represent the complex and diverse structures found in documents, SGML uses a grammar called a *DTD* (*Document Type Definition*) rather than a schema with a fixed number of attributes as usual in conventional database systems. SIM uses the SGML DTD directly as the database schema to specify the structure of documents. As far as the query language is concerned, SIM provides comprehensive Boolean querying capabilities for searching text, while supporting queries by providing automatic relevance judgments.

More recently, the availability of a lightweight standard for the representation of content and structure such as XML revived the interest for the use of markup languages in association with multimedia information. As shown in Figure 9.8, XML readily lends itself to defining the hierarchical structure of video information:

```
<xml version="1.0">

<!DOCTYPE videodoc [

<!- hierarchical structure of videodoc -!>
<!ELEMENT videodoc (sequence*)>
<!ELEMENT sequence (scene*)>
<!ELEMENT scene (shot*)>
<!ELEMENT shot (frame*)>
<!ELEMENT frame(object*)>
<!ELEMENT object(object*)>
```

Figure 9.8. An XML DTD Fragment for Video Information

XML query languages have been proposed (Ceri et al. 1999) that could be used for querying multimedia data. However, XML provides neither inheritance hierarchies nor adequate data typing for representing multimedia information.

Indeed, it is a moot point whether existing languages, formats, and multimedia document models such as HTML, MHEG, SMIL, HyTime, SGML, and XML, provide the appropriate modeling primitives needed to support interaction, adaptation, and presentation-neutral description of content of multimedia documents.

Besides XML, the RDF standard (Berners-Lee 1998), (Powell 1998)) shows some promise RDF has been developed as a part of the W3C metadata activity in order to provide a generic metadata architecture for Web sites that can be expressed in XML. RDF descriptions are based on assertions, i.e. triples made up of a `resource`, a `propertyType` and a `value`. Though `propertyTypes` can be thought of as attributes in traditional attribute-value pairs, RDF's choice of using the URI namespace for resources makes the assertion-based model very different from the standard Entity-Relationship model.

In fact, in conventional Entity-Relationship models each entity has its own set of attributes. In the RDF approach, the site publisher is completely in charge of consistent use of attributes to describe multimedia data content. A RDF schema definition for video information adapted from (Hunter and Armstrong 1999) is reported in Figure 9.9:

The technique seems indeed appealing, as it is both human- and machine-readable and simple to understand and customize.

However, query languages for RDF are not yet available, and many problems remain to be solved, as it appears to be difficult to use RDF to build a hierarchical structure based on part-of relationships and the limited data typing available within RDF.

Since each of these models seems to lack some significant concepts, Boll and Klas (Boll and Klas 1999) proposed a new approach for the semantic modeling of multimedia content, the *ZYX* model, later implemented on the basis of an object-relational database system. The ZYX approach allows for fine-grained representation and retrieval of structures and layout of multimedia material, providing flexible on-the-fly composition of multimedia fragments in order to create individualized multimedia documents. This allows for the realization of adaptation and personalization of multimedia presentations depending on the user environment specified by means of user profiles.

Much in the same line, John R. Smith and Shih-Fu Chang of the Columbia University proposed WebSeek, a content-based image and video search and catalog tool for the Web, which is fully operational at the URL http://disney.ctr.columbia.edu/WebSEEk/.

Another approach, conceived specifically for Web-based information, is based on the *WG-log* representation and query language (Damiani and Tanca 1997).

```
<rdfs:Class ID="MM_document">
<rdfs:comment>Class for representing a generic multimedia
  document</rdfs:comment>
</rdfs:Class>
<rdfs:comment>Define all of the DC elements for MM_document
  </rdfs:comment>
<rdf:PropertyType ID="Title">
<rdfs:comment>This is the DC Title element </rdfs:comment>
<rdfs:domain rdf:resource="#MM_document">
<rdfs:range
  rdf:resource="http://purl.org/metadata/dublin_core#Title"/>
</rdf:PropertyType>
<rdf:PropertyType ID="Creator">
<rdfs:comment>This is the DC Creator element </rdfs:comment>
<rdfs:domain rdf:resource="#MM_document">
<rdfs:range
  rdf:resource="http://purl.org/metadata/dublin_core#Creator"/>
</rdf:PropertyType>
etc.
<rdfs:Class ID="Video">
<rdfs:comment>Class for representing a video document. It is a
  subclass of MM_document</rdfs:comment>
<rdfs:subClassOf rdf:resource="#MM_document"/>
</rdfs:Class>
<rdfs:Class ID="Sequence">
<rdfs:comment>Class for representing a sequence from a video
  document. It is a subclass of MM_document</rdfs:comment>
<rdfs:subClassOf rdf:resource="#MM_document"/>
</rdfs:Class>
<rdfs:Class ID="Scene">
<rdfs:comment>Class for representing a scene. It is a subclass of
  MM_document</rdfs:comment>
<rdfs:subClassOf rdf:resource="#MM_document"/>
</rdfs:Class>
<rdfs:Class ID="Shot">
<rdfs:comment> Class representing a shot</rdfs:comment>
<rdfs:subClassOf rdf:resource="#MM_document"/>
</rdfs:Class>
<rdfs:Class ID="Frame">
<rdfs:comment> Represents a single frame. It is a subclass of
  #MM_document</rdfs:comment>
<rdfs:subClassOf rdf:resource="#MM_document"/>
</rdfs:Class>
<rdfs:Class ID="Object">
<rdfs:comment> Represents an object within a frame. It is a
  subclass of #MM_document</rdfs:comment>
<rdfs:subClassOf rdf:resource="#MM_document"/>
</rdfs:Class>
```

Figure 9.9. A Sample RDF Schema Definition for Video Information

9.8. Flexible Queries to Web-based Multimedia Sources

As we have seen in the previous sections, descriptors of multimedia data stored on the World Wide Web are a particularly interesting example of semi-structured information. At a typical Web site, data is varied and irregular, and the overall structure of the site changes often. Of course, Web site data may sometimes be fully unstructured, consisting in collections of images, sound and text. Other times, especially on Intranets, WWW data are extracted from traditional relational or object-oriented databases, whose completely specified semantics is exactly mirrored by the site structure.

Often, however, WWW information lies somewhere in between these extremes. In fact, the availability of effective design tools for hypermedia has profoundly affected the design of Web sites. Several methodologies such as HDM (Garzotto et al. 1995) or RMM (Isakowitz et al. 1995) provide to the Web site designer some more or less formal means to express data semantics. Based on these methodologies, research prototypes of Web site generators and many commercial Web authoring environments are currently available and widely used.

Today, very few Web sites store all their available information in a database system; it is clear, however, that Web users could take advantage of database support, e.g., by having the ability to pose queries involving logical data relationships. In order to design a complete Web Query System for multimedia data the following basic requirements must be taken into account:

- Semantics aware representation and indexing of semi structured and multimedia data sources
- Efficient querying
- Data presentation depending on the user's profile.

The WG-Log language (Damiani and Tanca 1997), (Comai et al. 1998) deals with these requirements, exploiting a schema-based representation of WWW semi-structured data. Wg-log is a graph-oriented language supporting the representation of logical as well as navigation and presentation aspects of hypermedia. The WG-log approach is also closely related to the XML-GL proposal for a query language for XML (Ceri et al. 1999).

A data description and manipulation language for the Web must gracefully support schemata that are huge or subject to change. Moreover, it easily retains the representation of semantics created during the design process, allowing Web users to exploit this semantics by means of a database-like querying facility. In our opinion, the availability of a schema is the key for providing effective query execution and view construction mechanisms. While expressing clearly Web site semantics, WG-log schemata take also into account the semi-structured nature of WWW information, providing:

- Graceful tolerance of data dynamics, i.e., an easy mechanism for schema updates resulting from instance evolution over time
- Efficient checking of instance correctness with respect to a given schema
- Efficient checking (i.e. at the schema level) of query applicability to a certain instance

- Direct semantics-aware querying, without forcing the conversion of data to a fully structured format.

The advantages of this approach are many, the first one being the immediate availability of a uniform mechanism for query and view formulation. Uniform graph-based representations of schemata and instances are used for querying, while Web site data remain in their original, semi-structured form. Thus, WG-log does not require Web site conversion to a fully structured format to allow for database-style querying.

Schema availability supports adaptability allowing users to build queries that are partially specified with respect to the schema; the expressive power of WG-log, which is fully equipped with recursion, is another powerful resource to this effect.

Adaptability is also supported by WG-log's ability of building custom views over semi-structured information. When a new view over a site is created, clients can formulate queries to include additional information and a good deal of restructuring to the result. In fact, this mechanism can be exploited to reuse existing sites' content as well as schemata to produce new sites.

9.9. Open Problems

As we have seen in previous sections, multimedia researchers have long since begun to address the issues of developing frameworks and facilities to support both integration and analysis of multimedia data, as well as processing. (Griffioen et al. 1995) describes *MOODS*, a framework for developing content-based retrieval applications that allow users to go beyond searching for features such as colors and textures by combining a database, processing engine, and knowledge base. Their framework is illustrated in its application to music note recognition and ancient manuscript analysis. In addition to tools and frameworks to control and coordinate processing, multimedia systems also require multimedia query and analysis tools that go beyond the visual query reported in earlier sections.

The more recent *Mirror* project (deVries 1998) investigates the implications of multimedia information retrieval on database design. It assumes a modern extensible database system with extensions for feature based search techniques. In the Mirror approach, a multimedia query processor has to bridge the gap between the user's high level information need and the search techniques available in the database. We therefore propose an iterative query process using relevance feedback. The query processor identifies which of the available representations are most promising for answering the query. In addition, it combines evidence from different sources. Mirror multimedia retrieval model turns out to be a generalization of a well-known text retrieval model, based on Bayesian reasoning over a concept space of automatically generated clusters. The experimentation platform uses structural object-orientation to model the data and its meta-data flexibly, without compromising efficiency and scalability. While this approach is indeed interesting, several problems must be solved in order to enable multimedia indexing, search, and navigation for large scale, heterogeneous collections. The most significant challenges at system level (see also section 9.3 of this chapter) include:

- *Scalability and Performance*: Dealing with orders of magnitude larger volumes of multimedia information, that support (storage and time) efficient indexing and real-time retrieval.
- *Portability*: Creating algorithms that rely minimally on domain-specific knowledge and can rapidly be applied to new multimedia collections.
- *Robustness*: Dealing with increasing levels of irrelevance, dirty data and unstructured data.
- *Extensibility*: As new processing algorithms emerge (e.g., for indexing, search, extraction, summarization), system architectures should be sufficiently flexible to support seamless integration with existing approaches (e.g., augmentation a video indexing system with a speaker identification or face identification algorithm).
- *Usability*: A mechanism that can mitigate complexity to ensure user-friendly and learnable interfaces that ameliorate task accomplishment.

9.10. Summary

Intelligent multimedia information retrieval is improving information access in three principal ways. First, it ensures more effective access to content: getting the right data and tailoring it to the context of the user, their task, and their environment. The goal of achieving context sensitivity is limited only by the richness of models that can be created. Second, by providing only the most relevant information, together with high performance browsing and search tools. Finally, enabling the user to ask for and receive information in a natural manner (e.g., by speaking, drawing, or pointing to similar artifacts) providing a less stressful and more pleasant interaction.

However, as we have seen, many problems are still to be solved, especially those involved with relating the symbolic and sub-symbolic content related parameters for indexing and retrieval. Users see the world in terms of objects and complex shapes, while database indexing and retrieval is often available in terms of low-level parameters as color and elementary shapes. Many approaches have been proposed to solve this fundamental problem, in order to put the user's perception of the information space at the center of the search activity.

Acknowledgements

The authors wish to thank Gianni Degli Antoni and Aaron Sood for some useful discussions on multimedia information processing. Thanks are also due to Patrick Bosc for his valuable comments on the perspectives of fuzzy queries to multimedia databases.

References

Aberer K. and Klas J. (1994), "Supporting Temporal Multimedia Operations In Object-Oriented Database Systems. *Proceedings of the IEEE International Conference on Multimedia Computing and Systems*, Los Alamitos, CA (US) pp. 352-361

Aberer K. and Hollfelder S. (1999), "Resource Prediction and Admission Control for Video Browsing", In R. Meersman, Z. Tari, S. Stevens (eds). *Proceedings of the IFIP Working Conference on Database Semantics DS8*, Kluwer, 1999

Adali S., Candan K.S., Su-Shing Chen, Erol K. and Subrahmanian V.S (1999), "Advanced Video Information System: Data Structures and Query Processing *ACM-Springer Multimedia Systems Journal* vol. 2 no.3 pp. 187-210

Bach M. (1986) *The Design of the Unix Operating System*, Prentice-Hall, 1986

Berson, S., Muntz, R., Ghandeharizadeh, S. and Xiangyu, J. (1994) "Staggered Striping in Multimedia Information Systems", *Proceedings of SIGMOD-94*, pp.321-335

Berners-Lee T. (1998) "Introduction to RDF MetaData" available at http://www.w3c.org

Black P.M. (1997), "Development and Implementation of Intraoperative Magnetic Resonance Imaging and Its Neurosurgical Applications", *Trans. on Medical Imaging*, vol. 15 no. 4 pp. 67-75

Blum T., Keislar D., Wheaton J. and Wold E. (1995), "Audio Databases with Content Based Retrieval", *Proceedings of the International Joint Conference on Intelligent Multimedia Information Retrieval*, Montreal (Canada) pp.95-102

Boll S. and Klas W. (1999), "ZYX: A Semantic Model for Multimedia Documents and Presentations", *Proceedings of the IFIP Working Conference on Database Semantics DS8*, Kluwer

Cattell R. (1991), *Object Data Management: Object-Oriented and Extended Relational Database Systems,* Addison-Wesley.

Cattell, R. (1994) "The Object Database Standard: *ODMG-93* (release 1.1)", Morgan Kaufmann

Cattell R. (1996), "The Object Database Standard: ODMG-93 (release 1.2)", Morgan Kaufmann

Candan K. and Li W.S. (2000), "Similarity Based Ranking and Query Processing in Multimedia Databases", to appear in *Data Knowledge Engineering*

Ceri S., Comai S., Damiani E., Fraternali P., Paraboschi S. and Tanca L. (1999), "XML-GL: A Graphical Query Language for XML", Proceedings of the 8th World Wide Web Intl. Conference, Toronto, Canada pp.93-110

Chang W., Murphy D., Zhang A. and Syeda-Mahmood T. (1997), "Metadatabase and Search Agent for Multimedia Database Access over internet. *Proceedings of the 4th IEEE International Conference on Multimedia Computing and Systems (ICMCS '97)*, Ottawa, Canada

Chen C., Pau L. and Wang P. (1993), "Handbook of Pattern Recognition and Computer Vision", *Singapore World Scientific*

Christodoulakis S. (1986), "The Multimedia Object Presentation Manager of MINOS: A Symmetric Approach *Proceedings of the SIGMOD '86 Conference*, pp. 118-126

Comai S., Damiani E., Posenato R. and Tanca L. (1998) "A Schema-based Approach to Modeling and Querying WWW Data" *Proceedings of FQAS'98,* Roskilde, Denmark, LNAI 1495, Springer

Damiani E. and Tanca L. (1997), "Semantic Approaches to Structuring and Querying Web Sites" *Proceedings of 7th IFIP Work. Conf. on Database Semantics (DS-7),* Chapman & Hall

Dublin Core (1998), "Dublin Core Metadata Element Set", available at http://www.purl.oclc.org/metadata/dublin_core

Dubois D., Prade H., Sedes F. (1999), "Fuzzy Logic Techniques in Multimedia Databases", *Proceedings of 8th IFIP Work. Conf. on Database Semantics (DS-8),* Kluwer

Fagin R. (1998), "Fuzzy Queries in Multimedia Database Systems", *Proceedings of the 17th ACM Symp. On Principles of Database Systems (PODS '98)*

Flickner M., Sawhney H. and Niblack W. (1995) "Query by Image and Video Content: the QBIC System", *IEEE Computer,* vol.23 no. 31

Florescu D., Levy A. and Mendelzon A. (1998), "Database Techniques for the WWW: A Survey", *SIGMOD Record*, vol. 3 no. 2.

Fuhrt B., Smoliar S. and Zhang H (1996), *Video Image Processing in Multimedia Systems,* Kluwer, 1996

Garzotto F., Mainetti L. and Paolini P. (1995),``Hypermedia Design Languages Evaluation Issues", *Communications of the ACM* vol. 38 no. 8 pp. 23-30

Gemmell, D.J., Harrick, M.V., Kandlur, D.D. Rangan, P.V. and Rowe, L.A. (1995). "Multimedia Storage Servers: a Tutorial". *IEEE Computer,* vol. 40 no.2, pp.40-49.

Gibbs, S., Breiteneder C. and Tsichritzis D. (1994). "Data Modeling of Time-Based Media" *Proceedings of SIGMOD-94,* pp.81-90

Griffioen J., Yavatikar R. and Adams R (1995), *A Framework for Developing Content-Based Retrieval Systems,* in M.Maybury, (ed.) *Intelligent Multimedia Information Retrieval,* MIT Press

Grosz B., Sparck-Jones B and Webbers B. (eds.) (1986), *Readings in Natural Language Processing, Morgan*-Kaufmann, 1986

Hunter J. and Armstrong L. "A Comparison of Schemas for Video Metadata Representation", *Proceedings of the 8th World Wide Web Conference*, Toronto (Canada) pp.353-374

Isakowitz T., Stohr A., Edward D. and Balasubramanian P. (1995), ``RMM: a Language for Structured Hypermedia Design", *Communications of the ACM* vol. 38 no.8 pp. 74-80

Jolesz F.A. (1997) "Image-Guided Procedures and the Operating Room of the Future", *Neurosurgery*, vol. 41 No. 4, pp. 18-30

Kanehara F., Satoh S. and Hamada T. (1995), "A Flexible Image Retrieval Using Explicit Visual Instruction", *Proceedings of the Third International Conference on Document Analysis Recognition*, Montreal, Canada pp. 20-27

Khoshafian S. and Brad Baker A. (1996), *Multimedia and Imaging Databases,* Morgan-Kaufmann

Kim W. (1990), *Introduction to Object-Oriented Databases,* The MIT Press

Konopnicki D. and Shmueli O. (1995) ``W3QL: A Query System for the World Wide Web", *Proceedings of the 21th Intl. Conf. on Very Large Databases*, Zurich, Switzerland pp. 27-35

Kunii T.L, Shinagawa Y., Paul R.M., Khan M.F. and Khokhar A.A (1995), "Issues In Storage and Retrieval of Multimedia Data". *Multimedia Systems*, vol.3 no.5-6, pp. 34-42

Liu Y. and Dellaert F. (1998), "A Classification-Based Euclidean Similarity Metric for 3D Image Retrieval", *Proceedings of the Computer Vision and Pattern Recognition Conference*, pp.22-30

Manola F. (1998), "Towards a Richer Web Object Model *SIGMOD Record*, Vol. 27 no.1 pp.25-30

Mendelzon A., Mihaila G. and Milo T. (1996), ``Querying the World Wide Web", *Proceedings of the Conf. on Parallel and Distributed Information Systems,* Toronto (Canada)

Niblack W. (1999) "SlideFinder: A Tool for Browsing Presentation Graphics Using Content-Based Retrieval" *Proceedings of the IEEE Workshop on Content-based Access of Image and Video Libraries (CBAIVL-99),* Fort Collins, CO, (US) pp. 181-190

Oomoto E. and Tanaka K. (1993), "OVID: Design and Implementation of a Video-Ohject Database System," *IEEE Trans. on Knowledge and Data Engineering*, vol. 5, no.4, pp.629-643

Oracle Corp, "The Oracle Media Server", available at http://www.oracle.com/products/media_server/html/

Powell A., ``Metadata for the Web: RDF and the Dublin Core", available from http://www.ukolno.uk/metadata/presentations/ukolug98/

Rakow T.C. and Lohr M. (1995) "Audio Support For An Object-Oriented Database Management System". *Multimedia Systems,* vol. 3 no. 5-6 pp. 118-130

Sakauchi M. (1999), "Towards the Construction of the Multimedia Mediation Mechanism" *Proceedings of the IFIP Working Conference on Database Semantics DS8,* Kluwer

Sheth A. and Kshah K. (1999), "Searching Distributed and Heterogeneous Digital Media", *Proceedings of the IFIP Working Conference on Database Semantics DS8,* Kluwer

Sheikholeslami W., Chang G., and Zhang A. (1998), "Semantic Clustering and Querying on Heterogeneous Features for Visual Data *Proceedings of the ACM Multimedia Conference (ACM-MM98)* pp. 234-240

Soffer A. and Samet H. (1999), "Two Data Organizations for Storing Symbolic Images in a Relational Database System", *Proceedings of the IFIP Working Conference on Database Semantics DS8,* Kluwer Academic Publishers

Staehli R., Walpole J. and Maier D. (1995), "A Quality-Of-Service Specification For Multimedia Presentations" *Multimedia Systems*, vol.3 no.5-6, pp.78-85

Stonebraker M. and Kemnitz G. "The POSTGRES Next Generation Database Management System" *Communications of the ACM* Vol.34, No. 10 pp. 37-45

Sterling W. (1996), "The Teradata Multimedia Database System: A Massively Parallel Solution for Object-Relational Databases", Proceedings of the DoD Database Colloquium '96, San Diego, CA (US) pp. 127-133

Vogel A., Kerherve B., Bochmann G. and Gecsei J., "Distributed Multimedia and QOS: a Survey". *IEEE Multimedia* vol. 2 no.2, pp.138-145

Waibel A. and Lee K. (1990), "Readings in Speech Recognition", Morgan-Kaufmann, 1990

Wang J., Chang W., and Acharya R. (1999), "Efficient and Effective Similar Shape Retrieval" Proceedings of the 6th *IEEE International Conference on Multimedia Computing and Systems (ICMCS'99),* Florence, Italy, pp. 340-352

Zhang A., Chang W., Sheikholeslami G. and Syeda-Mahmood T. (1998). "NetView: Integration of Large-Scale Distributed Visual Databases". *IEEE MultiMedia,* Vol. 5, No. 3, pp.62-78.

Zhao W., Wang J., Bhat D., Sakiewicz K., Nandhakumar N., and Chang W. (1999), "Improving Color Based Video Shot Detection". *Proceedings of the 6th IEEE International Conference on Multimedia Computing and Systems (ICMCS'99)* pp. 280-292

Maybury M. (ed.) (1995) *Intelligent Multimedia Information Retrieval,* AAI/MIT Press

MPEG Requirements Group, "MPEG-7Requirements Document", *Doc. ISO/MPEG N2461, MPEG* Atlantic City Meeting

MPEG Requirements Group, "Applications for MPEG-7, *Doc. ISO/MPEG N2462, MPEG* Atlantic City Meeting

MPEG Requirements Group, " MPEG-7Evaluation Procedure, *Doc. ISO/MPEG N2463, MPEG* Atlantic City Meeting

MPEG Requirements Group, " MPEG-7 Proposal Package Description (PPD*), Doc. ISO/MPEG N2464, MPEG* Atlantic City Meeting

Synchronized Multimedia Integration Language (SMIL) 1.0 Specification (1998), available at http://www.w3.org/TR/1998/REC-smil-19980615

de Vries A. P. (1998), "Mirror: Multimedia Query Processing in Extensible Databases", Proceedings of the *14th Workshop on Language Technology in Multimedia Information Retrieval*, Enschede, The Netherlands, pp.189-193

10 A BROKERAGE SYSTEM FOR ELECTRONIC COMMERCE BASED ON HCVM

10.1. Introduction

Electronic commerce (EC) can be broadly seen as the application of information technology and telecommunications to create trading networks where goods and services are sold and purchased, thus increasing the efficiency and the effectiveness of traditional commerce.

The internet has introduced many new ways of trading, allowing interaction between groups that, due to limited resources or to remoteness, previously could not economically afford to trade with one another.

Whereas traditional commercial data interchange involved the movement of data from one computer to another, without user interaction, the new model for Web-based commerce introduced by the internet is typically dependent on human intervention for EC transactions to take place.

As this new model gained acceptance, there has been a fundamental shift in how data used for commerce is processed. The original cycle of information processing, using individual application programs, is now being replaced by the concept of active agents. In recent years, software agents, both independently and through their interaction in multi-agent systems have been transforming the internet's character and mission. Indeed, as a consequence of the broad diffusion of electronic commerce we are experiencing a new "ecology" of global internet commerce with buyers, sellers, and intermediaries forming numerous industry-specific internet markets that in turn form extended trading communities.

Many software architectures have been proposed for supporting such trading networks in the past few years; in Hands et al. (1998) some of them are presented, addressing issues such as sales, ordering and delivery of products in the framework of the global internet.

Indeed, internet based electronic commerce is currently a driving force behind the evolution of many Web-based technologies such as HTTP, HTML, Java, CGI and others (Hamilton 1997), all of which were originally conceived for different applications.

Moreover, both academic and industrial research has developed topics related to money exchange in the context of Web-based architectures, such as support for security, cryptography, electronic currency and payments on the WWW (a complete survey on these subjects is presented in Lynch and Lundquist (1996).

Payment and security-related features are especially relevant to a business-to-consumer electronic commerce model, involving vendors, buyers and brokers who exploit the internet as a common marketplace.

Indeed, on the global Net commerce transactions are often occasional in nature, thus justifying a strong need for facilities supporting identification and mutual trust between buyers and vendors.

Internet-based trading networks present some major differences with respect to the linear supply chains of traditional commerce. In fact, traditional supply chains are often based on stable, long-lasting business relationships, while EC trading networks allow and indeed encourage a higher rate of change in commerce relations, causing customers to frequently change suppliers on the basis of short-term business opportunities.

Provisions for automatically ensuring mutual trust between EC partners (Fromkin 1995) are further increasing the number of occasional transactions.

In this setting, nearly every electronic commerce purchase is preceded by a network search or product brokering phase (Tenenbaum et al. 1998), when the customer navigates the trading network looking for the needed products or services.

However, the avalanche of on-line suppliers and multimedia information about goods and services currently available makes it difficult to locate, purchase and obtain the desired products at the best prices. General-purpose internet search engines seem wholly unfit to this task, though some work is being done to complement them with electronic commerce related capabilities. For instance, the well-known Excite search engine has a shopping guide to find products and prices on the Web, which is called Product Finder. Yahoo exploits a technology that aggregates information and prices for merchandise sold on the Web, enabling consumers to compare and shop for online products. More recently, Infoseek announced Express, which uses many search engines to execute multiple searches for products.

These developments notwithstanding, it is generally recognized that a general solution to the brokering problem requires the definition of an agent-based brokerage architecture, where mediation via on-line electronic brokers joins customers and suppliers, stimulating the market activities of both parties. Moreover, electronic brokers may prove useful in allowing small and medium size suppliers to compete with larger ones on a fair basis. In particular, software agent technology can be used on retail markets to carry out the most time-consuming stages of the buying process. Several descriptive models have been proposed in the framework of Consumer Buying Behavior (CBB) characterization (Maes et al. 1999), trying to model actions and decisions involved in purchasing and using goods and services.

As we shall see in this chapter, a *consumer-oriented market model* based on semi-structured metadata is needed in order to support a new generation of brokerage architectures. Using such a market model, broker agents must be able to locate all vendors carrying a specific product or service, and then query them in parallel to locate the best deals.

However sophisticated and powerful, modeling techniques are not enough: several architectural requirements must be taken into account as well. In order to be effective, electronic commerce brokerage architecture must also be scalable and applicable to the distributed multimedia nature of today's electronic commerce trading networks, handling the diverse nature of existing and future goods and services to be traded. In the past few years, several research groups have investigated requirements for electronic commerce brokering and some preliminary evaluation of agent-based technology was attempted with respect to those requirements (Connolly 1995; Maes et al. 1999). However, experience has brought to the forefront several problems with the existing technology.

In this chapter, we start by outlining the requirements for EC brokerage on the internet and evaluate them in the framework of the HCVM approach, investigating their feasibility and presenting a complete solution. The rest of the chapter is organized into six parts. Section 10.2 deals with basic requirements of agent-based systems for electronic brokerage, pointing out several open problems. section 10.3 briefly presents technology-centered approaches such as Electronic Data Interchange (EDI) standard systems and CommerceNet's *eCo* platform for electronic commerce, while section 10.4 outlines the role of the *eXtensible Markup Language* (XML) Electronic Commerce transaction. Section 10.5 contains an outline of our approach and an itemized usage scenario. In section 10.6 we describe the construction of user-centered market models by means of transformations on semi-structured documents via the XML. In section 10.7, a sample HCVM application to the domain of Hardware Adapters is presented in detail.

10.2. Basic Requirements

At the beginning of the last decade, in his seminal paper (Englebart 1990) David Englebart proposed a list of twelve requirements for electronic commerce systems, originally conceived in the framework of *Computer Supported Cooperative Work* (CSCW) research.

Engelbart's requirements mainly addressed the need for flexibility in the format of *hyperdocuments* used for information exchange between business partners.

More recent implementations of electronic commerce systems (Powley et al. 1997) emphasized architectural requirements, such as support for multiple communication paths and decentralized development of electronic commerce applications.

This brings us to revise Englebart's list and define the following six *basic requirements* for agent-based electronic commerce brokerage:

1. User-centered market model
2. Information interchange between users and agents based on flexible, standard object documents.
3. Document "back-link" capability, allowing users and agents to browse links from other objects to the one they are examining.

4. Access to object documents through multiple communication paths, including e-mail, X.12 Electronic Data Interchange (EDI), agent based interfaces and others
5. Encryption-based data security in user-to-agent communication
6. Access control on the basis of user profiles
7. Decentralized development of agent-based electronic commerce services and applications

After describing these requirements in some detail, we shall discuss how they have been addressed by current architectures for electronic commerce brokerage, underlining some open problems.

10.2.1. User-Centered Market Model

The basic assumption we rely on is that each customer organizes a *mental model* of the market. This model is only loosely related to supply-side product classification. Therefore, the first and main requirement for an agent-based electronic commerce broker is to be able to fully comprehend and utilize such a model. This requirement is not satisfied by many of the approaches proposed in the literature.

10.2.2. Flexible Object Documents

The first Web documents contained only text, but support for multimedia data types was gradually added from 1993 onwards through the use of a structuring and labeling mechanism called *Multipurpose Internet Mail Extension* (MIME). Besides processing multimedia data, agent-based brokers need to deal with *Electronic Data Interchange* (EDI) documents (section 10.3).

It is widely recognized that EDI lacks many of the knowledge representation capabilities required to support agent-based brokerage on the global Net. Moreover, deployment of EDI-based systems needs substantial investment that is regarded as too high by many small and medium enterprises. As a consequence, the *eXtensible Markup Language* (XML) (section 10.3) has been chosen as a standard representation technique, and organizations already using internal EDI formats for electronic commerce should now use XML translators to switch to a market-wide object standard. Recently, XML and agent-based technology have been jointly used to access such servers and exchange EDI documents between business partners. Moreover, agents have been used to implement translators between different EDI formats used in various organizations (Powley et al. 1997). Since these translations potentially require specialized knowledge, XML/EDI data manipulation agents (*DataBots*) may supplement translation rules (Bryan et al. 1999). DataBots ensure that users can express their requirements in high-level natural language terms, automatically create appropriate rule templates and XML syntax to match user requirements and broker the entire interchange.

10.2.3. "Back-Link" Capability

When accessing metadata and documents online, agents need to utilize information about links from other objects to the one they are examining. Moreover, back-links could lead to sensitive information that must not be accessible to all operators. This "back-link" capability is not ensured by current Web-based systems, where hypertext links are one-way and reverse links are not guaranteed to exist. However, XML more flexible link formats provides back-links as well as a number of other application oriented hypertext links. Exploiting XML markup, a mobile agent cooperating with a remote server can set up and maintain explicit back-links in a secure fashion.

10.2.4. Access through Multiple Communication Paths

Current agent-based technology does not allow for easy access to electronic commerce services through multi communication paths. While the Java language provides code portability across multiple platforms, including personal computers, network computers and workstations, the support for other communication interfaces is still limited..

Integration of fax and voice interfaces with XML is currently available via several Java-based third-party products, but it has been seldom if ever used in the framework of electronic brokerage. The high interoperability degree between Java code and some agent-based platforms such as IBM's Aglets (Lange and Oshima 1998) should make this requirement technologically viable.

10.2.5. Encryption-Based Data Security

S/MIME (*Secure Multipurpose Internet Mail Extension*) has been used since long in lieu of conventional MIME as a transport format for EDI documents (*secure EDI*). However, as XML-EDI solutions are gaining acceptance, more sophisticated techniques are becoming available. Since the advancement of public-key cryptography has solved most of the security problems in Web-based electronic commerce, it is widely recognized that authentication and encryption will play a large role in fine-grained XML security as well. Thanks to authentication, an encryption-based XML server knows what information can be sent to a user based on that user's access level, and employs element-wise encryption to prevent users without appropriate decryption keys to access the parts of the documents containing private information. However, encryption-based approaches unequally split security responsibilities between the connection protocol, the XML content, and the application processing the document, while the need for a standardization of access control is becoming well recognized for XML data (Rutgers Security Team 1999). Moreover, some of these techniques leave encrypted private information in the hand of unauthorized users, a design choice that may well prove unwise in the long run.

As far as digital signature technology is concerned, it must be noted that though a number of digital signature systems based on public-key cryptography exists (such as PGP, (Garfinkel 1995)), some of them providing a complete infrastructure, such as X.509 and EDIFACT standards, none has yet been globally adopted and deployed on the Web, partially as a consequence of United States' strict export control legislation. Though digital signature standards have been implemented in agent-based frameworks, this situation makes agent-based brokers to lack interoperability to some degree.

10.2.6. Access Control Based On User Profile

The distributed nature of the Web allows business partners to implement any access control policy they choose, down to the object level. Basic support for username/password authentication can be provided at the Web server level, allowing for custom access control based on user profiles. However, electronic commerce applications need a more sophisticated approach. Several mechanisms for strong client authentication and confidentiality have been proposed in the agent-based framework, with the explicit aim of developing electronic commerce applications (Lynch and Lundquist 1996). Moreover, in business transactions where not all participants are equally trusted, secure trading places can be located on special-purpose remote hosts. HTML-based Web sites transfer information one page at a time; thanks to XML, developers expect to deliver information with a finer level of granularity, making for a richer Web experience.

For this reason, security control at the tag level appears to be far more important with XML than it was with HTML, especially considering that the *XLink* standard for XML linking and the new XML query languages (Ceri et al. 1999) will enable users to retrieve portions of documents based on their tagging. Thanks to these new technologies, users will be able to directly control their Web experience by pulling out of an XML document the information that most interests them. In the same document, though, information that user should not see might be present, and the server will need to know whether the user should get specific data. (Damiani et al. 2000) describes an approach exploiting XML's own capabilities, defining an XML markup for a set of *security elements* describing the protection requirements of XML documents. This security markup can be used to provide authorizations with the granularity of XML elements.

Taken together with a user's identification and its associated group memberships, as well as with the support for both permissions and denials of access, this security markup allows expressing different protection requirements with support of exceptions. The enforcement of the requirements stated by the authorizations produces a view on the documents for each requester; the view includes only the information that the requester is entitled to see. A simple propagation algorithm ensures fast on-line computation of such a view on XML documents requested via an HTTP connection or a query. The proposed approach, while powerful enough to define sophisticated access to XML data, makes the design of a server-side *security processor* for XML rather straightforward.

10.2.7. Decentralized Development of Electronic Commerce Brokerage Services

Traditionally, brokerage applications require a high level of integration under the control of a central authority.

In the conventional view, these centralized applications typically map onto private or proprietary *Value Added Networks* (VAN). However, experience has shown that such applications are difficult to build and maintain, while agents should allow for parallel, modular development of individual component services and be effectively used in conjunction with pattern-based design techniques (Buschmann et al. 1996).

While it is recognized that agent-based components should be assembled to suit the needs of specific organizations by populating a skeleton system with selections from alternative component offerings (Neches 1994), no current system provides such a modular design.

10.3. Technology-Centered Approaches

While large-scale, internet based EC is relatively recent, standards for business-to-business applications have been available for over two decades. Experience with business-to-business EC systems as a part of information system design partially explains the hyphen on architectural and data format standards that characterized early approaches to internet-based EC. In this section we shall briefly outline some of these *technology-centered* approaches.

10.3.1. Electronic Data Interchange

Electronic Data Interchange (EDI) comprises a large set of standards developed by various organizations, including United States' ANSI and United Nations' EDIFACT committees. In particular, the recent ANSI X.12 standard [DATA INTERCHANGE STANDARDS ASSOC. 1996] specifies a general format for electronic interchange of business data, including request for quotes and purchase orders.

The structure of EDI messages exchanged between business partners can be hierarchically decomposed as follows:

Interchange Message: a routing and control header followed by transactions sets grouped by business functions, to allow for routing within the recipient organization.

Transaction set: Information needed for a single business transaction, possibly including several segments. There are hundreds of types of transactions sets specifying all possible interactions between vendors, brokers and buyers.

Segment: intermediate information unit, typically an address, in turn composed of several data elements

Data element: an atomic value, such as a qualifier or text, to be interpreted according to the segment it belongs to and its position inside that segment.

It is worth noting that, while ANSI X.12 EDI fully specifies messages' syntax, their meaning depends on the interpretation of the segment names and therefore requires a preliminary agreement between business partners on the interpretation of each field.

There are basically two approaches for such an agreement:

external data format, requiring all partners to implement translation to and from a conventional "universal" document format

transmitter makes it right, often used when a large corporation or a government agency exploits its contractual strength to require that all its business partners conform to its internal document format.

The X12 EDI standard does not specify the message transport mechanism, which can be a public network, Intranet/Extranet or a special Value Added Network (VAN).

In 1995, the Internet Engineering Task Force (IETF) defined the means for carrying EDI documents via the Net (Crocker 1995)

Thus, Web-based servers on Intranet/Extranet networks can straightforwardly be used to transport EDI traffic in substitution of traditional public X.25 networks and X.400 e-mail.

The drawback for this approach is however twofold. Firstly, companies usually refrain form standards that do not fully meet their business needs. Secondly, since the standards are pre-set there is no mechanism to transfer processing rules and associated information. It is assumed that the data meets the defined constraints and if not, it must be modified to conform. This means that companies must conduct exacting analysis to determine precisely how they are going to move their business data to and from the predefined EDI formats. The cost of these constraints resulted in excessively long, complex and expensive implementation cycles for traditional EDI systems.

Today, the interchange of electronic data (EDI) between multiple data sources on the Web is considered as a core technology for internet based business-to-business EC. Specifically, many suppliers rely on *lightweight* implementations of EDI based on XML. However, as we shall see in the following sections, XML adoption alone does not imply a modification of the technology-centered paradigm that characterized early implementations of EDI systems. As XML data proliferates on the Web, applications will need to integrate and aggregate data from multiple sources and clean and transform data to facilitate exchange.

10.3.2. The eCo System Approach to Electronic Commerce

A further step toward open markets standardization has been envisioned in eCo, the reference architecture for electronic commerce proposed by the CommerceNet Consortium (including Actra, Bank of America, Visigenic, World Wide Web Consortium, Mitsubishi, NEC and Oki), which exploits Web-based technologies such as HTML, and Java/CORBA (Tenenbaum et al. 1998).

The eCo platform was originally a framework of reusable software components based on CORBA middleware standard, (Yang and Duddy 1996) which can be used to build electronic commerce applications.

It includes a high-level domain specific language, the *Common Business Language* (CBL) allowing software modules to communicate much like humans involved in commercial transactions, but exchanging EDI-compliant object documents instead of traditional paper documents.

Industry-wide standardization of eCo objects should allow companies to build open markets for business-to-business electronic commerce. Objects can be created in all the main proprietary environments, including IBM, Oracle, JavaSoft and Netscape.

In 1997, eCo-system was entirely recast on an XML foundation, due to XML adoption by all key vendors of the CommerceNet Consortium.

It should be noted that organizations already using internal EDI formats for electronic commerce should use translators in order to fully exploit eCo system capabilities.

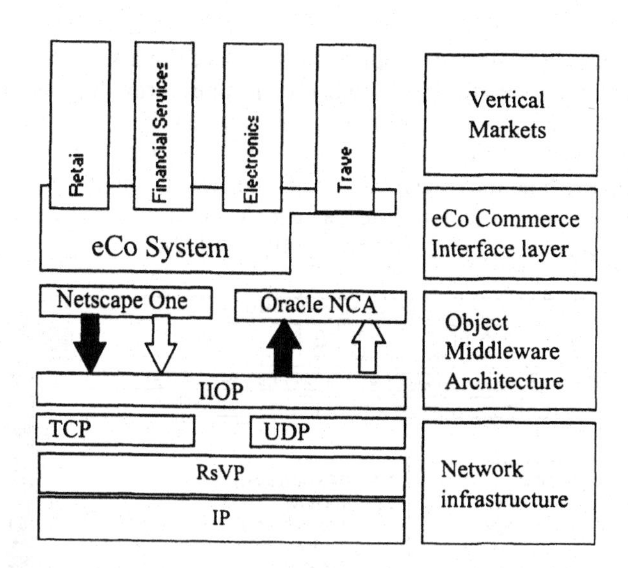

Figure 10.1. An Overview of the eCo Framework for Electronic Commerce

Figure 10.1 presents a layered view of the eCo architecture, showing how this EC framework relies on Internet/Intranet network infrastructure: transport is provided by standard TCP/IP layers (the RsVP layer is interposed to guarantee Quality-of-Service control), while the IIOP layer provides a CORBA-compliant *software bus* (Orfali and Harkey 1997) allowing object-oriented components to communicate via *remote method invocation* rather than via standard connection-oriented protocols such as HTTP.

The eCo system itself is provided as a vendor-independent application layer, supporting several specialized Application Objects for various application fields such as Retail, Travel and others. The Application Objects interfaces were initially written and published on the network in CORBA *Interface Definition Language* (IDL). By adopting XML descriptions, the eCo system framework overcame a long-standing barrier to the development of electronic commerce, as XML documents provide, at least in principle, an incremental path to business automation, whereby browser-based tasks are gradually transferred to software agents.

This development might allow traditional supply chains to evolve into open markets, while agents interact with business services through object documents. However, as we shall see in the remainder of the chapter, the human readability of the XML language, while an obvious advantage over a technology-centered standard such as CORBA Interface Definition Language (Glushko 1999), does not eliminate the drawbacks of a *supply-side* market model, where the structure and content of metadata are modeled according to the needs of vendors and distributors, leaving it to the brokers to transform them into a form more suitable for the buyers.

10.3.3. CommerceNet Evolvable Taxonomies

Besides dealing with the need for a standard architectural design pattern for EC systems, the electronic commerce research community has since long recognized the need to provide a systematic and complete view of the EC market. The CommerceNet consortium, including nearly 250 member companies providing solutions to EC technology issues, has been sponsoring some industry pilot projects in order to develop domain representation based on evolvable taxonomies (Hamilton 1997). An evolvable taxonomy functions much like a database schema and provides metadata for agent searches. Figure 10.2 illustrates a sample from CommerceNet's proposed Taxonomy of Everything, as presented in Hamilton (1997).

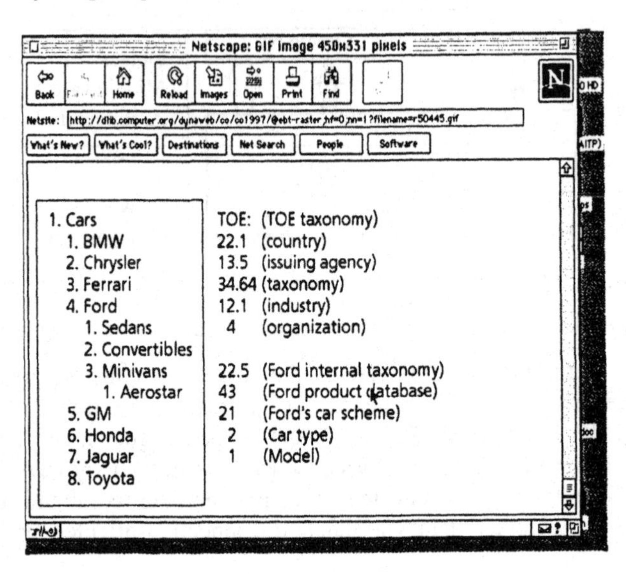

Figure 10.2. Sample evolvable taxonomy

Again, we observe that such taxonomies are *supply-side oriented*, being based on a classification of goods and services provided by a group of suppliers operating in a given application domain. Therefore, though they are useful in business-to-business transactions, in our opinion it is by no means guaranteed that they will present a view of the market that is satisfactory, or even comprehensible, for the casual user.

10.4. eXtensible Markup Language and Electronic Commerce

XML is a mark-up *metalanguage* originally designed to enable the use of SGML on the World Wide Web and standardized by World Wide Web Consortium (W3C) Bray et al. (1998).

A detailed introlduction to XML has been given in section 2.11 of chapter 2.

From the introduction in chapter 2 one can see that the advantages of XML support for EC transactions are indeed many. In fact, XML can be integrated with existing EC systems (for instance, in EDI-based solutions) by:

- Allowing data received in EDI format to be interpreted according to sets of predefined rules for display by the receiver on standardized browsers using a user-defined templates
- Providing application-specific forms that users can complete to generate EDI messages
- Generating EDI message formats for transmission over the internet

Moreover, XML can be used to extend existing EC applications by:

- Allowing message creators to add application-specific data to standardized message sets
- Allowing message receivers to display the contents of each field in conjunction with application-specific explanatory material
- Allowing field value checking to be integrated with checks on the validity of the data with respect to information stored on local databases.

The main advantage of XML data format, however, is that it lends itself to be transformed and extended to provide a vision of the electronic market more suited to the user' perception and needs. We shall deal with this problem in the following sections.

10.5. A Human-Centered View of the Electronic Market

As we have seen, though technology-centered support for electronic commerce brokerage requirements is still far from complete, current distributed agent-based architectures capabilities in no way prevent electronic commerce requirements from being realized. However, many problems are left unsolved by existing agent technology.

In this section, we outline a human-centered (as against *technology-centered*) brokerage architecture, showing how the *Human-centered Virtual Machine* (HCVM) layered architecture allows the broker design to proceed seamlessly from a human-centered representation of (a portion of) the electronic market, based on a nominal scale, to the ordinal and interval-based representations that are more suitable for intelligent search agents.

This will allow identification of the *decision classes* to be submitted to the user as the possible target for an EC transaction. As a by-product, we shall show how HCVM approach ensures generality and applicability of the brokerage system to a wide range of EC application domains.

10.5.1. The Electronic Brokerage Reference Model

In order to describe the functionality of our agent-based architecture for electronic brokerage, we briefly outline its reference model at the highest level of abstraction, in Table 10.1. It also defines some of the terminology used in the agent-based architecture. Based on the reference model, scope of various components involved is shown in Figure 10.3.

Table 10.1. Roles and Actions of the Reference Model

Abstract Model
Roles: Customer Broker Supplier Actions Search Order Deliver

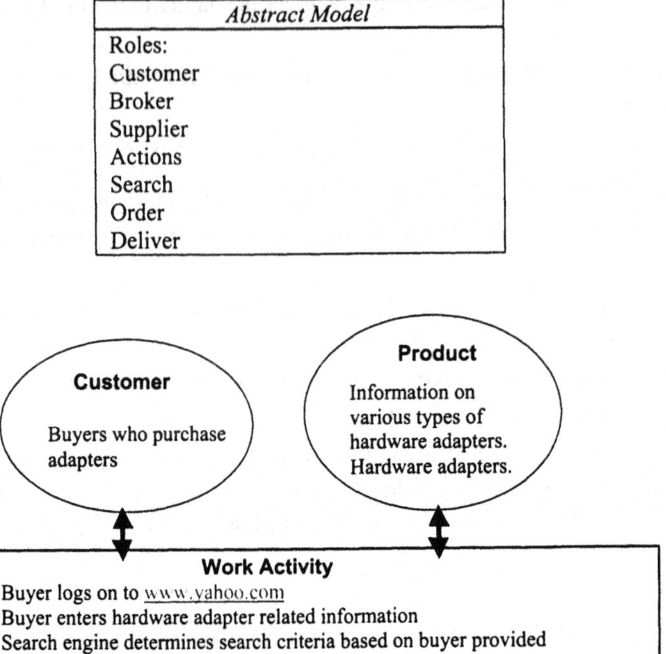

Figure 10.3. Scope of Six Components of Hardware Adapter Application

The core concepts of this reference model are the roles of customer, broker and supplier and the simple actions of search, order and deliver.

The broker acts as a supplier of information or services to the customers, and as a distribution mechanism for suppliers wishing to promote their products over the trading network.

A basic assumption of this model is that brokers should acquire and maintain information about online services, in order to be able to locate the product required by the customers. This information is multimedia in nature and can be arbitrarily specialized, with special brokers for particular domains or geographical regions.

The function of the broker is therefore providing a path whereby the customer may find and obtain a product offering the required characteristics (and, of course, price) to the highest possible degree.

In the remainder of this chapter, we shall focus on the broker's search functionality. As we will see, the use of both domain-specific and domain-independent metadata is central to the functioning of the broker, both for the customer to describe the needed product and for the broker itself to propagate the search to other brokers.

It should be noted that we take a line of different research to the technology-centered approached outlined in previous sections. Indeed, we focus on putting the user's, i.e. the customer's, perception of the market at the center of the knowledge organization process. The basic assumption we rely on is that the customer organizes a mental model of the market via a limited number of general features, and an agent-based broker should be able to fully comprehend and utilize such a model. Moreover, this model ought to be independent from any existing ontology.

Based on HCVM decomposition phase, we employ small number of coarse-grain input features to partition a global concept (for instance, the Hardware Adapters EC sub domain), in order to identify a hierarchy of abstract classes (refer Figure 10.4).

In order to deliver a user-centered market model, this decomposition relies on general features and structured features. General features are binary-valued and allow identification of orthogonal classes.

For instance, in the Hardware Adapters domain, `AT_bus_interface` is a general feature that can be used to identify the class `Old_PC_Cards`. In passing, we note that this class, though very unlikely to be used in a supply-side oriented taxonomy of products, is indeed crucial in the users' perception of the market. Structured features, in turn, can have multiple discrete values (for instance a connector-type feature could have values `DB-9`, `DB-25`,...) and represent basic domain structure.

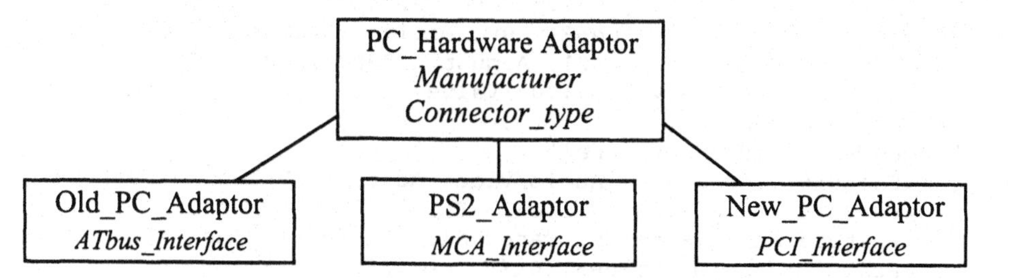

Figure 10.4. A Simplified Decomposition of the PC_Hardware Adaptors Sub-domain Using HCVM.

In the next sections, we shall describe how the HCVM-based methodology supports a standard metadata format such as XML.

10.5.2. A Human-centered Approach to XML-based Metadata

It is worth noting that as far as the involved metadata are concerned, the set of HCVM abstract classes identified by the decomposition phase (Figure 10.4) can be straightforwardly mapped into domain specific XML Document Type Definitions in order to obtain a user-centered market model (step 5 in section 10.5.3). In our approach, a general DTD can be seen as a *schema* for the market model and an XML document referencing the DTD as an *instance* of that schema. Obviously, it is possible for two instances of the same schema to have a different structure. In fact, some elements in the DTD can be optional and other elements can be included in an XML document zero, one, or multiple times.

```
<!ELEMENT HardwareAdaptor        (TaggedValue*)>
<!ATTLIST HardwareAdaptor
          SUBDOM (Multimedia, MassStorage, Networking)
                #REQUIRED
                >
<!ELEMENT TaggedValue      (GenFeature*,Feature*)>
<!ATTLIST TaggedValue
          TYP (General, Auxiliary, Functional, NonFunctional)
                #REQUIRED
                >
<!ELEMENT Feature          (#PCDATA, Value)>
<!ELEMENT GenFeature       (#PCDATA)>
<!ELEMENT Value            (#PCDATA)>

<?xml version="1.0"?>
<!DOCTYPE HardwareAdaptor SYSTEM "HA.DTD">
<HardwareAdaptor SUBDOM="Multimedia">
<TaggedValue TYP="General">
    <GenFeature>AT_Bus_Interface </GenFeature>
 </TaggedValue>
 <TaggedValue TYP="Auxiliary">
    <Feature>Title <Value>Sample PC Card</Value> </Feature>
    <Feature>Author <Value>Ernesto Damiani</Value> </Feature>
    <Feature>Date  <Value>21, August, 1999</Value></Feature>
    <Feature>Version<Value>1.0</Value>            </Feature>
  </TaggedValue>
<TaggedValue TYP="Functional">
    <Feature>Type <Value>Audio/Video Adapter</Value> </Feature>
    <Feature>Frame rate <Value>30 Fps</Value> </Feature>
    <Feature>Audio Sampling rate<Value>16 Khz</Value></Feature>
    <Feature>Version<Value>1.0</Value></Feature>
  </TaggedValue>
<TaggedValue TYP="NonFunctional">
    <Feature>Price <Value>30$</Value> </Feature>
    <Feature>VAT <Value>6$</Value> </Feature>
    <Feature>Date<Value>21, March, 1999</Value></Feature>
</TaggedValue>
</HardwareAdaptor>
```

Figure 10.5. A Sample XML DTD and a Tagged Document for the Hardware Adapters Class

Figure 10.7 illustrates a DTD for XML documents describing Hardware Adapters in the PC cards domain. Our DTD aims at being independent from existing supply-side ontology, as features' names are not included as tags but are part of the tagged document data (which can be constructed on-line, by means of an XML query language (Ceri et al. 1999)). Much of the DTD is self-explanatory; the elements definition and structure tell us that a PC hardware card is described by its features; each block of features composes a `TaggedValue`, whose type is denoted by an XML attribute.

The XML representation shown allows for dealing with both HCVM general and structured features. General features are mapped in XML `<GenFeature>` elements, while structured features, in turn, can be straightforwardly represented by XML `<Feature>` elements as shown in Figure 10.5.

10.5.3. The Brokerage Architecture

From the user point of view, our human-centered approach to Electronic Brokerage involves seven basic steps:

1. **Access to the Broker**, providing a set of keywords
2. **DTD Repository search**, in order to identify relevant sites
3. **Sites identification**, on the basis of their DTDs
4. **Document retrieval**, i.e. downloading a collection of XML documents from relevant sites,
5. **Market model construction**, that is the transformation of retrieved documents into a customized catalogue according to a user-centered DTD
6. **Query execution**, allowing for different user interaction styles
7. **Presentation of results**, providing to the user all available information about needed products and services.

These seven steps, together with the basic system modules used to perform them, will be briefly described in the present section.

A usage scenario of our system architecture is shown in Figure 10.6.

The sites identification and DTD retrieval (steps 1-3) involve a dialogue between the client module and a Broker. Users can activate the construction of a market model by providing a keyword list to be used against remote sites' DTDs or adopting more sophisticated hybrid styles of interaction. Essentially, our Broker is a Trader agent that identifies relevant Web sites' on the basis of their DTDs, i.e. of a synthetic description of the semantics of their content. After the user has approved the set of identified sites, the Broker downloads relevant XML documents (step 4) and uses them to build a market model based on a uniform, user-centered DTD (step 5). Once the market model has been created the user can either navigate it or adopt a query-language style of interaction (step 7-8).

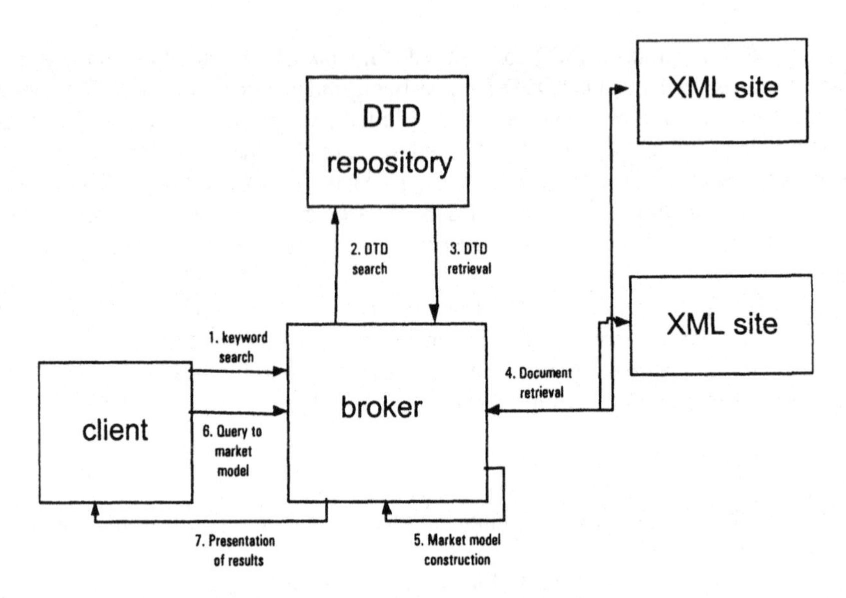

Figure 10.6. A Usage Scenario for the Human-Centered Broker

Let us now describe by a simple example how appropriately structured descriptors can be used in order to enhance the identification of relevant DTDs. The DTD fragment shown in the previous section provides the following data dictionary: {`adapter, audio, video, network, card, manufacturer`} from `%items` and {`order, deliverylocation, invoicing, order-no, item`} from the main DTD (from the sake of simplicity, here we do not consider the contribution of the `%address` entity to the data dictionary). The basic descriptor submitted by the user to the Broker could be an XML well-formed document containing a flat keyword list, as for example the following:

```
<?XML 1.0?>
<DESCRIPTOR> adapter, audio, video, order </DESCRIPTOR>
```

The matching of this descriptor (X) to the above data dictionary (Y) using a standard measure of lexical nearness is computed as follows:

$$|X \cap Y| \ / \ |X \cup Y| = 3/11$$

Giving a measure of the relevance of the supply-side DTD including `%items` to the user needs.

The user interface for DTD identification phase is shown in Figure 10.7.

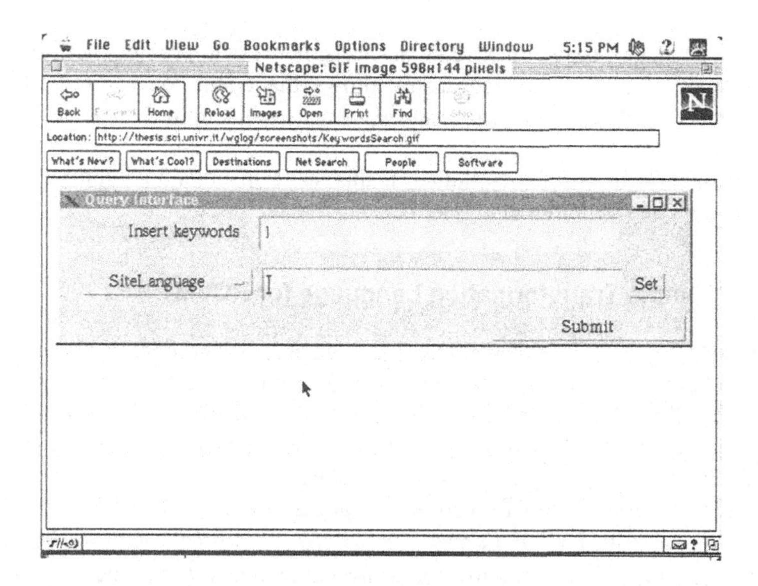

Figure 10.7. User Interface for DTD identification

The lexical nearness provides a quick-and-dirty filtering technique for assessing the relevance of candidate DTDs. However, it should be noted that it is easy to modify syntax in order to obtain a more semantics-aware descriptor. An example of such a descriptor is reported below:

```
<?XML 1.0?> <DESCRIPTOR> <GENERAL> audio video adapter
</GENERAL> <AUXILIARY> order </AUXILIARY> </DESCRIPTOR>
```

Separately computing lexical nearness for the two parts of the data dictionary, we get 1/2 + 1/5 = 6/10, which is roughly double of the 3/11 value obtained before.

Of course, a different syntax for descriptors should be provided for any supply-side DTD, but the computation of such syntax on the basis of the entities composing the target DTD is indeed not difficult and can be made transparently to the user. After helping the user with the identification of relevant DTDs via lexical nearness, the Broker will collect relevant sites' documents. This collection only involves XML documents, as multimedia data can be easily linked to and therefore need not be downloaded by the Broker. Then, the downloaded XML documents will be processed in order to provide a user-centered market model (step 5 in section 10.5.3) according to a uniform DTD. Finally, as we shall see in more detail in the next sections, users will browse or query this model exploiting a fuzzy query execution engine in order to deal with vagueness and imprecision in users' requirements.

As far as the presentation of results is concerned, there are two basic modes for our remote Broker to return results, i.e. documents belonging to the user-centered market model: as a list of handles (e.g. URLs), in order to keep network traffic at a minimum; and as a complete result instance, composed of XML documents that are valid with respect to the user-centered DTD. It should be noted that Brokers may well include a

Presentation Manager, based on XSL style sheets, in order to provide facilities for the presentation of results. This mechanism allows the client to require customized presentation styles on the basis, for instance, of constraints on the resources available at the user location.

User feedback can also be exploited at this stage in order to improve the DTD selection technique (Bellettini et al. 1999).

10.6. A Sample Transformation Language for HCVM

The aim of this section is describing how, once relevant documents have been collected, a human-centered market model can be built by means of a *transformation language* for XML documents. Through the use of this language, supply-side XML-based product descriptions are *transformed* by the Decomposition Agent of HCVM to support the user's perception of the market structure.

Informally speaking, transformation languages are specifications of *rewriting operations* to be performed on a string *S* belonging to a formal language L_G in order to produce a string *S'* belonging to another formal language $L_{G'}$, where grammars *G* and *G'* may not coincide. In the latter case, *G'* should be deducible from *G* and the rewriting procedure.

Several transformation languages were proposed in the past for applications to marked-up documents whose syntax specification is made by a SGML DTD.

Some older proposals such as *DSSL* (Prescod 1998) included a transformation language within a fully-fledged, general-purpose environment for rendering, management and processing of SGML documents. Several DSSL execution environments, called *processors*, are currently available, some of them as commercial products.

On the other hand, as mentioned in the previous sections, the recent *XSL (XML Stylesheet Language)* standard, while defining a complete language for document transformation, is a heavyweight specification also concerned with style sheet-based rendering capabilities for XML documents.

The transformation capabilities of XSL are now described by a separate specification, namely XSLT (XSL Transformation Language), available at http://www.w3.org/TR/1999/WD-xslt-19990421.

As we shall see in the next section, XSL allows sets of *actions* to be associated with particular XML *patterns*. Actions can be defined in terms of values to be assigned to a set of data presentation attributes (*styles*), or in terms of a data processing *script* that users can define using a *define-script* object.

Which actions are associated with which elements can be defined using XML element sets known as XSL *rules*. A simplified set of *style-rules* allows presentation properties to be *applied* to element classes.

10.6.1. XSL path expressions

A path expression on an XML document is a sequence of element names or predefined functions separated by the slash character (/). Path expressions may end with an attribute name, syntactically distinguished preceding it with the special character @.

For instance, path expression `CATALOG/BOOK/TITLE` denotes the nodes of the `TITLE` element, which are children of `BOOK` elements, which are children of `CATALOG` elements. Absolute path expressions start from the root of the document while relative path expression start with an element name. The path expression may also contain the operators dot (.), which represents the current node; double dot (..), which represents the parent node; and double slash (//), which represents an arbitrary descending path. For instance, relative path expression `BOOK//@isbn` retrieves all the attributes `isbn` belonging to element `BOOK` or to any of its sub-elements.

Finally, path expressions may also include functions in place of element names. Such functions perform various tasks, like the extraction of the text contained in an element and navigation in the document structure. Expression `CATALOG/ancestor(BOOK)` returns the catalog node which appears as an ancestor of the book element. Table 10.2 shows XSL main predefined functions.

Table 10.2. XSL Functions

function	*return value*
id	returns the element to which an IDREF or IDREFS attribute refers
ancestor	returns the first among the ancestors of the current node which satisfy a condition appearing as function argument
ancestor-or-self	same as ancestor, but includes the current node
comment	extracts the comment present in the current element
text	extracts the text contained in the current element
not	negates the predicate appearing as function argument
first-of-type	true if the current element is the first child of its type
last-of-type	true if the current element is the last child of its type
first-of-any	true if the current node is the first child of its parent element
last-of-any	true if the current node is the last child of its parent element

The syntax for XSL patterns also permits to associate conditions with the nodes of the path. The path expression identifies the nodes that satisfy all the conditions. The conditional expressions used to represent conditions may operate on elements content (i.e., the character data in the elements) or on names and values of attributes. Conditions are distinguished from navigation specification by enclosing them within square brackets. Given a path expression $l_1 ... l_n$ a condition may be defined on any label l_i, enclosing in square brackets a separate evaluation context in which a predicate appears, which compares the result of the evaluation of the relative path expression with a constant or another expression. Logical operators may be used to combine conditional expressions to build Boolean expressions. Note that conjunction is assumed in multiple conditional expressions appearing in the same path expression (i.e., all the conditions must be satisfied). For instance, expression `//PERSON/FLNAME./text()` = "John Doe" returns the `FLNAME` elements which contain "John Doe" as text.

10.6.2. A Sample Transformation Language

Due to the complexity of the position-based notation of the XSL language, in the remainder of this chapter we shall use a simpler transformation language for XML documents (*XML Transformation Language, XTL*), loosely based on XSL. In XTL, *pattern-based matching* is supported allowing for easy and effective processor implementation. Though having a limited expressive power (for instance, it does not provide recursion) XTL allows to express most of the computations needed in practice.

Our aim is showing through simple examples how HCVM-based Broker Agents and their associated XTL scripts can provide facilities that allow XML/EDI systems to:

- Automatically analyze and manipulate the data structures embodied in XML documents. The XTL program used to accomplish this is defined externally.
- Allow system developers to express transformations through an interface to a rule template that is clear and simple to follow. This allows users to focus on the business rules and interactions, without having to use programming language level tools.
- Exchange rule templates as a way to integrate their data interactions
- Extract and update data to and from heterogeneous supply-side documents, provided a transformation has been defined.
- Provide in-memory, single pass, document translation that frees suppliers and buyers from constraints of database and message structure conditions traditionally imposed by EDI systems
- Allow brokers to exchange data integration rules and database structures between systems on an ad hoc and rapid basis.
- Exchange data, update data and translate data as required as it processes the supply side XML documents.

Some additional considerations also need to be taken into account include Object-Oriented support for scripting languages. As we shall see, script support can be provided through W3C's *Document Object Model* (DOM), which provides a CORBA-compliant IDL definition for XML objects.

In the remainder of this section, after some explanations about the graphical representation of XML DTDs and documents, we outline a simple grammar for XTL; then we give some examples together with some observations on the implementation of its execution environment as part of the Decomposition Agent.

10.6.3. Representing XML DTD and Documents as Graphs

We represent XML DTD and documents as *E-R-like directed graphs* shown in Figure 10.8 according to the following simple convention:

- Each "non-terminal" (i.e. containing other elements) XML element is represented as a square node
- Each terminal (i.e. containing #PCDATA only) element or XML attribute is represented as a round node. To distinguish terminal elements from attributes, we shall sometimes paint in black the attributes leaf nodes.
- Arcs are drawn to represent the containment relationship between elements.

Consider for instance the following simple DTD fragment:

```
<!ELEMENT DOCUMENT> (ORDER)+
<!ELEMENT ORDER> (DESCRIPTION, PRICE)
<!ELEMENT PRICE> (#PCDATA)
<!ELEMENT DESCRIPTION> (#PCDATA)
```

and its graphical representation:

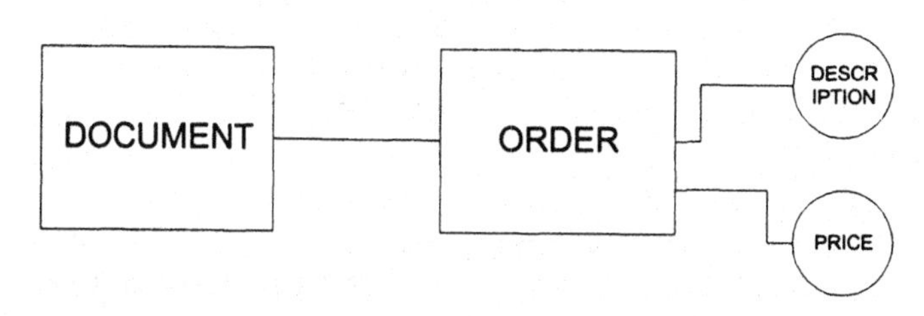

Figure 10.8. Graph Representation of a DTD

The following document is valid according to the DTD given above:

```
<DOCUMENT>
<ORDER>
<DESCRIPTION>Christmas decorations</DESCRIPTION>
<PRICE>$100</PRICE>
</ORDER>
<ORDER>
<DESCRIPTION>Christmas tree</DESCRIPTION>
<PRICE>$50</PRICE>
</ORDER>
</DOCUMENT>
```

The document will be represented by the following graph:

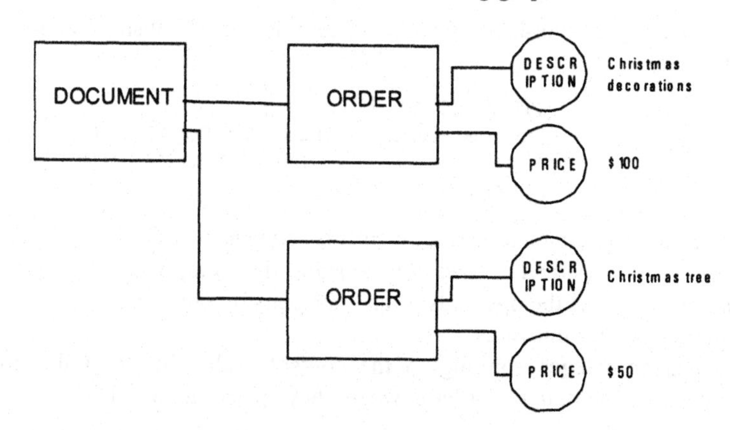

Figure 10.9. Graph Representation of a XML Document

10.6.4. XTL Grammar

We are now ready to give the grammar for our simple transformation language. The main productions of the XTL grammar are reported below in Figure 10.10:

```
PROGRAM         --->   TRANSFORM*

TRANSFORM       --->   PATTERN(I+CONDITION)
                              {make(newelement |newattribute)
                              LOCALITYSPECIFIER
                                     ELTAG
                                     EXPRESSION
                                     /ELTAG
                                     }
CONDITION       --->   BOOLEXPR

BOOLEXPR        --->   ((ATTRIBUTE+ELTAG)=
VALUE)+((ATTRIBUTE+ELTAG)= VALUE.OPR.(ATTRIBUTE+ELTAG)=VALUE)+

PATTERN                --->  (ELTAG+ATTRIBUTE)*
```

Figure 10.10. The XTL Grammar

As it is easy to see, TRANSFORM defines a transformation specifying a pattern and an *action* to be performed on the source document wherever an occurrence of a given *pattern* is found. It should be also noted that, as in DSSL, XTL patterns are sequences of tags and/or attributes.

Using pattern matching rather than tag matching allows one to extract the content of a tag (or attributes' values) selectively, i.e., only when the tag is used inside a suitable context.

CONDITION allows for specifying optional conditions on attributes values and element contents. Some auxiliary productions of the XTL grammar are listed below:

```
EXPRESSION    --->   RPN (OPERAND, EXPRESSION)
```

This production is shorthand for the classic Reverse Polish Notation (RPN) expression syntax.

```
OPERAND        --->   (TRANSFORM) |
                      'thiselement.FIELDNO.TYPEQUALIFIER |
                      'PATTERN.FIELDNO.LOCALITYQUALIFIER.TYP
                      EQUALIFIER |CONSTANT
```

The use of the ' operator to extract elements' *content* is a DSSL heritage. Using TRANSFORM as an operand allows for *nesting*, i.e. supplying the result of a transformation as one of the operands of another transformation. Execution of nested transformation proceeds outwards.

FIELDNO is an integer pointing to the content (value 0) or to the attributes associated to the element, in the order in which they appear in the DTD.

```
TYPEQUALIFIER         --->          AsInteger| AsString| AsReal
```

XML elements' content is not typed, while the XTL processor for expression evaluation requires information about types. `AsString` is a possible default.

```
LOCALITYQUALIFIER    --->    AsChild| AsSibling| AsFirst|
                             AsContent| AsSubstitute
```

Tag searching and generation can be done at the same or at different levels with respect to the matching pattern's level. `AsFirst` is the default and corresponds to searching from the last match.

10.6.5 Worked-out Examples

Let's start with the simplest unary transformation, generating a new tag (and its content) from an existing one. The new tag will compute the Value Added Tax (VAT) as shown in Figure 10.11..

Example 1 (Unary)

XTL transformation sheet:

```
<PRICE>
{make newelement AsSibling
<VAT>
'thiselement. AsInteger 0,2 *
</VAT>
}
```

Original XML document:

```
<PROD>
   <CODE>012<CODE>
   <PRICE>100.000</PRICE>
</PROD>
```

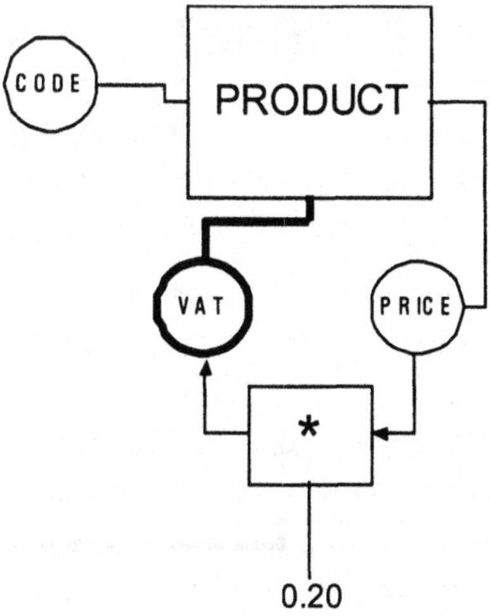

Figure 10.11. Dataflow Notation for the XTL Transformation in Example 1

```
Transformed document:
<PRODUCT><CODE>001
</CODE>
<PRICE>100.000</PRICE>
<VAT>20.000</VAT>
</PRODUCT>
```

Example 2 (binary)

A binary expression can now be used to compute the total gross price, including VAT, in the document obtained from Example 1.

```
XTL transformation sheet
<PRICE>
{make newelement AsSibling
<GROSSPRICE>
'thiselement. AsInteger
'<VAT> AsSibling AsInteger
+
</GROSSPRICE>
}
Transformed document
<PRODUCT><CODE>001
<'CODE>
<PRICE> 100.000 </PRICE>
<VAT>20.000</VAT>
<GROSSPRICE> 120.000 </GROSSPRICE>
</PRODUCT>
```

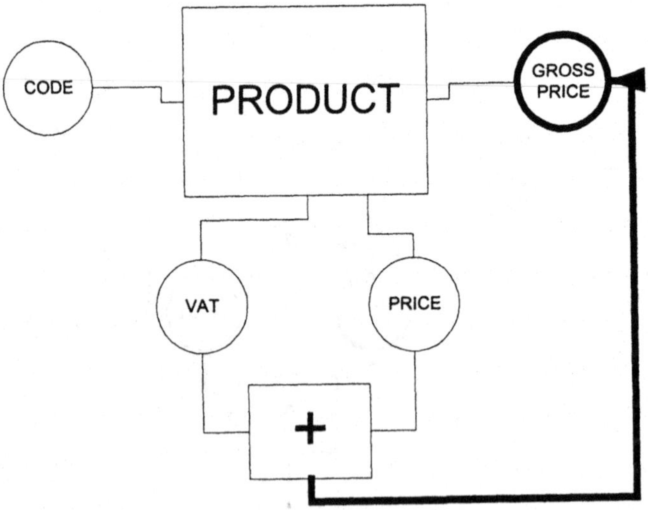

Figure 10.12. Dataflow Notation for the XTL Transformation in Example 3

Example 3 (nesting)

The same effect of applying the transformation defined in the previous two examples can be obtained by nesting the two transformations one inside the other (Figures 10.12 and 10.13. This example also hints at the limitations of XTL expressive power, as it makes it easy to see that the language does not support recursion.

```
XTL transformation sheet
<PRICE>
{make newelement AsSibling
'<GROSSPRICE> 'thiselement. AsInteger
'( <PRICE>
   {make newelement AsSibling
   <VAT>
   'thiselement. AsInteger 0,2 *
   </VAT>}
) AsSibling AsInteger
+
</GROSSPRICE>
}
```

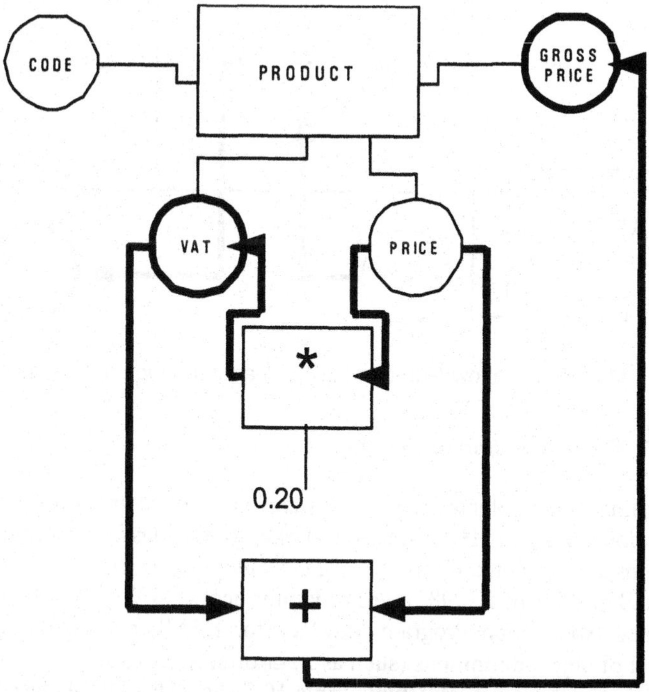

Figure 10.13. XML-TL dataflow Notation for the Nested Transformation in Example 3

Example 4 (pattern matching)

This example shows the computation of the price of a product starting from the prices of its two parts. Pattern (as opposed to tag) matching ensures that only parts of the same product are found. `AsFirst` (taken as default in the operands) ensures two different parts are used as operands. The graphical representation is shown in Figure 10.14.

```
XTL transformation sheet
<PRODUCT CODE=001><PRICE> /*will match the first part's price*/
{make newelement AsChild
<TOTALPRICE>
'<PART><PRICE> AsInteger
'<PART><PRICE> AsInteger +
</TOTALPRICE>
}
```

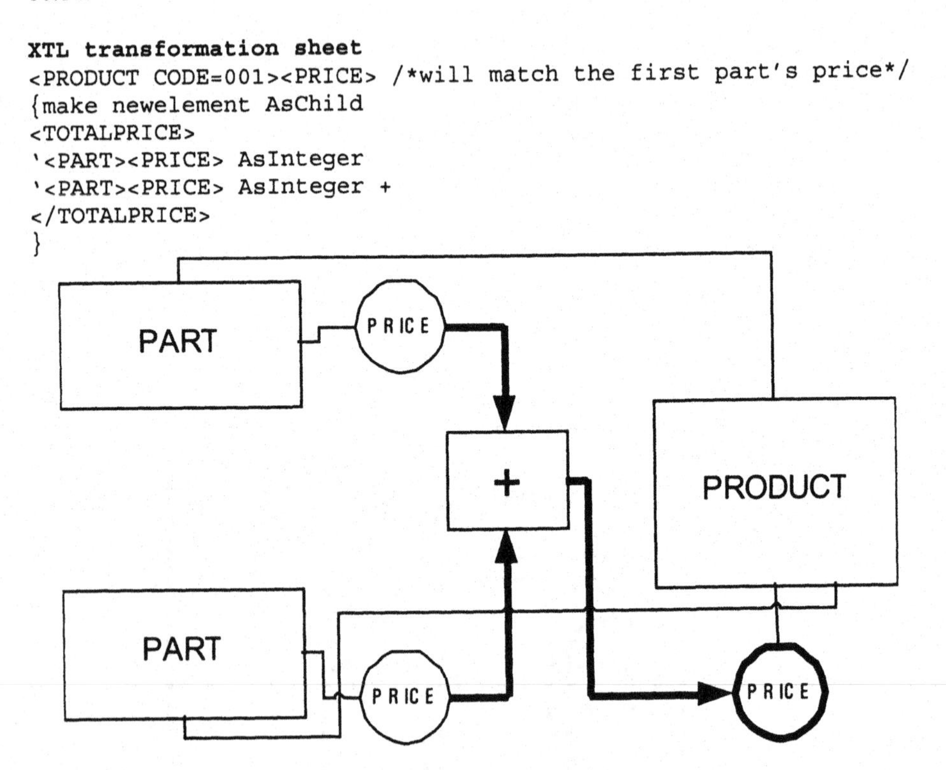

Figure 10.14. XML-TL Dataflow Notation for the Transformation in Example 4

10.6.6. XTL Processor Implementation

While designing and implementing the system, an explicit transformation between XML DTD, documents and HCVM based Market Model classes is defined by a set of XTL programs and executed by means of *Document Object Model Level One* (Core) specification. DOM (Wood 1998) gives programs access to Web data by providing an object-oriented Application Program Interface for HTML and XML documents. It employs a set of object definitions (such as `Element`, `Attribute`, `Text`) that closely model the tagged documents' structure. Table 10.3 shows the DOM class hierarchy.

Table 10.3. DOM Level 1 Class Hierarchy

Level 1	*Level 2*	*Level 3*	*Level 4*
Node (Root)	DocumentFragment		
	Document		
	{Attr		
	Element		
	DocumentType		
	Notation		
	Entity		
	Entry Reference		
	Processing Instruction		
	Character Data	Comment	
			CDATA section

The above class hierarchy defines the object types that appear in the memory representation of a validated XML document after parsing. Each object provides a standard DOM interface, which can be invoked using Java or JavaScript languages. The main classes are `Node`, `Document`, `Element`, `Attr` and `Text`, corresponding to the main parts of XMl documents. Specifically, `Node` corresponds to the generic element in a XML document and provides basic methods for inserting, deleting and substituting an object, performing a transformation on the underlying XML document. Via the inheritance mechanism, these functions are available throughout the DOM hierarchy. `Document` back-links the DOM object structure to the text document that was used to construct it; its `ownerDocument` property contains the XML document URL. Each `Element` object represents a XML tag, while `Attr` represent XML attributes associated to elements. Moreover, `Node` provides some navigation primitives such as `parentNode()`, `firstChild()` and `NextSibling()`allowing for visiting the document graph. Fast indexed access is also available via the `NodeList()` method, returning all the children of the current node. . Note that Attr is not a subclass of `Document`, and its interface does not inherit the navigation methods. The DOM standard specification includes the CORBA-IDL interfaces. The `Node` class interface is shown in Figure 10.15.

```
interface Element: Node{
readonly attribute DOMString tagName;
DOMString getAttribute(in DOMString name);
void   setAttribute   (in   DOMString   name,in   DOMString   value)
  raises(DOMException);
void removeAttribute(in DOMStringName) raises(DOMException);
Attr getAttributeNode(in DOMStringName);
Attr setAttributeNode(in Attr NewAttr) raises(DOMException);
Attr removeAttributeNode4 in Attr NewAttr) raises(DOMException);
NodeList GetElementsByMarcatoreName(in DOMStringName);
void normalize();
}
```

Figure 10.15. Node IDL Interface

The objects and application interfaces defined by DOM Level 1 are sufficient to allow software agents and other applications to access and manipulate individual parts of a document (including all mark-up and, according to DOM Level-2 specification, DTDs) without having to parse the document. Consider the sample XML document in Figure 10.16.

```
<?xml version="1.0" encoding="ISO-8859-1" standalone="no"?>
<!DOCTYPE HardwareCatalog SYSTEM "ct.dtd">
<Catalog>
<Product><Title>ATM-PCI Card</Title>
<Hardware-Adapter SupplierCode="4012345000951" Order-
from="http:\\www.acme.com"
Interface="PCI">
<Model-No>967634</Model-No>
<Order-Line Reference-No="0528835">
<Manufacturer>Adapters Inc.</Manufacturer>
</Hardware-Adapter>
</Product>
<Product><Title>Ethernet ISA Card</Title>
<Hardware-Adapter SupplierCode="4012345000952" Order-
from="http:\\www.acme.com"
Interface="ISA">
<Model-No>967642</Model-No>
<Order-Line Reference-No="0528835">
<Manufacturer>Adapters Inc.</Manufacturer>
</Hardware-Adapter>
</Product>
</Catalog>
```

Figure 10.16. Sample XML document

The following code fragment performs a simple visit to the DOM representation of the above document:

```
var catalog = document.getElementsbyTagName("catalog")[0];
var product = document.documentElement.firstChild;
while (product) {
   var title=product.firstChild;
   var titledata=product.firstChild.data;
       if (titledata=="ATM-PCI Card ")
       {var code=title.nextSibling;
       // do something with code's attributes}
   else if (titledata=="Ethernet ISA Card ")
       {var code=title.nextSibling;
       // do something else}
product=product.nextSibling;}
}
```

Figure 10.17. Visiting the Document Tree

We are now ready to briefly discuss the XTL processor implementation. The broker agent loads supply side XML documents and parses them according to their original DTDs. This operation can be performed using any standard XML validating parser. Then, Java code is used to execute the XTL program, modifying the document to conform it to the user-centered DTD. This code uses the DOM interface to access the original document structure and to build a new one. If needed, the user-centered document can be reconverted in text format via the un-parsing operation.

The interface of the DOM-based XTL processor is shown in Figure 10.18(a)-(c).

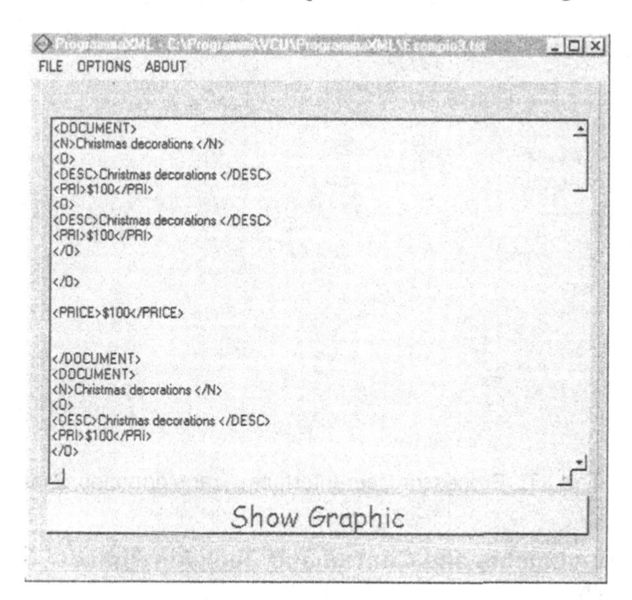

Figure 10.18(a). The XTL Processor User Interface: Textual Representation

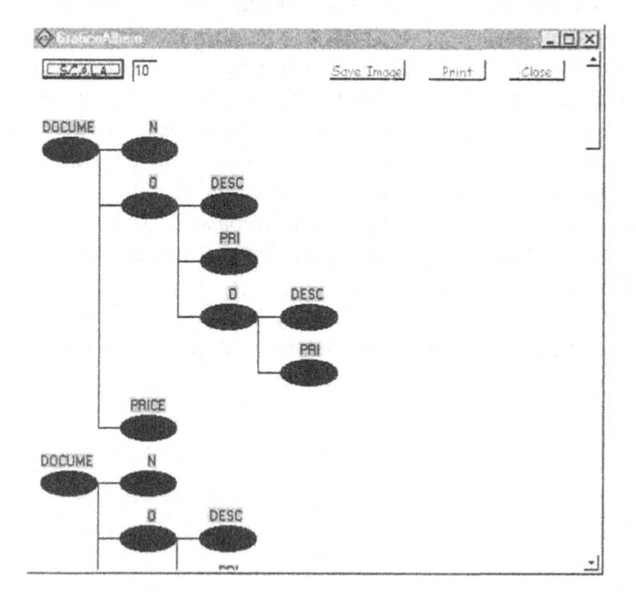

Figure 10.18(b). The XTL Processor User Interface: Graphical Representation

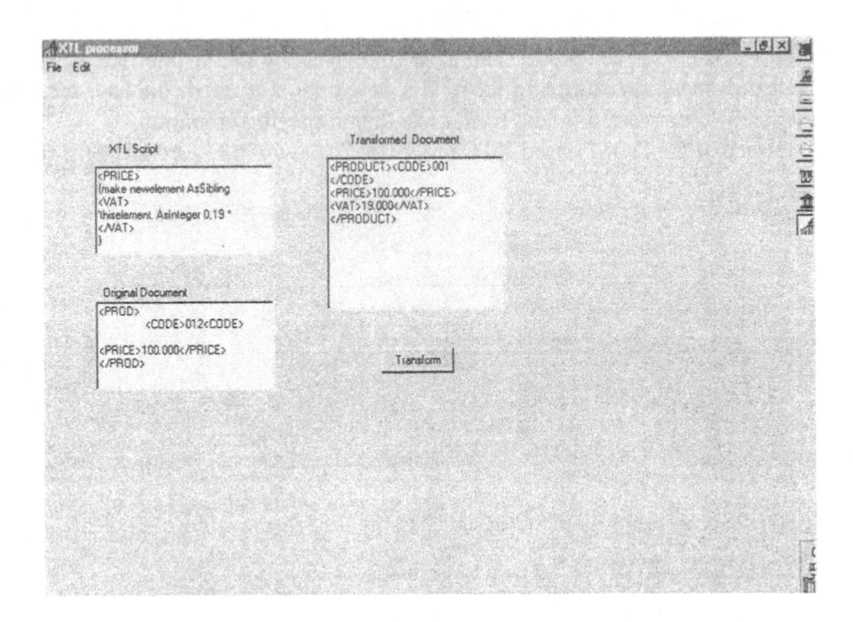

Figure 10.18(c). The XTL Processor User Interface: Transformation Sheet

10.7. XML Documents and Control and Decision Phases of HCVM

In Figure 10.5 we have shown XML representation of the decomposition phase of the HCVM. This section outlines how XML documents can be integrated with the control and decision phases of HCVM. The abstract classes identified by the decomposition phase and the corresponding transformed XML documents present a model of the market that is familiar both to the user and the supplier. However, the decomposition classes are not directly related to the solution of any particular search problem. The control phase uses a control agent to execute a further specialization of each of the XML documents corresponding to the items identified in the decomposition phase. The control agent, on the basis of fine grain features defines decision classes. The values of the fine grain features can be drawn on an ordinal or interval-based scale

In the first case, values will belong to any ordinal domain such as the integers, while in the second case each feature will be associated to a fuzzy linguistic variable.

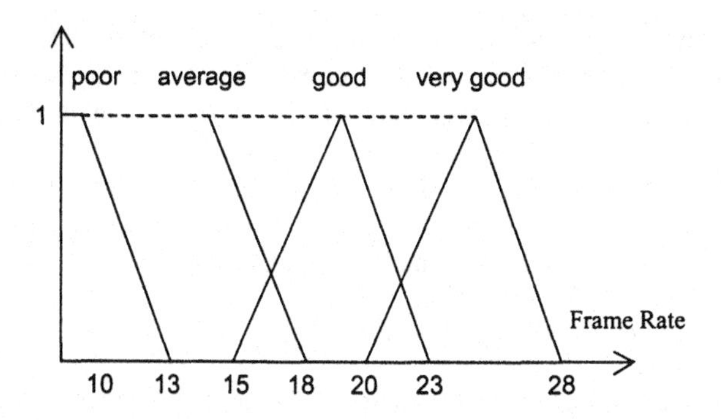

Figure 10.19. Sample Fuzzy Feature for a Video Adapter

The classes obtained in this phase are directly involved in the decision support process of the broker system.

Class construction is executed by taking into account the Market Model classes computed from the XML representation and the fuzzy knowledge base of the Decision Agent, mapping each feature of the market model in a fuzzy linguistic variable. Mapping is executed simply by identifying feature with the associated linguistic variable and using the feature value tag content to compute the membership value for each of the fuzzy elements of the linguistic variable.

It is interesting to observe that the model and metadata developed in the previous phase allows us to easily deal at this level with the user's implicit knowledge about the (sub) domain.

For instance, the user's linguistic knowledge about a modem card may include the fact that it has a "fast" data rate. In this case, the meaning of "fast" depends on the modem card being compared with other cards of the same kind and not, says, with a digital communication adapter.

In our model, being a class feature, the fuzzy linguistic variable `data_rate` may well have different definition intervals for different classes of adapters, thus dealing with the implicit different meanings of `fast` in different sub domains.

The broker, in association with user input, to compute the decision instance for a specific user query (e.g., "a frame rate of 30 frames per second"), will then use decision level classes defined by the control phase. This step involves a decision agent using the search and identification technique, (e.g. rule based, fuzzy, neural networks, and others) that is more suited to the specific multimedia information available from the supply sites.

The next section describes the control and decision agents of the hardware adapter application.

10.8. A Sample Application

The user-centered design of our architecture requires the Broker Agent to compute and retain information about the available products in the format of a class hierarchy.

Classes at the higher levels exhibit general and structured features based on a nominal scale, while lower level classes may present ordinal and interval-based features.

To decompose the chosen sub domain on the basis of the nominal scale, the Decomposition agent exploits the available body of domain knowledge to obtain decomposition such as the one shown in Figure 10.4. In this section, a detailed description of an application is given with regard to the retail (business-to-consumer) segment of the Hardware Adapters in the Personal Computers domain.

10.8.1. The Hardware Adapters Domain

Recent findings (Almeida et al. 1999) clearly indicate that electronic commerce sites' success is strongly tied to local language, national customs and regulations, currency conversion and logistics, and internet infrastructure. Several domains exist, however, where such cultural differences can be safely neglected, among which automotive, software and hardware products domains. In this section we focus on the Hardware Adapters for Personal Computers domain, as many successful EC sites for this domain are already in place.

Personal computers have been on the market since the second half of the seventies, about three decades after the development of the first digital computer. Early market-oriented classifications (see for instance [Barna, Porat 1976]) mentioned personal computers as the second segment of the computer systems market, after home computers and before workstations, minicomputers, mainframes and supercomputers. While time somewhat blurred the original distinction between personal and home computers, the other segments of this classification remain valid to this day, mainly for architectural reasons. In fact, the architecture of a modern personal computer can be described by means of the well-known *adapter-based* model. For our purposes, a brief recapitulation of this architectural model will suffice; for further details, the interested reader may consult the classical book (Patterson and Hennessy 1994) and its later editions.

An adapter-based system includes a fast communication link (the *system bus*) to which adapters or boards, each providing a different functionality are connected. The personal computer's *main board* hosts a CPU (Central Processing Unit) and the main RAM memory (connected to the CPU via an on-board fast *memory bus*); the *graphic adapter* hosts a MPU (Micro Processor Unit) designed for controlling the screen, while the disk controller cards ensure an interface to mass memory devices such as hard drives and CD-ROMs.

Finally the *network card*, also equipped with a special purpose MPU, allows for interconnecting the personal computer to an external network.

Over the years, many standards have been proposed for system buses (ISA, EISA, PCI to mention but a few) and for the adapters' internal architecture (for instance, SCSI and SCSI-2 for disk controllers). From the electronic commerce point of view, however, it is important to remark that this model has been carefully implemented by companies producing both personal computers and adapters, to the point that often buyers are presented by Web sites with a set of adapters they can choose from in order to assemble a personal computer suited to their specific needs.

Several EC Web sites, such as `shopping.com`, present a complete catalog of adapters that can be browsed and queried by users in order to locate the desired components. However, HTML-based representation of catalog pages, including scripts and presentation tags as well as product information, does not lend itself to agent-based automatic processing.

A simplified HTML catalog page (obtained omitting scripts and most presentation tagging) is reproduced in Figure 10.20.

```
<html>
<head>
       <title>Sample.com - Computer Hardware</title>
</head>
<body bgcolor="#ffffff" link="#000000" vlink="#000000" alink="#000000"
leftmargin=0 topmargin=0 marginwidth=0 marginheight=0>
<!-- toolbar.asp start -->
<!-- toolbar.asp end -->
<br>  3Com Fast Etherlink ISA 10Base-T/100Base-TX Network Interface Card
</UL><B>Features:</B>
<UL><LI> Lifetime warranty
<LI>Top performance using 3Com's patented Parallel Tasking technology
<LI>Easy migration - supports both 10Base-T and 100Base-TX networks -
autosenses hub speed
</UL><B>System and Other Requirements:</B>
<UL><LI> Ethernet 10Base-T or 100Base-TX network (for 100Base-TX, at
least 2-pair Category 5 UTP wiring is required)
 </UL><B>
Physical Dimensions:</B>
<UL><LI>Length 6.15
<LI>Width/Depth 4.15
</UL><B>Warranty:</B>
<UL><LI> Lifetime
 </UL><B>Product Specifications:</B>
<UL><LI>Device type Ethernet/Fast Ethernet
<LI>Network Port(s) 10/100Base-TX (RJ-45) Autosensing
<LI>Number of connectors 1
<LI>Speed (Mbits/sec) 10/100
<LI>Full duplex capability? No
<LI>Bus
<LI> ISA16
<LI>NE2000 compatiblity? No
<LI>Configuration Software
<LI>Promiscuous? No
<LI>Drivers included Windws NT,Windows 3.x,Windows 95,ODI,NDIS
<LI>Buffer size (kbyte) 64
<LI>Interrupts supported 3,5,7,9,10,11,12,15
<LI>Available I/O Addresses 11 possible addresses, 280 to 3EO in steps
of 20.
</font> </td> </tr></table>
                      </td>
               </tr>
       </table>
</body>
</html>
```

Figure 10.20. An HTML Catalogue Page for a Hardware Adapter

To allow for agent based processing of catalogue information, several supply-side XML namespaces and DTD definitions for hardware adapters have been suggested by the industry. *Namespaces* are domain-specific glossaries designed to provide DTD designers with a domain specific controlled lexicon.

However, switching from HTML to XML is just a first step towards a user-centered market model. In fact product information in semi-structured form is more often than not extracted on the fly from databases at the server side, rather than being stored in static documents. While promising from the technological point of view, this mapping between XML documents and entries in vendors' databases must be made flexible in order to build an effective user-centered market model.

This point can be better understood through an example. A fragment of a supply-side XML document describing a hardware adapter is reproduced below:

```
<?xml version="1.0" encoding="ISO-8859-1" standalone="no"?>
<!DOCTYPE Hardware-Adapter SYSTEM "hw.dtd">
<Hardware-Adapter SupplierCode="4012345000951" Order-
   from="http:\\www.acme.com"
Interface="PCI">
<Title>ATM-PCI Card</Title>
<Model-No>967634</Model-No>
<Order-Line Reference-No="0528835">
<Manufacturer>Adapters Inc.</Manufacturer>
<MinimumQuantity>1</MinimumQuantity>
<Warranty>Multi Year</Warranty>
<Availability>Temporarily No Stock<\Availability>
<Shipping Charge>4.95</Shipping Charge>
<Description>ATM fiber adapter PCI operating at 155 Mbps
   <\Description>
</Hardware-Adapter>
```

Figure 10.21. A Supply-side XML Document

The comparison with the HTML page of Figure 10.20 should further clarify how XML allows for a *semantics-aware* mark-up, switching the content-representation burden from free text to XML tags. As mentioned above, tags in the above fragment comply to a domain specific controlled vocabulary, allowing for a uniform representation across sites which is far more suitable than HTML to agent-based data collection and processing. However, this XML representation is still linked to a supply-side rather than to a user-centered market model. Luckily enough, transformations can be defined on XML information, preserving well-formedness and even ensuring validity with respect to special purpose DTDs. Such transformations, which can be seen as *intelligent filters*, aim to putting the user perspective at the center of the market model, while retaining all the relevant information provided by suppliers.

In our application, the Broker Agent applies such transformations, expressed as sets of XTL programs to XML documents. It should be noted, however, that transformations could also be expressed using XSL or a XML query language like XQL, XML-QL (Deutsch et al. 1999) or XML-GL (Ceri et al. 1999)

The following XTL program fragment computes a general feature and some auxiliary features belonging to our user-centered market model:

```
<Hardware-Adapter SupplierCode=* Order-from=*>
/*will match any attribute value*/
{make newelement AsSubstitute
<GenFeature>
'<Hardware-Adapter>.3.AsString
/* Takes the content of the 3rd tag from pattern <Hardware-
   Adapter> */
</GenFeature>
}

........

{make newelement AsSubstitute
<Feature>
'<ShippingCharge>.0.AsString
/* Takes the content of the 0-th tag from < ShippingCharge >,
   i.e. the tag name itself */
</Feature>
}
{make newelement AsSubstitute
<Value>
'<ShippingCharge>.AsString
/* Takes the content of the < ShippingCharge > tag name */
</Value>
}
```

Figure 10.22. The XTL Transformation Program

The resulting document, complying with the user-centered DTD of Figure 10.5, is reproduced below:

```
<?xml version="1.0"?>
<!DOCTYPE HardwareAdaptor SYSTEM "HA.DTD">
<HardwareAdaptor SUBDOM="Networking">
<TaggedValue TYP="General">
    <GenFeature>PCI_Bus_Interface </GenFeature>
 </TaggedValue>
 <TaggedValue TYP="Auxiliary">
    <Feature>Title <Value>Networking PC Card</Value> </Feature>
    <Feature> Model-No <Value>967634</Value> </Feature>
       <Feature> Manufacturer <Value>Adapters Inc.</Value>
</Feature>
       <Feature> Manufacturer <Value>Adapters Inc.</Value>
</Feature>
       <Feature> SupplierCode <Value>0528835</Value> </Feature>
<Feature> Order-from <Value>http:\\www.acme.com </Value>
</Feature>
    <Feature>Date  <Value>21, August, 1999</Value></Feature>
    <Feature>Version<Value>1.0</Value></Feature>
```

```
  </TaggedValue>
<TaggedValue TYP="Functional">
    <Feature>Type <Value> ATM Card </Value> </Feature>
    <Feature>Speed<Value>155 Mbps </Value> </Feature>
       <Feature>Bus Interface<Value>PCI </Value> </Feature>
</TaggedValue>
<TaggedValue TYP="NonFunctional">
    <Feature> Shipping Charge <Value>4.95$</Value> </Feature>
    <Feature> Availability <Value>Temporarily No Stock</Value>
</Feature>
    <Feature> MinimumQuantity <Value>1</Value></Feature>
</TaggedValue>
</HardwareAdaptor>
```

Figure 10.23. The Transformed Document

It should be noted that the XTL transformation, while producing more verbose XML documents, has several important effects:

- It cleanly separates features according to type, for instance distinguishing between functional and non-functional ones. As we will see in the next section, some values are later to be mapped from ordinal to interval-based scales. Search based on these features will also be outlined in the following.

- It reinstates content representation via text instead of relying on XML tags for that purpose. This choice, analogously to the one made in Suzuki and Yamamoto (1998), produces a *flattened* representation of contents; besides being easy to manage, such representation is inherently uniform across different sites, thanks to the fact that our user-centered XML DTD is independent from supply-side DTDs, which in turn usually depend on the schemata of vendors' databases.

- Benefits coming from standard namespaces are not lost, since in the user-centred market model the same controlled vocabulary formerly used for tags is used for document contents.

10.8.2. HCVM Based Adapter Definitions and Computations

Table 10.4 outlines the decomposition, control and decision phase adapter definitions of the harware adapter application. The shaded rows in Table 10.4 show the mapping of HCVM adapter terms and attributes with those of the hardware adapter application. In the rest of the section we describe the fuzzy computations used by the old-_PC_adapter control agent to activate the decision agents.

In the control phase, unlike the decomposition phase, interval and ordinal values need to be computed on the basis of data published by supplier sites, which is multimedia in nature and whose format is likely to change across different vendors. In order to illustrate this we have used the old_PC_adapter control agent.

In the next paragraph, the control knowledge for selecting and activating the decision agents is based on interval based *features*, in turn based on a domain specific knowledge base including the definition of interval-based fuzzy linguistic variables like `FrameRate {l, a, h}` and `DataRate {l, a, h}`, where `{l, a, h}` stand for `{low, average, high}`, that is the fuzzy elements of the linguistic variable definition. Such interval-based features compose *fuzzy descriptors* to be used in the decision phase, as will be explained in detail in section 10.8.3.

A price band variable from the market model has been used as means for naming the decision level classes where price appears as a fuzzy feature. The decomposition phase employs nominal features to determine the control classes and each control class employs ordinal/interval features to determine decision classes.

Table10.4: HCVM Phase Adapter Definitions of the Hardware Adapter Broker Agent

Adapter / Attribute	Decomposition	Control	Decision
Goal (Generic)	Restrict input context, reduce complexity, enhance reliability	Establish domain decision control constructs for orthogonal concepts based on desired outcomes from the system	Provide user/stakeholder defined outcomes from the system
Goal (PC H/W)	*Restrict input by connector-type of PC hardware adapter.*	*Define control knowledge for price-based adapter decision classes.*	*Provide a list of products to satisfy the requested properties (e.g. a multimedia adapter featuring 30 fps and 16KHz audio sampling rate*
Task (Generic)	Determine abstract orthogonal concepts	Determine decision level concepts	Determine decision instance
Task (PC H/W)	*Partition a global concept of hardware adapter into abstract classes Old_PC_Adapter, PS2_Adapter, New_PC_Adapter)*	*Determine price based decision classes that can be used with Old_PC_Adapter*	*Retrieve adapter specifications for the selected price based old PC adapter decision class*
Task Constraint (Generic)	Orthogonality, reliability, scalability	Imprecise and incomplete data, scalability, reliability, and other domain dependent	Imprecise, noisy data, generalization, adaptability, and other domain dependent
Task Constraint (PC H/W)	*Orthogonal types of adapter, scalable adapter classes*	Imprecise, fuzzy features	Imprecise, fuzzy features
Precondition (Generic)	Filtered/conditioned domain data	Orthogonal concept defined, concept related data available	Decision class case data, decision level concepts defined (optional)
Precondition (PCCC H/W)	*Domain of PC platform, filtered user query adapter data*	*Old_PC_Adapter class, fuzzy membership functions for various features*	*Price band,frame rate , data rate defined by user*
Postcondition (Generic)	Orthogonal abstract classes	Decision concepts/classes defined, control knowledge defined	Specific decision instance computed

Table10.4 (Cont'd): Various Phases of the Broker Hardware Adapter Broker Agent.

Adapter / Attribute	Decomposition	Control	Decision
Postcondition (PC H/W)	*Different types of adapters to select from*	*Price based decision classes defined and decision control knowledge for decision classes defined*	*Retrieved adapters based on user specifications and price, data scale, etc.*
Technology Artifacts (Generic)	Symbolic rule based, neuran networks, agent, multimedia	Agent neural networks, fuzzy logic,	Agent, fuzzy logic
Technology Artifacts (PC H/W)	*Perceptual (multimedia)*	*Fuzzy logic*	*Fuzzy logic/Division*
Domain Model (Generic)	Structural, functional, causal, geometric, heuristic, spatial, color, shape, etc.	Structural, functional, casual, geometric, heuristic, spatial, shape, color, etc.	Structural, functional, causal, geometric, heuristic, spatial, shape, color, etc.
Domain Model (PC H/W)	*Adapter Structure model.*	*Fuzzy price model*	*Functional specification model*
Represented Features (Generic)	Qualitative/linguistic – binary, structured Non-linguistic – continuous features	Qualitative/linguistic – binary, structured, fuzzy data Non-linguistic – continuous data related to an orthogonal concept	Qualitative/linguistic – binary, fine grain fuzzy decision concept data Non-linguistic – continuous decision concept data
Represented Features (PC H/W)	*Binary-valued(e.g. AT_bus_interface) Structure (e.g. connector type DB-9, DB-25)*	*Fuzzy (frame rate, data rate, price band), weight (refer Table 10.5)*	*Fuzzy (user selected frame rate, data rate. See Table 10.6), fuzzy division data,weight*
Psychological Scales (Generic)	Nominal; Formal property; category	Nominal, ordinal, interval, ratio	Nominal, ordinal, interval, ratio, or none
Psychological Scales (PC H/W)	*Nominal (e.g. different manufacturer card products)*	*Interval (e.g. price)*	*Interval (user selected selected frame rate, data rate*
Representing Dimensions (Generic)	Shape, location, size, position, etc. on the nominal scale	Shape, size, length, distance, density, location, position, orientation, color, texture	Shape, size, length, distance, density, location, position, orientation, color, texture
Representing Dimensions (PC H/W)	*Shape, color*	*Orientation* *Price band* 50 75 25 100	*Shape, length, color of retrieved adapter. Functional specification sheets*

Moreover, it should be noted that while in this application features are mainly functional in nature, other kinds of non-functional (e.g. structural) features could be more suitable to other applications.

In the remainder of this section, we outline how the decision computation is carried out in our application through an example, exploiting a repository storing interval-based *descriptors* of the products available on the network. The repository is a structured collection of simplified descriptions of products properties; in the line of (Damiani and Khosla 1999), such a repository can be promptly built from the user-centered market model defined in the previous subsection. We consider that such a repository to be associated with an application subdomain of electronic commerce or to a specific theme, such as, for instance, `Houseware`, `PC_Hardware Adapters`, `Fashion and Clothing`.

In such a repository, the O-O control level classes computed from the XML documents are stored as a set of *fuzzy relations*, which are defined by applying an imprecise criterion through a fuzzy predicate on a crisp relation. For such a fuzzy relation, built on a set of domains D_i, every t-uple is supplied with a membership degree μ_R, from 0 to 1, interpreting how this t-uple satisfies a fuzzy predicate **P** applied to the relation **R**.

In the simplest case, the repository is a single fuzzy relation whose attributes are: *OID, feature, fuzzy element, weight.* Each feature corresponds to a linguistic variable having several fuzzy elements. To each fuzzy element of each feature a weight is associated, describing to which extent the object offers the corresponding property. From a syntactic point of view, *nouns* express features whereas *adjectives* describe fuzzy elements.

Table 10.5 shows a fuzzy relation describing the properties of two audio/video adapters.

Table 10.5. An Example of Fuzzy Descriptor Relation

OID	*Feature*	*Fuzzy Element*	*Weight*
1	frame rate	good	.8
1	frame rate	average	0.
1	audio sampling rate	good	.4
2	frame rate	good	1.
2	audio sampling rate	average	.5

The data in this table has been computed by the Broker's Control Agent from crisp values associated to functional features by applying the definition of the `Frame_Rate` linguistic variable shown in Figure 10.19. The linguistic variables' definitions are a part of the *domain knowledge* stored by the broker, their names complying to the domain specific namespace. It is important to observe that this computation can take place both when the suppliers' servers *sign up*, i.e. communicate to the broker the availability of their products and services, and as they are periodically *polled* allowing the broker to take into account new prices or delivery conditions. Moreover, we remark that the nature of the linguistic variables representing features, as well as the number and shape of their fuzzy elements, are domain dependent and user-centered. Indeed, it is a well-known empirical fact received in several psychological studies that the number of linguistic terms used by humans to describe an intensity of some feature reflects a compromise between a loss of information as a result of granulation and a cost of manipulating those terms. In our approach, the number of linguistic values on a given scale stems from ideas such as Likert's scale and Osgood's semantic differential technique (Osgood et al. 1957), where, for example, visual stimuli allow using more values than the auditory ones.

The broker agent's role is to help the client to choose among the available objects. User interaction is used to select a part of the domain model built in the Decomposition phase, for instance `Old_PC_Adaptors`; this selection is performed by means of the general and structured features. In a second step, user input will detail the features of the desired product (For instance `price=100$` or `data rate=64kbps`). This input is filtered to compute a fuzzy predicate for each interval/based or ordinal feature, as well as an absolute value in the definition universe of the selected predicate.

As we will see in the following, in this example the control agent of our broker will use a simple fuzzy technique to deal with values supplied by the user as *reference values* or as *thresholds*, rather than as absolute according to predicates definition, the input filter transforms crisp values provided by the user into weights.

In this application, we exploit the availability of a namespace allowing both functional and non-functional information about products to be uniformed through a naming discipline, in order to deal with a standard context-dependent vocabulary [Dam97]. As we have seen, this allows both brokers and clients to use a domain-specific language to express features, without any explicit reference to the fuzzy model. Fuzzy elements and membership values i.e. the internal knowledge representation, are only computed and dealt with inside the broker.

10.8.3 User Query and Computation of Decision Instance

In the example used in the preceding section (10.8.2), a user could request to the broker a multimedia adapter having the following features: a frame rate of 30 frames per second and a audio sampling rate of 16KHz. User input filtering by preprocessing adapter computes a list of properties, each one associated with a certain fuzzy predicate and weighted by a value between 0 and 1. These values are obtained by transformation of crisp values according to the linguistic variable definition determined in the control phase. The processed input defines a *fuzzy request* to the broker, which is nothing but another fuzzy relation shown in Figure 10.19.

Table 10.6. A Fuzzy Request

frame rate	good	1
audio sampling rate	average	0.5

In order to get the list of products satisfying the requested properties, the Decision Agent can simply compute the fuzzy relational division by the query table (Damiani and Fugini 1997). As we shortly outline below, various fuzzy operators are available to compute this division, among which the broker can perform a choice in order to obtain the semantics desired by the user.

The Division Operation

Let us consider two relations R(X,A) and S(Y,A) where A, X, and Y point out sets of attributes. The *division* of R by S, denoted R[A/A]S is a relation on X, which can be defined as follows:

$$x \in R[A/A]S \text{ if } \forall a \in S[A], (x,a) \in R$$

Following (Bosc et al. 1997), we examine the extension of the division to fuzzy relations.

The operation of division of R by S can be considered as a set inclusion:

$$x \in R[A/A]S \Leftrightarrow S[A] \subseteq \Gamma^{-1}(x), \text{ with } \Gamma^{-1}(x) = \{a, (x,a) \in R\}$$

This inclusion, in the case of the extension of the division to fuzzy relations, can be interpreted either using the concept of cardinality of a fuzzy set or using a fuzzy implication, as follows:

$$Inc(S \subseteq R) = min_S (\mu_S(a) \rightarrow \mu_R(x,a))$$

The second type of division operation based on fuzzy implications is more appropriate in our case since it retains the logical aspect we are interested in. Among the main families of fuzzy implication connectives, only three are appropriate for our Decision Agent:

- *R-implications*, denoted $a \rightarrow b = \sup \{c \in [0,1], a^*c \leq b\}$. In this family, we get Goguen implication $a \rightarrow b = 1$ if $a \leq b$, b/a otherwise if we associate T with the multiplication operation, and Goedel implication $a \rightarrow b = 1$ if $a \leq b$, b otherwise, if we associate T with minimum.
- *S-implications*, namely $a \rightarrow b = n(T(a, n(b)))$, where n is an involutive order reversing negation operation, and T a conjunction operation modeled by a triangular norm. This norm has to respect several properties such as associativity, commutativity, monotonicity and 1 as neutral element. We get: Dienes implication $a \rightarrow b = \max(1-a,b)$, if we associate T with the minimum, and Goedel reciprocal $n(b) \rightarrow n(a) = 1$ if $a \leq b$, $1-a = n(a)$ otherwise
- *R&S-implications* such as Lukasiewicz one, defined by: $a \rightarrow b = 1$ if $a \leq b$, $1-a+b$ otherwise, obtained with Lukasiewicz norm $T = \max(a+b-1, 0)$.

10.8.4. User-dependent Decision Semantics

By selecting an implication, the user assigns the intended meaning of μ_S degrees in the fuzzy division R[A/A]S, i.e. the semantics of the query submitted to the broker.

If $\mu_S(a)$ values are considered as weights (i.e., we are interested in their importance), any element x will completely satisfy the query if, for each element a of S different from 0, we have a maximum membership degree for the corresponding t-uple (x,a) of R.

$$\mu_{R[A/A]S}(x) = 1 \Leftrightarrow (\forall a, \mu_S(a) > 0 \Rightarrow \mu_R(x,a) = 1)$$

In the same way, an element x will not satisfy at all a condition if there exists any element a of S which is completely important, or the t-uple (x,a) has membership degree equal to 0.

$$\mu_{R[A/A]S}(x) = 0 \Leftrightarrow (\exists a, \mu_S(a) = 1 \wedge \mu_R(x,a) = 0)$$

This desired behavior leads to define the quotient operation by using Dienes implication. Then, we have:

$$\mu_{R[A/A]S}(x) = \min_S \mu_S(a) \rightarrow \mu_R(x,a) = \min_S \max(1\text{-}\mu_S(a), \mu_R(x,a)).$$

where S is a fuzzy normalized operation (i.e. $\exists a \in S$, $\mu_S(a) = 1$) in order to have a complete scale of importance levels.

10.8.5. An Algorithm For The Decision Agent

Whatever implication is chosen to perform the division, we can give the following naive algorithm (Figure 10.24) for the Decision Agent in our example. The algorithm sequentially seeks for each element x of the divided relation R, the t-uple (x,a) for each element a of the relation S.

```
for each x of R do
mR/S(x) := 1.0;
for each a of S do
seek sequentially (x,a) in R;
if found then
mcurrent(x) := mS(a) → mR(x,a);

else

mcurrent(x) := mS(a) → 0;

end;
mR/S(x) := min (mR/S(x), mcurrent(x));
done;
done;
```

Figure 10.24. A Naïve Algorithm for the Decision Agent

This algorithm is very costly in terms of memory accesses (when the t-uple (x,a) does not exist the algorithm examines the whole relation R).

Improvements, based on heuristics and indexes, are necessary. For example, supposing the existence of a threshold l that the products final weights must reach in order to be selected, the following heuristics can be used:

A *heuristic of failure* valid for any implication: element x will not be retrieved if $\exists a \in S, \mu_S(a) \rightarrow \mu_R(x,a) < l$, since the division compute a minimum value.

The second heuristic concerns the implications of Dienes and Lukasiewicz, as well as the reciprocal of G^del implication. If we assume that S is sorted on decreasing μ_S degrees, one can stop the computation as soon as the current degree $\mu_{R/S}(x)$ is lower than $1 - \mu_S(a)$. Indeed, in this case, if the values $1 - \mu_S(a)$ are increasing then the degree of satisfaction for the considered element x can not decrease anymore. The element x will only be included in the division if $\mu_{current}(x) \geq l$. Finally, dealing with Gödel and Goguen implications, we note that for a given element x, if there exists an element a in S such that the t-uple (x,a) does not exist in R, then we have $\mu_R(x,a) = 0$, and $\mu_S(a) \rightarrow \mu_R(x,a) = 0$.

This heuristic can be used whenever the number of t-uples of any partition of the relation R is inferior to the number of t-uples of the relation S.

10.9. Summary

In this chapter we have seen, the HCVM approach allows for providing a seamless connection between a human-centred domain decomposition, based on nominal scale features expressed in XML, to lower level classes based on ordinal and interval-based features suitable to be processed by intelligent agents.

We are currently exploring a number of applications of this approach in the framework of electronic commerce.

Acknowledgements

We wish to thank Piero Bonatti, Piero Fraternali and Letizia Tanca for many stimulating discussions on the role of XML in the framework of EC agent-based systems.

References

Adler S. (1998), "Initial Proposal for XSL", available from: http://www.w3.org/TR/NOTE-XSL.html

Almeida V., Ribeiro V. and Ziviani N. (1999), "Efficiency Analysis of Brokers in the Electronic Marketplace", *Proceedings of the WWW8 Intl. Conference*, Toronto, Canada, pp.1-12

Data Interchange Standards Assoc. (1996) "Electronic Data Interchange X12 Standards, *Rel. 3070,* Alexandria,VA, 1996

Barna A. and Porat L. (1976) *Introduction to Microcomputers and Microprocessors,* Wiley Interscience

Bellettini C., Damiani E. and Fugini M.G. (1999) "User Opinions and Rewards in a Reuse-Based Development System" *Proceedings of the International Symposium on Software Reuse (SSR '99)* Los Angeles, CA (US), pp.98-110

Blair B. and Boyer J. (1999), "XFDL: Creating Electronic Commerce Transaction Records Using XML", *Proceedings of the WWW8 Intl. Conference,* Toronto, Canada, pp. 533-544

Bosc P., Dubois D., Pivert O. and Prade H. (1997), "Flexible Queries In Relational Databases - The Example of The Division Operator", *Theoretical Computer Science*, vol.171, pp.45-57

Bray T. et al. (ed.) (1998), "Extensible Markup Language (XML) 1.0", available at http://www.w3.org/TR/1998/REC-xml-19980210

Bryan M., Marchal, B., Mikula, N., Peat, B. and Webber, D. "Guidelines for using XML for Electronic Data Interchange", available from: http://www.xmledi.net/

Buschmann F., Meurier R., Rohnert H., Sommerlad P. and Stal M. (1996) *A System of Patterns,* J. Wiley

Ceri S., Comai S., Damiani E., Fraternali P., Paraboschi S. and Tanca L. (1999) "XML-GL: A Graphical Query Language for XML, *Proceedings of the WWW8 Intl. Conference*, Toronto, Canada, pp.93-110

Connolly D. (1995), "An Evaluation of the WWW as a Platform for Electronic commerce, *Proceedings of the Sixth Computer, Coordination and Collaboration Conference,* Austin, TX (US), pp. 55-70

Crocker D., "MIME Incapsulation of EDI Objects", RFC 1767 2/3/1995 http://www.ietf.org

Damiani E. and Fugini M.G. (1997) "Fuzzy Identification Of Distributed Components, *Proceedings of the 5th Intl. Conference On Computational Intelligence*, Dortmund, Lecture Notes in Computer Science 1226, pp.95-98

Damiani E. and Khosla R. (1999) "A Human Centered Approach to Electronic Brokerage, *Proceedings. of the ACM Symposium. on Applied Computing* (SAC '99) San Antonio, TX, February 1999, pp.243-249

Damiani, E. De Capitani S., Paraboschi S. and Samarati P. (2000),"Securing XML Documents" *Proceedings of the Seventh International Conference on Extending Database Technology*, Kostanza, Germany, LNCS 1777 pp. 121-135

Deutsch A., Fernandez M., Florescu D., Levy A. and Suciu D. (1999) "A query language for XML" *Proceedings of the WWW8 Intl. Conference,* Toronto, Canada pp.77-92

Englebart D. (1990), "Knowledge Domain Interoperability and an Open Hyperdocument System", *Proceedings of the 1990 Conference on Computer Supported Cooperative Work*, pp.45-58

Finin T., Fritzson R., MacKay D. and MacEntire R. (1994), "KQML as an Agent Communication Language, *Proceedings of the Third International Conference on Information and Knowledge Management*, pp.112-124

Fromkin, A.M., (1995) *The Essential Role of Trusted Third Parties in Electronic Commerce*, in Kalakota R., Whinston A. (eds.) *Readings in Electronic Commerce*, Addison-Wesley

Garfinkel S. (1995), *Pretty Good Privacy*, O'Reilly

Glushko R., Tenenbaum J. and Meltzer B. (1999), An XML-framework for Agent-Based Electronic commerce, *Communications of the ACM*, vol. 42 no. 3

Hands J., Patel A., Bessonov M. and Smith R. (1998), An Inclusive and Extensible Architecture for Electronic Brokerage", Proc. of the Hawai Intl. Conf. on System Sciences, Minitrack on Electronic Commerce, pp.332-339

Hamilton S. (1997), Electronic Commerce for the 21st Century, *IEEE Computer*, vol. 30, no. 5 pp. 37-41

Lange D. and Oshima M. (1998), *Programming and Deploying Java Mobile Agents With Aglets,* Addison-Wesley

Lynch D., Lundquist L (1996), *Digital Money: The New Era of Electronic Commerce,* John Wiley

Maes, P., Guttman, R., and Moukas, A.G. Ágents that Buy and Sell", *Communications of the ACM*, Vol. 42, n. 3, pp. 81-85

Neches R., "Electronic Commerce on the Internet, *white paper to the Federal Electronic Commerce Action Team,* http://www.isi.edu/dasher/Internet-Commerce.html

Orfali, R. and Harkey, D., *Client/Server Programming with Java and COBRA*, John Wiley Computer Publishing

Osgood E., Suci G. and Tannenbaum P. (1957) *The Measurement of Meaning,* Oxford University Press

Patterson D. and Hennessy J. (1994), *Computer Organization and Design,* Morgan-Kaufmann,
Powley C., Benjamin D. and Grossman D. (1997) "DASHER: A Prototype for Federated Electronic commerce Services, *IEEE Internet Computing,* vol. 1, no. 6
Prescod P., "An Introduction to DSSL", available from http://cito.uwaterloo.ca/~papresco/dsssl/tutorial.html
Rutgers Security Team, "WWW Security. A Survey", available from http://www-ns.rutgers.edu/www-security/
Suzuki, J., Yamamoto, Y., "Making UML Models Exchangeable ovre the Internet with XML: the UXF Approach", *Proceedings of the UML'98 Conference,* S. Antonio, TX US, pp. 134-143.
Tenenbaum J., Chowdhry T. and Hughes K. "eCo System: CommerceNet's Architectural Framework for Internet Commerce", http://www.commercenet.org
Yang Z. and Duddy K. (1996), "CORBA: A Platform for Distributed Object Computing, *ACM Operating Systems Review,* vol. 30, no. 2.
Wood L. (1998), "Document Object Model Level 1 Specification", available from http://www.w3.org/pub/WWW/REC-DOM-Level-1/

11 A USER-CENTERED APPROACH TO CONTENT-BASED RETRIEVAL OF MEDICAL IMAGES

11.1. Introduction

In chapter 9 we discussed that the challenge of managing huge multimedia *digital libraries* has turned attention of researchers toward the problem of efficient and effective access to structured data and textual sources as well as media with spatial and temporal properties (e.g., sound, maps, images, video) (Maybury 1995)

In the field of image databases, while older systems relied on the straightforward technique of providing textual *descriptions* to be stored together with the images, a number of more recent approaches focuses on using *color*, *texture* or *shape* (Faloutsos et al. 1994) as the basis for image indexing and querying. Such systems try to exploit directly perceived pictorial features to model and reproduce the human evaluation of similarity between relevant features; the outcome of this similarity evaluation is then used for retrieving images from a multimedia archive.

However, while promising from the purely technological point of view, many of these systems (see for instance Binaghi et al. (1994)) explicitly renounce incorporating in the query language naming or coding methods already familiar to user communities in specific application domains, as they turn out very difficult to be mapped to mathematically satisfying definitions of similarity.

Some other approaches to image databases, in order to attain effective context-based indexing, manage *feature data*, which can be seen as points in a feature space (Chiueh 1994) These features are classically described as a probability distribution function, rather than as a single point. However, such systems need a probabilistic *similarity measure* to support image classification, requiring complex trade-offs between efficiency and effectiveness of retrieval (Barros et al. 1995).

Moreover, all systems mentioned above are *monolithic*, i.e. they integrate indexing, search and storage facilities at the same location, requiring the whole multimedia database to be stored in the same place and available for indexing at the same time.

In this chapter we follow a different approach, describing query support to a distributed collection of medical images as an example of *human-centered* retrieval architecture for multimedia data. Following this user-centered approach, we shall describe an application of HCVM to *on-line indexing and retrieval* of medical images, investigating its feasibility and presenting a complete solution.

Most conventional multimedia databases for medical applications execute content-based indexing *off-line*, independently from their query execution mechanism. In our user-centered approach, image data are interpreted immediately before they are queried, generating suitable content descriptions on-line. It should be noted that such a system is a tool for collecting and retrieving images designed to highlight certain pathologic processes and from which students learn about human disease information, and not a diagnostic help. Since the requirement of on-line operation prevents costly computations, only rough estimates of content related indexes are computed; this imprecision is dealt with by exploiting fuzzy query support, in the line of the application of fuzzy query languages to multimedia databases proposed in Dubois et al. (1999).

Telehealth applications, like all other applications involving distributed components, require standardization both at the logical and at the physical interface levels to ensure modularity and cooperation among different systems. The application presented in this chapter relies on the MASIF standard (OMG 1999) for CORBA-compliant agent-based systems, proposed by the Object Management Group.

The remainder of this chapter is organized into six parts. In the following section, the characteristics of the problem are first outlined. section 11.3 describes a reference model for distributed image retrieval, while the fourth section proposes HCVM based solutions to some of the classification and retrieval problems presented in the chapter.

In section 11.5 a sample application for distributed classification and retrieval of medical images is presented in some detail, and section 11.6 briefly comments on the implementation of an agent-based system complying to the MASIF standard.

11.2. Characteristics of the Problem

According to a standard definition agreed in 1990 among the Ministries of Health of the European Union, *Telematics for Health Care* is "the integration, monitoring and management of medical data, including those used for personnel education, by means of communication systems providing ready access to medical databases and/or the advice of remote experts" (Telemedicine Research Center 1998).

In the framework of such *telehealth* systems, searching and querying image collections is considered a routine activity in the medical field and constitutes an important support to differential diagnosis and clinical research. This application field lies at the borderline between telehealth and multimedia data analysis and processing, and several layered standards for image representation and management have been developed, originally aimed at radiological images transmission, and later extended to other fields such as dermatology and dentistry. Examples of such domain-specific layered standards for storage and transmission of medical images are DICOM/3 (CEN Project Report 1998) and SPI (Siemens AG 1998).

In this chapter we shall focus on the area of human anatomy and *pathology*, the latter addressing illnesses involving human cellular tissues. While a broad range of laboratory techniques are available in this field, pathologists are well aware of the increasing importance of consulting remote image libraries illustrating specific pathologies in the framework of distributed telehealth systems.

11.2.1. Image Libraries and Diagnostic Systems

Many image-based diagnostic applications are currently being developed, such as tumor detection systems for full-digital images. In this field, even simple techniques for content-based indexing of images are expected to have a considerable impact on the accuracy and efficiency of the diagnostic process. Image processing for tumor detection poses two fundamental problems:

- *Shape enhancement* It is not rare that tumors have very weak contrast against their background and how to detect "suspicious regions" in digital images is one of the key points to establish reliable system.
- *Feature extraction* Expert knowledge must be exploited in order to extract features that best characterize pathologies such as malignant tumors.

The former problem has been often tackled by means of *adaptive filters*. Such filters (Kobatake and Murakami 1996) can be very effective in enhancing approximately rounded opacities no matter what their contrasts are. Clues for discriminating between malignant tumors and other pathologies are believed to be mostly in their boundary areas. Therefore, the detection of boundary is an important preprocessing for feature extraction. The boundary of malignant tumor is usually fuzzy, and fuzzy-based filters are often adopted to estimate boundary of suspicious region. In Kobatake and Murakami (1996) the optimal region of support of a filter was shown to coincide with the boundary of an approximately rounded convex region if the pixel of interest is in its inside. By applying it to real mammograms, it was possible to extract probability of the existence of the tumor boundary. In other words, the fuzziness of the boundary was reflected on the boundary probability obtained by the filter.

Several feature parameters can be adopted to identify malignant tumors from suspicious regions. It is widely recognized that the most effective of them reflect boundary characteristics; the shape parameter "ellipticity" was adopted as a feature in Kobatake and Murakami (1996) and will be used as a reference in the remainder of the paper.

Many experiments are documented in the literature to test the effectiveness of feature parameters. Classification experiments were conducted for instance in Kobatake et al. (1998) using 1313 suspicious regions, which were detected by an adaptive filter from 354 images. The number of suspicious regions corresponding to malignant tumors is 71. Adjusting the system parameters so that 96% of malignant tumors could be identified correctly, the average number of false positives was 0.9 per image. These results in cytopathology represent the closest anyone has come to automate diagnosis, i.e. to design of screening machines that can help to sort out abnormal from normal images.

However, a word of caution is mandatory: there are still many difficulties in extending the automated diagnosis approach to other fields of the medical practice.

First of all, even if some images of interest can be obtained from surgical pathology specimens, cytopathology slides, and clinical laboratory specimens, many other images cannot be obtained except through autopsy.

Moreover, while such medical images can be easily categorized by components, obtaining descriptions of image content like "dots in the upper right" or "pale smudge in the center", performing such a classification does not mean to be able to make a diagnosis and correlate it with the patient's condition, obtaining a word description such as "Hodgkin's disease, nodular sclerosis type" which conveys much more information.

Regardless of all the attempts to utilize morphometric analysis for diagnosis, many physicians maintain that the gold standard remains an experienced pathologist looking at a glass slide. This widely shared opinion has not diminished the interest of the computer science and medical research communities in designing applications and systems for content-based indexing of medical images that can be used for teaching and research, as will be clarified later in the chapter.

11.2.2. The SNOMED language

While a global framework for medical images sharing and processing is still in its infancy, a number of well-known medical images repertoires illustrating various pathologies are currently available in GIF or JPEG format over the internet, though few of them are stored in fully-fledged image databases. Indeed, most of these collections are only accessible through standard Web sites, each of them maintained by a different medical institution.

Moreover, while the medical community has since long agreed on a basic textual coding for image contents (namely *SNOMED*, the *Systematized Nomenclature of User and Veterinary Medicine*), no standard technique for indexing or retrieval of medical images based on color or shape is currently available or planned.

SNOMED is a concept-based *reference terminology* including 110,000 concepts linked to clinical knowledge; its lexicon includes more than 180,000 terms, each with unique computer readable codes, reflecting contemporary clinical practices. The SNOMED terminology also includes about 260,000 relationships between terms enabling retrieval based on any of a variety of criteria (e.g., disease hierarchy, anatomic relationships)

By functioning as a reference terminology, SNOMED allows easy transmission of patient-related data across diverse and otherwise incompatible information systems.

SNOMED is applied to a wide variety of applications, including clinical documentation and decision support; cancer research, literature retrieval; cardiovascular case findings and data warehousing. SNOMED has already received significant acceptance internationally, as it is has been translated into 16 languages and is being used in more than 25 countries.

Medical users are therefore accustomed to using SNOMED patterns as search and classification codes for images depicting pathologies.

Indeed, retrieval of images tagged by a SNOMED code is easily and efficiently performed through string *pattern* matching; Figure 11.1 presents some sample queries that can be posed using SNOMED pattern-based language (Mauri et al. 1996).

11.2.3. SNOMED
(Systematized Nomenclature of Human and Veterinary Medicine)

```
                    T: Topography

T=2*    Breathing Apparatus
T=28*   Lungs
T=281*  Right Lung
```

Figure 11.1. Sample SNOMED patterns

In the following sections, a technique for transparently superimposing hybrid agent-based search to SNOMED compliant Web-based collections of medical images will be presented as a user-centered alternative to storing such images in a conventional, monolithic multimedia database.

11.3. A Reference Model for Distributed Image Retrieval

Following the same presentation technique used in the previous chapter, before describing the functionality of our agent-based architecture for medical image indexing and retrieval, we briefly outline its reference model at the highest level of abstraction, in order to clarify the use of the terminology.

Table 11.1. Roles and Actions of the Image Classification and Retrieval Reference Model

Abstract Model	
Roles	Customer Broker Supplier
Actions	Classify Search Deliver

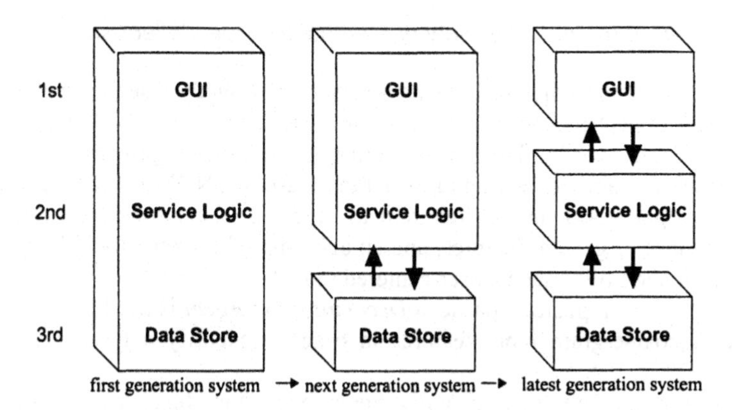

Figure 11.2. An architectural view of roles

The core concepts of our simple reference model are the roles of *customer* (the medical researcher querying the system), *broker* (our multimedia retrieval agent) and *suppliers* (the sites storing collections of medical images). To these roles, the actions of *search*, *classify* and *deliver* are associated.

In Figure 11.2, an architectural view of this distinction is given, outlining how all service logic can be located in the broker agent, while customer and supplier respectively provide a graphical user interface and fast conventional data storage, possibly made accessible through a Web site.

In fact, in our system the Broker Agent acts as a supplier of images to the customers, and as a classification and distribution mechanism for suppliers wishing to make their images accessible over the internet.

A basic assumption of our design is that suppliers *do not index* images, other than providing their standard SNOMED codes; it is left to brokers to compute and store content representations of online images, in order to be able to locate the image required by the customers.

Therefore the main function of our Broker Agent is providing a path whereby the customer may find and obtain from a supplier a set of images offering the required characteristics to the highest possible degree.

11.4. User-Centered Image Retrieval

In this section we shall outline how HCVM's layered architecture allows the design of our image classification and retrieval system to proceed seamlessly from a human-centered representation of image content, based on a *nominal scale*, to the *ordinal* and *interval-based* representations that are more suitable for intelligent search agents.

This will allow identification of the *decision classes*, corresponding to candidate images to be submitted to the user.

As a by-product of our case study, we shall show how HCVM approach ensures generality and applicability of the retrieval system to a wide range of multimedia application domains.

11.4.1. A Human-Centered View of Medical Images Collections

The basic idea underlying HCVM approach is putting the user's (in this case, the medical researcher) perception of the information space (the collections of medical images available over the Net) at the center of the knowledge organization process. The basic assumption we rely on turns out to be use of SNOMED coding: the medical researcher organizes a mental hierarchical model of the information domain via a limited number of general features, and an agent-based broker should be able to fully comprehend and utilize such a user-centered model.

In HCVM decomposition phase, a *Decomposition Agent* is used employing a small number of coarse-grain input features in order to identify a hierarchy of abstract classes.

In the present setting, SNOMED hierarchical encoding, in addition to being familiar to the user, provides an easy and effective way to decompose the medical images domain. Moreover it has the additional advantage of being familiar to the user community. Therefore, we use SNOMED domain decomposition as the guideline for designing our system, while intervals and ordinal indexes are computed *on-line* on the basis of image content.

11.4.2. From the Domain Model to Decision Support Classes

The SNOMED-based taxonomy identified by the decomposition phase presents a model of the whole medical images information space that is both simple and familiar to the user. However, this user-centered model is not related to the solution of any particular search problem. HCVM Control Phase uses a *Control Agent* to determine decision level concepts on the basis of finer-grain features whose values can be drawn on an *ordinal* or an *interval-based* scale.

In the first case, values will belong to any ordinal domain such as the integers, while in the second case the feature will be associated to a fuzzy linguistic variable. In our present application, *fuzzy linguistic variables* with trapezoidal elements such as the one depicted in Figure 11.3 will be used to represent features of image content.

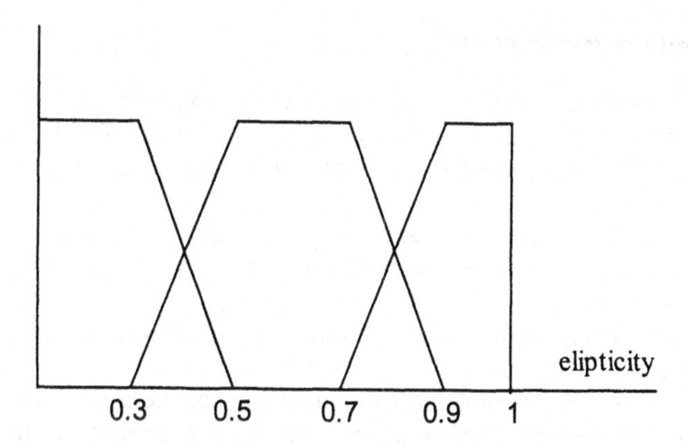

Figure 11.3. Fuzzy Elements of the Ellipticity Linguistic Variable

A basic assumption of our design is that such content-related features, though an integral part of the domain model, need not be stored together with the images, but can be computed on-line by the broker agent. However, as we shall see in the next section, the classes obtained in this phase will be directly involved in the decision support process of our image classification and retrieval system. It is interesting to observe that the SNOMED-based domain model developed in the previous phase allows us to easily deal at this level with the user's implicit knowledge about the (sub) domain.

For instance, the medical user's linguistic knowledge about a pathology of the uterus may well include the fact that it involves an "abnormal" ellipticity of black *blobs* in the image corresponding to pathologically suspect cells aggregates. In this case, the meaning of "abnormal" depends on the fact that the image is being compared with other images of the same kind and not, say, with the image of a lung.

In our model, being a class feature, the fuzzy linguistic variable `ellipticity` may well have different definition intervals for different classes of images, thus dealing with the implicit different meanings of `abnormal`.

11.4.3. The Decision Phase

The aim of the decision phase is to compute user specified outcomes or decision instances in each decision class. In this application it means computing decision instances based on specific user query. This step involves a hybrid Decision Agent, using the search and identification technique (Rule based decision system, Fuzzy matching, Neural Networks) which is more suited to the specific multimedia information available from the supply sites.

In the medical image retrieval application described in the next setion, we will exploit the fuzzy techniques similar to those used in chapter 10 and fuzzy queries techniques developed by Bosc et al (1997).

11.5. A Sample Application

The medical user-centered design of our architecture requires the broker to compute and retain information about the available images in the format of a hierarchy. Classes at the higher levels exhibit general and structured features based on a nominal scale, while lower level classes present ordinal and interval-based features.

To decompose the chosen sub domain on the basis of the nominal scale, the Decomposition agent exploits the available body of domain knowledge, in this case the SNOMED encoding.

At the Control level, however, intervals and ordinal values must be computed on the basis of image content.

In this section, we outline how such a computation can be carried out exploiting a broker-managed repository storing *descriptors* of the images available on the network.

11.5.1. Descriptor-based Image Classification

The repository holding image descriptors is a structured collection of simplified descriptions of image properties, in the line of (Damiani and Fugini 1997). We consider that such a hierarchical repository to be associated with each application sub domain, i.e. to each specific prefix of SNOMED coding, such as, for instance `T=3*` for the cardiac system.

In such a repository, the O-O control level classes are stored as a set of *fuzzy relations*, which are defined by applying an imprecise classification criterion on a crisp relation through a fuzzy predicate. For such a fuzzy relation, built on a set of domains D_i, every t-uple is supplied with a membership degree μ_R, from 0 to 1, interpreting how this t-uple satisfies a fuzzy predicate **P** applied to the relation **R**.

In the simplest case, the repository is a single fuzzy relation whose attributes are: *OID, feature, fuzzy element, weight*. Each feature corresponds to a linguistic variable, such as `max_blob_ellipticity`, having several fuzzy elements.

To each fuzzy element of each feature a *weight* is associated, describing to which extent the object offers the corresponding property. From a syntactic point of view, *nouns* express features whereas *adjectives* describe fuzzy elements.

Figure 11.4 shows a simplified fuzzy relation describing the properties of two sample medical images. The actual data model currently employed in our system will be discussed in detail in the next section.

Object Id	*Feature*	*Fuzzy Element*	*Weight*
1	Max_blob_ellipticity	slightly_abnormal	.3
1	Max_blob_ellipticity	normal	.8
1	Max_blob_magnification	high	0.7
2	Max_blob_magnification	high	0.8
2	Max_blob_ellipticity	normal	1

Figure 11.4. Sample Fuzzy Descriptor Relation

The Broker's Control Agent computed the above table on-line after executing a standard noise reduction based on fuzzy techniques Stephanakis et al. (1998), Russo and Ramponi (1995) by applying to two images the definition of the `max_blob_ellipticity` and `max_blob_magnification` linguistic variables (see Figure 11.4).

Our classification procedure involves, for each image, five main steps:

- First, the image is converted to a standard RGB representation
- Secondly, RGB encoding is converted to the well-known HSL color space representation. Hue is a linear scale expressing the darkness of each pixel as a function of its RGB values.
- Thirdly, a simple hue threshold is applied to identify the main "dark" blobs in the image.
- Then, a deterministic convex-hull algorithm is used to determine the crisp ellipticity value associated to the biggest black blob in the image (Castelletti et al. 1999) This kind of *ellipse fitting* algorithms fit an ellipse to a set of points, which are all roughly associated with the ellipse while allowing for some tolerance to scatter. Current implementation of the algorithm allows us to set a maximum amount of time for index calculation per image (namely, 5 seconds).
- Finally, applying the linguistic variable definition corresponding to the SNOMED code associated to the image fuzzifies the crisp value obtained in the previous step.

It should be noted that step 1 and 2 could in principle be avoided, as GIF or JPEG images collected from Web sites could be easily converted to a gray-scale representation before computing the threshold. After conversion to gray-scale, however, images to be indexed are large, rectangular arrays of pixels with different gray-levels. Within this array of pixels there may be several blobs. In order to apply an ellipse-fitting algorithm to such blobs, it would be necessary to segment out points belonging to each of them. It is well known that, in principle, it is not possible to identify points belonging to a blob on the basis of gray levels, if the blobs are superimposed on a background of much smaller rectangles of varying gray level.

Several *Shell-clustering* techniques are currently available (Dave 1990), providing shape finding approaches that use fuzzy, possibilistic, and robust methods to find unknown numbers of parameterized curves from sparse and noisy edge data. However, due to time constraints imposed by on-line indexing, in this application a simple threshold must be applied to identify black areas in the image, and ellipse fitting must be applied to these areas only. Hue representation proved to be more effective than others for the available images; the availability of fast conversion software allowed for choosing a representation more sensible to the chromatic property of the original image.

It should also be noted that the linguistic variables used for classification are part of the system's knowledge base for a certain value of the SNOMED code.

The linguistic variables' definitions for a given sub domain are a part of the domain knowledge stored by the corresponding broker. It is also important to observe that this computation takes place as the suppliers' sites are periodically *polled* with simple HTTP connections, allowing the Broker Agent to take updates into account.

Finally, it should be noted that, while the simple convex-hull deterministic algorithm (see the next section) proved to be reasonably fast and accurate, other techniques could be evaluated in order to be put at the Broker Agent's disposal as alternative tools to be selected according to time or accuracy constraints. A back-propagation neural network (Yiming et al. 1999) with a single hidden layer can usually map inputs to outputs for arbitrarily complex non-linear systems, and could be used in the framework of our system. Application of Kohonen's self-organizing maps (Kohonen 1997) is another interesting possibility.

11.5.2. Ellipticity Evaluation

In order to satisfy the time constraints imposed by on-line indexing, the computation of blob ellipticity is performed in a "quick and dirty" manner, leaving it to the query execution engine to deal with data imprecision introduced by the indexing procedure. Figure 11.5 shows the pseudo-code of the naive algorithm employed for computing ellipticity (Damiani et al. 1999):

```
Point:=ComputeBlobCenter(Blob)
MAxEll:=0;
InitialSlope:=0;
For step:=0 to maxstep do
Ratio:= (GetSegmentLength (Point, InitialSlope+step)/
GetSegmentLength (Point, InitialSlope+step+π/2));
if MaxEll<Ratio then MaxEll:=Ratio;
od;
Return MaxEll;
```

Figure 11.5. Pseudo Code of the Indexing Algorithm

The employed naive algorithm simply computes the center of the blob and then samples pairs of orthogonal straight lines passing through it with angular coefficients from 0° to 360° degrees (measured in the image local coordinate system).

The sampling step is a parameter of the algorithm as it can be tuned according to time constraints and/or image characteristics. It is usually set at $\pi/10$. On each line, the length of the segment belonging to the blob (i.e., whose hue values are all above threshold) is computed. Then the highest ratio between a pair of orthogonal segments is returned as a rough estimate of the blob's ellipticity.

Figure 11.6 shows the image of a human aorta (source: WebPath, courtesy of E. Klatt). The white arrow indicates a lipidic streak. It should be noted that, while the streak has poor contrast with respect to the background in RGB representation, it is easily identified in the HSL scale. Moreover, the high ellipticity of the streak makes it particularly suitable for index computation. Figure 11.7 shows a cervical Pap smear in which dysplastic cells are present that have much larger and darker nuclei than the normal squamous cells with small nuclei and large amounts of cytoplasm.

Figure 11.6 A Human Aorta with Lipidic Streaks (source: WebPath, courtesy of E.G. Klatt)

Finally, we remark once again that our procedure does not take into account blob identification, which is obtained by applying the threshold; this of course could mean "losing" some blobs whose hue is not different enough from the image background, or computing some "false positives" for the opposite reasons. For this reason, user feedback is collected: the user is given the possibility of manually modifying the index value for any of the images returned in the query result.

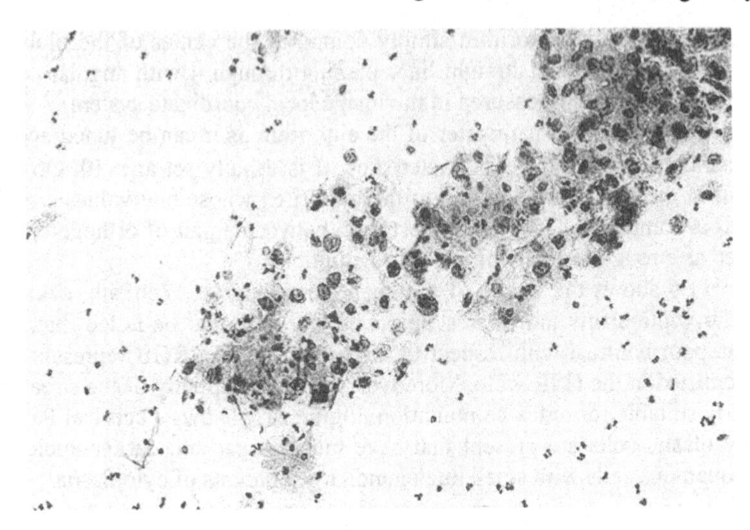

Figure 11.7 A cervical Pap smear with dysplastic cells *(source:* WebPath, *courtesy of E.G. Klatt)*

11.5.3. Querying the Broker Agent

User interaction is made through a media agent, whose graphical interface is shown in Figure 11.8.

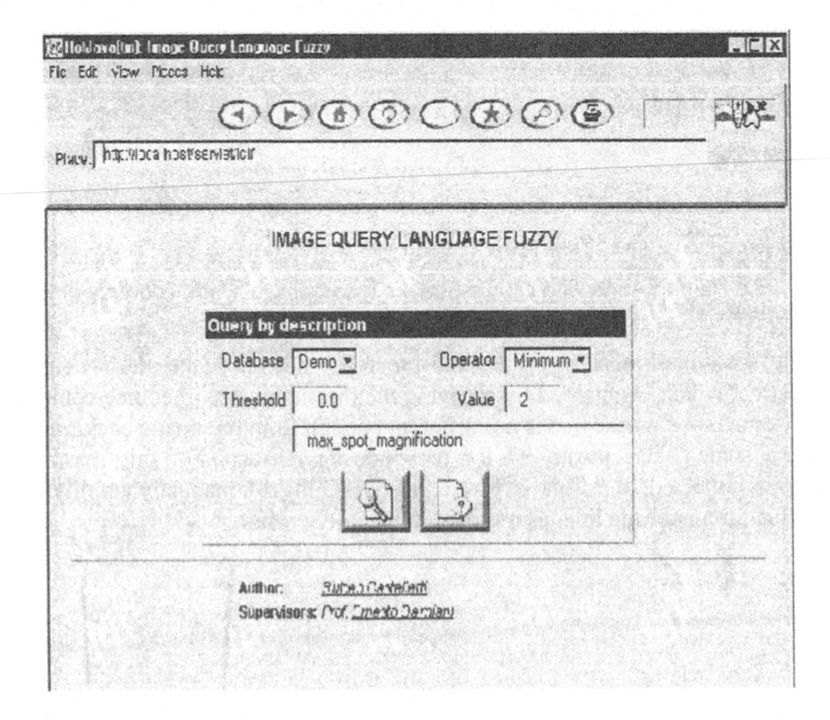

Figure 11.8. Sample User Interaction

From the user point of view, four main elements are involved in the interaction:

- The **user connected to internet** from a certain machine, which requires a service from the Broker Agent system. We assume that the client has previously contacted a "Master Broker", which contains lists of various URL of known Broker in the SNOMED domain under consideration.
- A **Broker site** where all classes defining the GUI interface are stored, together with the Broker Agent itself and its relational repository, including the characteristic values of the various fuzzy predicates. The Broker receives fuzzy requests from users who obtained the GUI connecting to the Broker Web site. On the basis of this information, performs the division (according to the fuzzy predicates) on the repository and returns the best matching images to the user.
- An **Adapter site** that supports a fuzzy adapter system whose role is to dialogue with Web servers storing medical images and maintain a coherent view of fuzzy predicates according to some images' content-related properties.
- Several **Web servers** that actually provide the images whose content is described in the Trader base. No fuzzy issue characterizes these sites: the server programmer or installer only specifies the SNOMED code for each image as a part of the HTML or XML markup of the page holding it.

The first interaction step requires the medical user to select a specific Broker agent, i.e. a part of the domain model built in the Decomposition phase, again via SNOMED codes. SNOMED provides a complete decomposition of the information space: a hierarchy of Brokers could well rely on the domain decomposition provided by SNOMED. In a second step, the user must detail the features of the desired image.

User input is collected through a graphical interface in order to identify a fuzzy predicate for each interval-based or ordinal feature specified by the user.

When provided by the user, an absolute value in the definition universe of the selected predicate is also selected (for instance `max_blob_magnification =2`).

This results in a query language having the simple grammar shown in Figure 11.9 in *Backus-Naur Form*:

Productions:

```
statement:= noun [':' adjective] ['=' value]
noun:= TERM |REGX
adjective:=TERM
value:=REAL
```

Terminal symbols:
TERM: a string coded in ISO-LATIN-1
REGX: a string coded in ISO-LATIN-1, including wildcards '?' and '*'
REAL: a real number

Figure 11.9. The Query Language Grammar

While the simple grammar given above is self-explanatory, a comment should be made on the fact that it does not explicitly involve any fuzzy number or linguistic modifier. Indeed, in a user-centered approach to image querying it would be rather unrealistic to require the medical researcher to be conversant in, or even knowledgeable about, fuzzy query techniques.

Here, all the user needs to know is that since classification based on content related parameters is performed on-line, it is forcedly raw, i.e. values are to be taken as approximate.

Therefore, while the query interface requires the user to specify crisp values for content-related queries, query result will automatically include also images whose indexes are reasonably "near" to the value specified by the user.

Of course, many content-based indexes other than `ellipticity` and `magnification` of black blobs could be used for querying; to this aim, we assume the availability of a Thesaurus, i.e. a medical *controlled vocabulary* allowing features about images in a given sub domain to be uniformed through a naming discipline, in order to deal with a standard context-dependent vocabulary [Dam97].

SNOMED itself provides a fully featured glossary (Mauri et al. 1996) that allows both Brokers and clients to use a sub domain-specific language to express features, without any explicit reference to the fuzzy model. Fuzzy elements and their membership values, i.e. the internal knowledge representation, are only computed and dealt with inside the broker.

11.5.4. Computation of the Decision Classes

With reference to the previous example, a medical user could request to the Broker Agent an image having the following features: a `max_blob_magnification` of 2 and a `max_blob_ellipticity` of 0.8 (the reader should note that this latter value is not a fuzzy membership value, but a crisp geometrical parameter).

User input filtering computes a list of properties, each one associated with a certain fuzzy predicate and weighted by a value between 0 and 1.

Once again, we remark that these values are obtained inside the Broker Agent, by transformation of crisp values specified by the user according to the linguistic variable definition determined at the Control level.

The processed input defines a *fuzzy request* to the Broker, which is nothing but another fuzzy relation shown in Figure 11.10.

max_blob_magnification	average	1
max_blob_ellipticity	abnormal	0.9

Figure 11.10. A Fuzzy Request

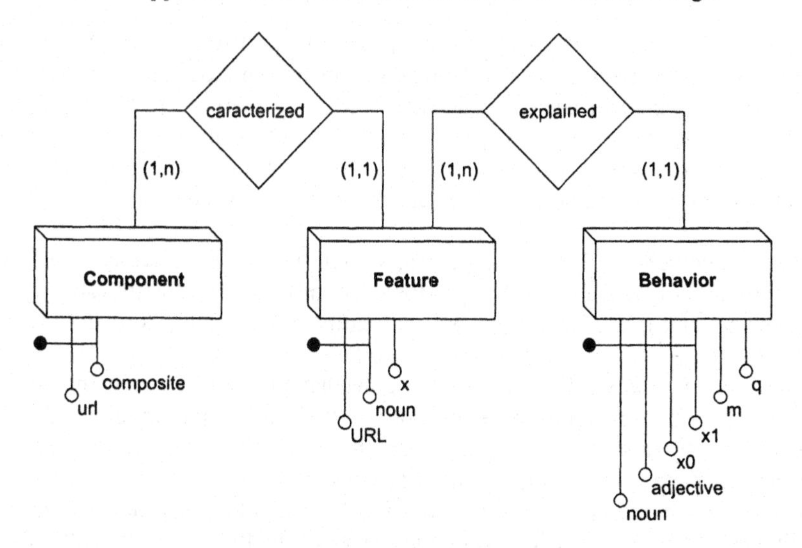

Figure 11.11. Relational Schema of the Broker's Repository

Executing a *fuzzy division* (Bosc, P., et al. 1997) between the user query and the Feature table in the broker's repository allows the Decision Agent to compute a ranked list of candidate images, whose URLs is returned to the user.

11.5.5. The Division Operation

In order to describe in detail the internal operation of the Broker, we now recall some basic definitions about relational division. Let us consider two relations R(X,A) and S(Y,A) where A, X, and Y point out sets of attributes. The division of R by S, denoted R[A/A]S is a relation on X, which can be defined as follows:

$$x \in \mathrm{R[A/A]S} \text{ if } \forall a \in \mathrm{S[A]}, (x,a) \in \mathrm{R}.$$

Following (Bosco et al 1997) we now examine the extension of the division to fuzzy relations.

The operation of division of R by S can be considered as a set inclusion:

$$x \in \mathrm{R[A/A]S} \Leftrightarrow \mathrm{S[A]} \subseteq \Gamma^{-1}(x) \text{, with } \Gamma^{-1}(x) = \{a, (x,a) \in \mathrm{R}\}$$

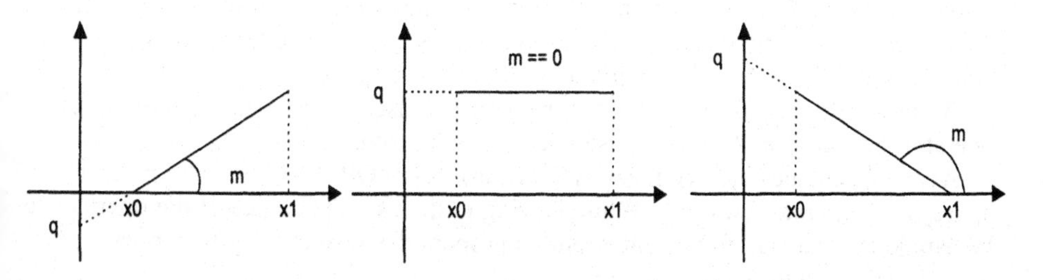

Figure 11.12. Membership Function Storage Format Used By The Broker Agent

In order to compute such a list of images, our Decision Agent exploits the Broker Agent's internal repository, whose relational schema is depicted in Figure 11.11.

The entity *Component* has two attributes: *URL* is a reference to the image on the Web and *Composite* is the HTML document holding the image. This entity gives access to the Web pages holding the images composing a query result. Note that the URL alone would not be sufficient, as an image may well appear in multiple Web pages.

As mentioned above, image features are stored in the entity *Feature*, much in the same way as shown in Figure 11.12. However, instead of explicitly storing fuzzy membership values, in our system we use entity *Behavior* to store definitions of trapezoidal fuzzy elements as shown in Figure 11.12.

In our current release, the linguistic variables have three fuzzy elements each, but (as is usually the case) the superposition in their definition involves the fuzzy elements two by two. Thus, in this case the maximum amount of fuzzy offer properties having non-zero membership for each server is 6.

In order to compute membership values for the content-related indexes the Broker actually uses the parameters describing the shapes of the fuzzy elements' membership functions.

Thus, if we call *inf* the inferior boundary, *sup* the superior boundary and *max* the value for which the membership function yields 1, we obtain for the first fuzzy element of the `ellipticity` feature the representation shown in Figure 11.13.

μ_{low}: $[0 .. 1] \rightarrow [0,1]$
$x \rightarrow 0$ if $x \geq \text{sup}$ or $x \leq \text{inf}$
$x \rightarrow 1$ if $x = \text{max}$
$x \rightarrow (x - \text{inf}) / (\text{max} - \text{inf})$ if $x \in [\text{inf,max}]$
$x \rightarrow (\text{sup} - x) / (\text{sup} - \text{max})$ if $x \in [\text{max,sup}]$

Figure 11.13. Fuzzy Elements Internal Representation

However, there is no need to explicitly evaluate such functions, as the broker agent can compute fuzzy membership values at run time as simple SQL queries.

With reference to the membership functions storage technique described in Figure 11.12, a sample query to the Broker Agent repository is reported below:

```
SELECT (b.m*f.x+b.q) AS weight
From Feature As f, Behaviour as b
WHERE b.x0<=f.x and f.x <=b.x1
```

The above SQL query computes the membership values (`weight`) exploiting the definition of linguistic variables held by the Broker Agent in order to transform crisp values into the fuzzy format required to perform the division.

Besides selecting the target site, the medical user can choose the aggregation operator the Decision Agent will use to compute the division.

Available choices include fuzzy AND (min), fuzzy OR (max) and any mean-based aggregation operator (AVG). While leaving to the user the choice of the operator to be employed may at first sight seem improper, the resulting system operation is indeed easy to understand.

The choice of the AND operator results in retrieving images possessing at the highest possible degree *all* the features required by the user. Images lacking even one of the requested features will be left out. The OR operator selects images possessing (at the highest possible degree) at least one of the features required by the user, regardless if they possess the other features or not.

Finally, the AVG operator corresponds to an intermediate behavior between AND and OR.

While the present choice of aggregation operators could in principle be extended to widen the spectrum of system behavior (Bosco et al. 1997) in order to fully support any user-selected retrieval semantics, it should be noted that currently supported operators are easily executable using standard SQL queries.

11.5.6. Collecting User Feedback

As the on-line indexing system described above is admittedly prone to errors, user feedback plays an important role in the framework of our system, corresponding to our methodology's validation or postprocessing phase.

Following the approach described in Bellettini et al. (1999), the user is given the possibility of manually modifying the index value for any of the images returned in the query result, in order to eliminate invalid data.

Users are required to input an approximate value, to be cached by the Broker and used in future query sessions involving the same image.

As a help to providing this estimate, users are shown the wrongly indexed images in a window where they can point out four segment endpoints to be used by a client applet to compute the ellipticity value to be transmitted to the Broker Agent.

11.6. System Architecture

Having presented the design of a general and flexible agent system based on the HCVM approach, we are ready to proceed to describe its implementation architecture. The system includes procedures for classification and retrieval as a part of the Broker Agent, as well as a User Agent based on a graphical user interface. The architecture of the current implementation of our image retrieval system is depicted in Figure 11.14.

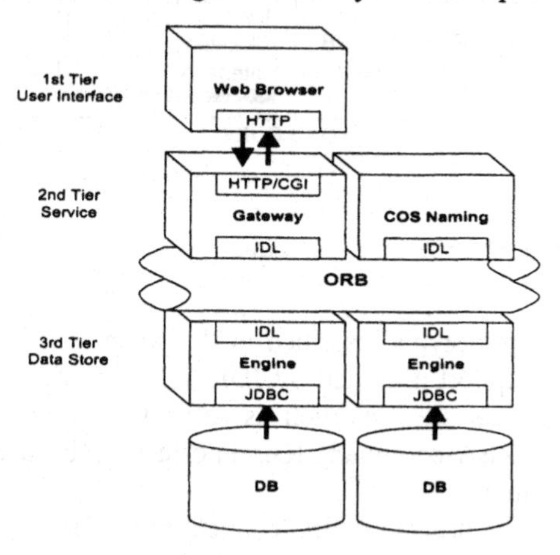

Figure 11.14. System Architecture

Both design and implementation are based on CORBA-based *Mobile Agent System Interoperability Facility (MASIF)* proposed by General Magic, IBM, GMD and the Open Group (OMG 1999). In order to standardize design and implementation of O-O agent-based distributed execution environments, the Object Management Group (OMG) decided to promote MASIF in the framework of the *CORBA* (*Common Object Request Broker Architecture*) software architecture (Yang and Duddy 1996).

CORBA reference architecture, called the *Object Management Architecture (OMA)* upon which applications can be constructed. OMA attempts to define at a high level of abstraction all facilities needed for distributed object-oriented computing. It consists of four components: an *Object Request Broker (ORB)*, *Object Services (OS)*, *Common Facilities (CF)*, and *Application Objects (AO)*.

The core of the OMA is the ORB component, a transparent communication bus for objects that let them transparently make requests and receive responses from other objects. In other words, the ORB allows client and server objects, possibly written in different languages, to interoperate. It intercepts calls and is responsible for finding an object that can execute them, pass it the parameters, invoke its methods and return the results. Invocations can be done either statically at compile time or dynamically at run time with a *late binding* of servers.

Objects Services specifications define a set of objects that perform fundamental functions such as naming services, life cycle services, transaction services or trader services. Generally speaking, they augment and complement the functionality of the ORB, whereas *CORBA Common Facilities* provide services of direct use to application objects. Finally, *Application Objects* implement distributed services to be invoked by clients through the ORB bus. The overall CORBA architecture is depicted in Figure 11.15.

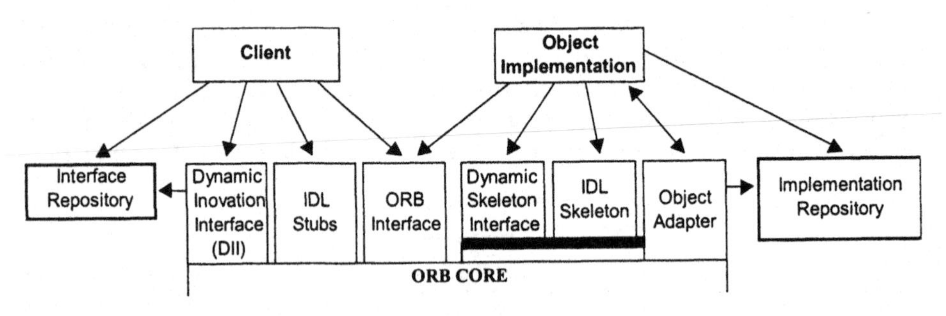

Figure 11.15. Structure of ORB Communication

The HCVM-based prototype described in this chapter is entirely implemented in Java, using CORBA support provided by JDK 1.2. It currently indexes images from several European medical sites, as well as some local demo databases; all Broker Agents provide a CORBA-compliant IDL interface to be used by User Agents in order to query the system.

On the client side, the user must first connect to the Broker site. His browser then downloads an HTML page including two Java applets (`TraderInitApplet.java` and `BestResult.java`), that is, Java classes implementing the predefined interface java.applet.Applet, together with the other

MASIF classes directly called by these two applets. Such classes are used as interfaces for exchanging messages with the Broker Agent.

Since the purpose of the Broker is to help the client to select a picture through a GUI interface, the applet `TraderInitApplet` first displays the window enabling the user to select the SNOMED code he or she is interested in. In Figure 11.8, the user can specify a SNOMED pattern.

Then the GUI displays the second frame (corresponding to the file `PropertiesFrame.java`) which lets the user make his selection over available features. All this information (i.e., the SNOMED pattern, the semantics and the corresponding operator, the selected features and the crisp values) compose a fuzzy request, that the `TraderInitApplet` passes on to the Broker to compute the division, after having contacted it using the naming service of the MASIF/CORBA architecture. For each picture regarding the subject, the Trader computes its degree of satisfaction relatively to the fuzzy query (using the `FuzzyCalc` class). Finally, it returns the `BestResult` applet information on the best matching service: the address of the server where the picture can be found, its name, and its final degree (see Figure 11.16). The client browser then connects to the target site to see the retrieved picture, by means of a third applet, `TraderImageViewer`.

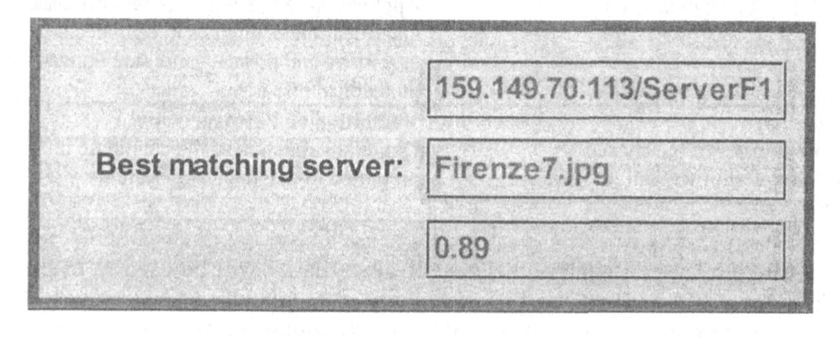

Figure 11.16. The Broker Agent Indicates the Best Target Site

11.7. Summary

In this chapter, we have presented a MASIF-compliant system for medical images retrieval based on the HCVM approach. Table 11.2 summarizes the mapping of our *Medical Images Retrieval Model to* HCVM layered architecture that has been throughout the chapter:

As we have seen, the HCVM approach allows for providing a seamless connection between a human-centered domain decomposition, based on nominal scale features, to ordinal and interval-based features suitable to be processed by intelligent agents. Besides developing a full-featured system for on-line indexing of pathology images (Catelletti et al. 1999), we explored a number of other applications of the HCVM approach, including on line analysis and retrieval of satellite images for crop control (Ahmed et al. 1999). Many measures of comparison of descriptions of objects have been proposed and studied in given frameworks or domains of applications. B. Bouchon-Meunier, M. Rifqi and S. Bothorel (Bouchon-Meunier et al. 1996) proposed a general classification consisting in four main kinds of measures of comparison, depending on the purpose of their utilization: they define measures of *satisfiability*, *resemblance* and *comparison* that on one hand, can be considered as measures of similarity, and on the other hand measures of dissimilarity.

Table 11.2. Retrieval Model to HCVM Mapping

Medical Image Retrieval Reference Model	*HCVM*
A *image description* to be supplied by customer/user	**Preprocessing** Input filtering
Domain Specific Knowledge SNOMED-based classification of images in the domain of interest	**Decomposition** *Domain model* as a hierarchy of abstract orthogonal classes. Domain specific metadata.
System Functional Units Fuzzy query to the broker's database.	**Control** Determining decision classes based on content-related features like ellipticity.
Candidate Selection Presentation of candidate images	**Decision** Application of fuzzy/neural network based identification techniques in each decision class to retrieve and present candidate images based on medical researcher's query.
Transaction Network transfer of the chosen images/Request of validation of choice process	**Validation or Postprocessing** Optional validation of the candidate images presented to the medical researcher

In a future release, a general measure of dissimilarity will be used to evaluate to which extent an object is different from a reference, helping the user to choose objects that are closer to the reference than the others. Secondly, as far as the design of the architecture is concerned, we are currently working on the servers-to-Broker dialogue in order to support intelligent load balancing techniques.

However, several other research directions could be also pursued. First of all, the mechanism of retrieval of servers can be enlarged and complemented on the basis of features the images should *not* have, in order to eventually decide between those that would have a same level of fitness after the division. Indeed, the phase of division searches in the repository for images that have as many as possible of the desired requirements. On the contrary, the optional phase of differentiation, through a measure of comparison, should compute a similarity or a dissimilarity measure to eliminate images that have features the user did not ask for.

Acknowledgements

We wish to thank Maurizio Lestani, M.D., Maurizio Pea, M.D. and their colleagues at the Institute of Anatomy and Pathological Histology, Medical University of Verona, Italy, for valuable cooperation and precious assistance on medical issues. Prof. Edward C. Klatt, M.D. of the Department of Pathology of University of Utah, besides granting permission for the publication of the medical images from his *Web Path* service (`http://www-medlib.med.utah.edu/WebPath/`, provided valuable insight on the problem (see Sect. 2) that helped us to set the goals for this application. While we thankfully acknowledge their help, we would like to clarify that any mistake or inaccuracy on medical matters are our own. Thanks are also due to Letizia Tanca and Patrick Bosc for their encouragement and valuable suggestions. Ruben Castelletti implemented the first prototype of the system, while Giovanni Righini provided the convex-hull algorithm used for indexing.

References

Ahmed M., Damiani E., Tettamanzi A. (1999) "A Fuzzy System for Crop Control, *Proceedings of the 6th Intl. Conf. On Computational Intelligence*, Dortmund, Germany, LNCS 1625 pp.473-481

Barros J., French J., Martin W., Kelly P. (1995) "A System for Indexing Multi-spectral Satellite Images for Content-Based Retrieval", *Proceedings of IS&T/SPIE: Storage and Retrieval for Image and Video Databases III,* San Josè, CA,USA

Bellettini C., Damiani E., Fugini M.G. (1999) "Collecting User Feedback in Software Reuse Systems", *Proceedings of the ACM-SSR Workshop on Software Reuse*, Los Angeles, CA, USA

Binaghi E., Gagliardi I., Schettini R. (1994) "Image Retrieval Using Fuzzy Evaluation of Color Similarity, *International Journal of Pattern Recognition and Artificial Intelligence* vol.8, no.4, pp.123-135

Bosc P., Dubois D., Pivert O. and Prade H. (1997), "Flexible Queries In Relational Databases - The Example of The Division Operator", *Theoretical Computer Science,* vol.171, no. 4

Bouchon-Meunier B., Rifqi M., Bothorel S. (1996) "Towards General Measures of Comparison of Objects", *Fuzzy Sets and Systems,* vol.84, no.7

Castelletti R., Damiani E., Righini G., Khosla R. (1999) *A Human Centered Approach to Image Retrieval*, Proc. of *the 6th Intl. Conf. on Computational Intelligence*, Dortmund, Germany, LNCS 1625

CEN/TC251 Project Report (1998), available at http://miginfo.rug.ac.be:8001/index.htm

Chiueh T. (1994), "Content-based Image Indexing, *Proceedings of the 20th VLDB Conference*, Santiago,Chile, pp.188-197

Dave R. N. (1990), "Fuzzy Shell-Clustering and Applications to Circle Detection in Digital Images," *International Journal of General Systems* vol.16 no.4

Damiani E., Fugini M.G. (1997), "Fuzzy Identification Of Distributed Components", *Proceedings of the 5th Intl. Conf. On Computational Intelligence,* Dortmund, Germany, LNCS 1226

Dubois D., Prade H., Sedes F. (1999), "Fuzzy Logic Techniques in Multimedia Databases Querying: A Preliminary Investigation of the Potentials", *Proceedings of IFIP Working Conference on Database Semantics (DS-8) Roturoa, New Zealand*

Faloutsos C., Equitz W., Flickner M., Niblack W., Petrovic D., Barber R. (1994), "Efficient and Effective Querying by Image Conten"t, *Journal of Intelligent Information Systems* vol. 3 no. ¾

Kohonen T. (1997) "Self-organizing maps", *Series in Information Sciences*, Vol. 30, Springer 1995, 2nd ed. 1997

Kobatake H., Murakami M. (1996), "Adaptive Filter to Detect Rounded Convex Regions: Iris Filter", *Proceedings of the. Intl. Conf. on Pattern Recognition*, pp.340-344.

Kobatake H., Takeo H, Nawano S. (1998) "Tumor Detection System for Full-digital Mammography" *Proceedings of the 4th International Workshop On Digital Mammography* Nijmegen, the Netherlands

Mauri F., Beltrami A., Della Mea V. (1996) "Telepathology Using Internet Multimedia Electronic Mail", *Journal of Telemedicine and Telecare*, vol.6 n.3 pp.56-69

OMG (1999) "MASIF: A CORBA-based Mobile Agent System Interoperability Facility" available at http://www.omg.org/masif

Maybury M. (ed.) (1995) *Intelligent Multimedia Information Retrieval*, MIT Press

Russo F., Ramponi G. (1995), "A Fuzzy Operator for Enhancement of Blurred and Noisy Images", *IEEE Trans. on Image Processing*, vol. 4 (8), pp. 128-140

Siemens AG (1998) "The SPI standard: a White Paper", available at http://www.siemens.de/med

Stephanakis I., Stamou G., Kollias K. (1998), "A Fuzzy Associative Operator for Enhancement of Noisy Images By Detection of Local Multiresolution Features", *Proceedings of JCIS '98,* Rayleigh, NC, USA

Telemedicine Research Center (1998), "An Introduction to Telemedicine", available from http://tie.telemed.org/tiemap.html

Yang Z., Duddy K. (1996), "CORBA: A Platform for Distributed Object Computing", *ACM Operating Systems Review*, vol. 30 n.4.

Yiming T., Ping F., Yong Z., Siqian Y. "Evaluating Nugget Sizes of Spot Welds by Using Artificial Neural Networks, *Proceedings of the 6th Intl. Conf. on Computational Intelligence*

12 HCVM REVISITED

12.1. Introduction

In the last eleven chapters we have come a fair distance in realizing our original goal of adopting a human-centered approach towards development of intelligent multimedia multi-agent systems. These systems cover a range of areas including medicine, image processing, internet games, human resource management, multimedia retrieval, e-commerce, telehealth/e-health. A summary of the applications in these areas and in some others (not described in this book) is shown in Table 12.1. In these application areas we have illustrated a number of facets of the Human-Centered Virtual Machine (HCVM). It is instructive to revisit these facets as they help us to take stock of some of the contributions made by this book.

12.2. Successful Systems as Against Successful Technologies

In this book we have proposed the development of successful systems instead of successful technologies. The need for such systems has been outlined in chapter 1. In successful systems, among other aspects, human actors, instead of technology, are the prime driving force behind development of computer-based artifact. These systems are centered around three human-centered criteria, namely, human-centered research and design is problem/need driven, human-centered research and design is context bound, and human-centered research and design is activity centered are used as the guidelines for the development of successful systems in the book.

12.3. Pragmatic Considerations, Enabling Theories and HCVM

The book sees the human-centered approach as an evolutionary one rather than a revolutionary one. It considers various pragmatic issues and problems in areas like intelligent systems, software engineering, multimedia databases, electronic commerce, data mining, enterprise modeling and human-computer interaction. The book takes stock of these pragmatic issues which are driving these areas towards a human-centered approach and answers the question, "What needs to be done?" It then considers various enabling theories from philosophy (e.g., semiotic theory), cognitive science (e.g., situated cognition, distributed cognition, etc.), psychology (e.g., activity theory), and workplace (e.g., work-centered analysis, socio-technical framework) in order to answer the question "How it can be done?" The book integrates concepts from these enabling theories to lay down the conceptual and computational foundations of the HCVM.

12.4. Software Components, Adapters and HCVM

Software components and adapters are two emerging concepts in software development today. These two concepts have been realized by HCVM at a conceptual level and a computational level. At the conceptual level, from a human-centered viewpoint it integrates the external plane or context represented by the physical, social and organizational reality with the subjective reality based on organizational culture, stakeholder goals and incentives, their problem solving strategies, etc. Through this integration it develops four components, namely, activity-centered analysis component, problem solving ontology component, transformation agent component and multimedia interpretation component. It defines a detailed ontology of all these components in order to facilitate a seamless integration between the conceptual and computational levels of the HCVM.

The activity-centered analysis component analyzes the physical reality of a system in terms of six components or elements, namely, work activity, product, participant, customer, data and tool, social reality in terms of division of labor/tasks between human and the computer-based artifact, and organizational reality in terms of performance of the six components. It then analyzes the subjective reality by undertaking a context analysis of these components. In particular it analyzes stakeholder goals and incentives and organizational culture of participant, customer and work activity components respectively. It identifies the goals and tasks for a computer-based system by marrying the physical, social and organizational reality with the subjective reality of stakeholder incentives and organizational culture.

The problem-solving ontology component is then used to develop a human-centered domain model. It does that by systematizing and structuring the stakeholder goals and tasks model, stakeholder external and internal representation model and the domain model for various tasks by employing five problem solving adapters. These are preprocessing, decomposition, control, decision and postprocessing.

The transformation agent component is defined at the computational level. Its purpose is to transform a human-centered domain model defined with the help of problem solving adapters into a computer-based artifact. It does that by integrating problem solving ontology with technological tools like intelligent technology model, agent and object-oriented model, multimedia and distributed processing and communication models, etc.. The integration results into component based agent definitions of generic agents in four agent layers of the HCVM. These are the problem solving agent layer, intelligent hybrid technology layer, intelligent technology layer and software agent layer. These component-based agent definitions are used to model various software components of the human-centered domain model. In addition, an object layer is also defined to capture the structural and association relationships between different data entities.

Finally, the multimedia interpretation component is defined at the conceptual level to improve the representational efficiency and effectiveness of the computer-based artifact at the human-computer interface for various defined by the problem solving adapters. It is also used to model perceptual problem solving tasks. At the computational level a media agents are defined for presentation and coordination of the multimedia artifacts.

Table 12.1: Brief Summary of Some Applications of HCVM

Application	Area	Repres-entation Model	Some HCVM Features	Domain Model	Technology Tool
Medical Diagnosis and Treatment Support	Health Informatics: Diagnosis and Treatment Decision Support	External and Internal	Application of Activity-centered analysis, Adaptive problem solving ontology, Multiple practitioner based decision sequences, Multimedia based perceptual modeling of symptoms and problem solving tasks	Heuristic, Self-Organized Diagnosis Cluster, Inductive and Deductive Treatment Models	Multi-Agents (problem solving, intelligent technology, media) objects, self-organizing maps, back propagation neural networks, rule based techniques, databases, media objects (e.g., graphic and audio)
Face Detection and Annotation	Image Processing	External and internal	Human perception based problem solving ontology for face detection and annotation	Color, Shape, and Area Models	Agents, morphological techniques, Area and shape based filtering algorithms, etc.
Recruitment and Bench-marking	Sales Management, Human Resource management	External and Internal	Modeling human breakdowns, multiple users	Selling Behavioral, Heuristic profiling & benchmarking Models	Agents, behavior evaluation objects, rule based techniques, neural networks, databases, media objects (e.g. graphic)
Net Euchre Card Game	Internet games	External and Internal	Distributed human-computer interaction, multiple users, backup support agents	Heuristic model	Agents, objects, rule based techniques, internet technologies, media objects, net sprocket software
Buying PC hardware adapters on the internet	Electronic Commerce	External and Internal	User-centered Brokerage Architecture based on HCVM problem solving ontology, Integration of problem solving ontology with XML	User-Centered Market Model	Agents, Internet technologies, multimedia objects, Fuzzy logic and neural network techniques

Table 12.1 (Cont'd): Brief Summary of Some Applications of HCVM

Application	Area	Repres-entation Model	Some HCVM Features	Domain Model	Technology Tool
On line Medical image query & retrieval	Telehealth /E-Health	External and Internal	Medical Researcher Centered Image Processing Architecture	SNOMED Model;	Agents, Internet technologies, multimedia objects, Fuzzy logic and image processing techniques
Obstacle Avoidance (not described in the book)	Distributed Control	Internal	Open component based distributed software modeling environment	Fuzzy Control , Neural Network Control & PID Control Models	Agents, objects, fuzzy rule based systems, neural networks, genetic algorithms, fuzzy integral, media objects
Genome Classification (not described in the book)	Bioinformatics /Data mining	External and Internal	App. of problem solving ontology, and multimedia interpretation component for information visualization	Gene Functional and Location Models	Agents, Genome objects, Self-organizing maps
Buying home on the internet (not described in the book)	Internet/Image Processing	External and Internal	Direct manipulation of House Objects (e.g. roof line), problem solving ontology	Buyer-Centered functional and spatial location models	Agents, house objects, 2D Strings, neural network

12.5. Human Decision Ladder and Problem Solving Adapters

The five problem solving adapters of the HCVM can be used in a number of different sequences depending upon the decision ladder or path of the user/s. In chapter 6 we show in a medical diagnosis and treatment support application that transformation agent definitions of the problem solving adapters can be executed in three different sequences. Further, one problem solving adapter can also be used as a part of or within another problem solving adapter. For example, in chapter 7 on face detection and annotation, we show that preprocessing adapter is used within decomposition and control adapter definitions. Similarly, in chapter 6 we show that preprocessing adapter is used within control and decision adapter definitions.

12.6. HCVM and Multimedia

Multimedia is used in HCVM in two contexts. Firstly, we look into the human-centered modeling of data at the human-computer interface using multimedia artifacts. This is done by the multimedia interpretation component of the HCVM. The application of the multimedia interpretation component is described in the intelligent multimedia multi-agent diagnosis and treatment support application in chapter 6. Secondly, we look into multimedia information management and retrieval issues on the internet and role of user-centered models in addressing these issues. Chapter 10 of this book describes the application of HCVM in developing a user-centered model for querying and retrieval of medical images on the internet.

12.7. HCVM, Human Dynamics and Breakdowns

The role of computers as back up support agents has gained ground in recent times. Intelligent distributed Net Euchre game agents are developed in chapter 8 to provide solicited advice to human players playing Net Euchre on different computers on the internet. It also outlines the nature of distributed communication between different game agents supporting the human players.

In order to motivate stakeholders to use computers as an integral part of their work activity, it is important to use computers to model human breakdowns in decision making. Intelligent agents of the sales recruitment system described in chapter 8 help hiring managers in behavior profiling and benchmarking of incoming sales candidates with existing successful salespersons in a organization. The modeling of the breakdown situations is one of the reasons that the sales recruitment system has been in commercial use for the last three years.

12.8. HCVM, Internet, E-Commerce and XML

Intelligent agent applications on the internet have become a necessity because of information explosion, and huge amount of time being spent by the internet users to get the information they need on the internet. A need for developing user-centered models for searching on the internet has been clearly identified. Chapters 9 and 10 in this book have described internet based e-commerce (buying hardware adapters) and on-line medical image query and retrieval applications respectively. In both of these applications user-centered models based on HCVM are developed. Further, in the e-commerce application we also show how the HCVM based user-centered market model is integrated with XML, the new emerging internet standard.

12.9. Dynamic Analysis of Agents in HCVM

From a software design and implementation perspective it is useful to do a dynamic analysis of the agents in HCVM. Although not described in this book, in Khosla and Dillon (1997) we describe how State Controlled Petri Nets (SCPN) shown in Figure 12.1 can be used for dynamic behavior analysis of multi-agent systems.

Further, at a conceptual level, the SCPN model can also be used for analysis of the task-product-network of the activity-centred analysis component. The places in the SCPN model can be used to represent the task and product of the task-product-network. The action tokens of the SCPN model can be replaced by task and product tokens, respectively. Finally, the transition inscription structure (square box in Figure 12.1) can be used to represent precondition and postcondition in a task-product-network.

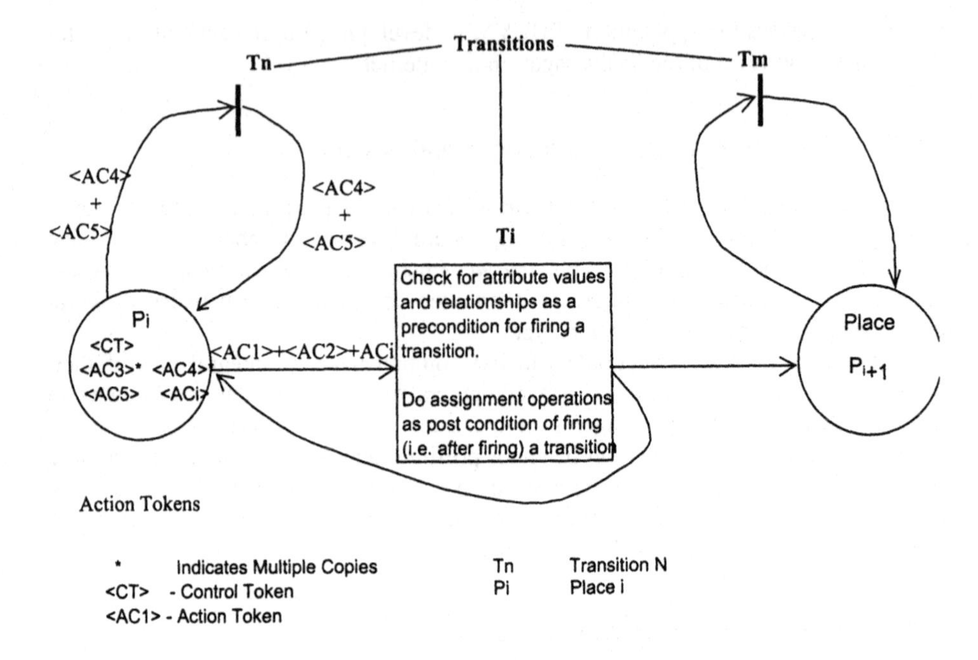

Figure 12.1: SPCN Model with Inscribed Transition Structure

References

Khosla, R. and Dillon, T. (1997) *Engineering Intelligent Hybrid Multi-Agent Systems*, Kluwer Academic Publishers, MA, USA.

F